TIME-DOMAIN SCATTERING

The wave equation, a classical partial differential equation, has been studied and applied since the eighteenth century. Solving it in the presence of an obstacle, the scatterer, can be achieved using a variety of techniques and has a multitude of applications. This book explains clearly the fundamental ideas of time-domain scattering, including in-depth discussions of separation of variables and integral equations. The author covers both theoretical and computational aspects, and describes applications coming from acoustics (sound waves), elastodynamics (waves in solids), electromagnetics (Maxwell's equations) and hydrodynamics (water waves). The detailed bibliography of papers and books from the last 100 years cement the position of this work as an essential reference on the topic for applied mathematicians, physicists and engineers.

Encyclopedia of Mathematics and Its Applications

This series is devoted to significant topics or themes that have wide application in mathematics or mathematical science and for which a detailed development of the abstract theory is less important than a thorough and concrete exploration of the implications and applications.

Books in the **Encyclopedia of Mathematics and Its Applications** cover their subjects comprehensively. Less important results may be summarised as exercises at the ends of chapters. For technicalities, readers can be referred to the bibliography, which is expected to be comprehensive. As a result, volumes are encyclopedic references or manageable guides to major subjects.

All the titles listed below can be obtained from good booksellers or from Cambridge University Press. For a complete series listing visit

www.cambridge.org/mathematics

130 T. G. Faticoni *Modules over Endomorphism Rings*
131 H. Morimoto *Stochastic Control and Mathematical Modeling*
132 G. Schmidt *Relational Mathematics*
133 P. Kornerup and D. W. Matula *Finite Precision Number Systems and Arithmetic*
134 Y. Crama and P. L. Hammer (eds.) *Boolean Models and Methods in Mathematics, Computer Science, and Engineering*
135 V. Berthé and M. Rigo (eds.) *Combinatorics, Automata and Number Theory*
136 A. Kristály, V. D. Rădulescu and C. Varga *Variational Principles in Mathematical Physics, Geometry, and Economics*
137 J. Berstel and C. Reutenauer *Noncommutative Rational Series with Applications*
138 B. Courcelle and J. Engelfriet *Graph Structure and Monadic Second-Order Logic*
139 M. Fiedler *Matrices and Graphs in Geometry*
140 N. Vakil *Real Analysis through Modern Infinitesimals*
141 R. B. Paris *Hadamard Expansions and Hyperasymptotic Evaluation*
142 Y. Crama and P. L. Hammer *Boolean Functions*
143 A. Arapostathis, V. S. Borkar and M. K. Ghosh *Ergodic Control of Diffusion Processes*
144 N. Caspard, B. Leclerc and B. Monjardet *Finite Ordered Sets*
145 D. Z. Arov and H. Dym *Bitangential Direct and Inverse Problems for Systems of Integral and Differential Equations*
146 G. Dassios *Ellipsoidal Harmonics*
147 L. W. Beineke and R. J. Wilson (eds.) with O. R. Oellermann *Topics in Structural Graph Theory*
148 L. Berlyand, A. G. Kolpakov and A. Novikov *Introduction to the Network Approximation Method for Materials Modeling*
149 M. Baake and U. Grimm *Aperiodic Order I: A Mathematical Invitation*
150 J. Borwein et al. *Lattice Sums Then and Now*
151 R. Schneider *Convex Bodies: The Brunn–Minkowski Theory (Second Edition)*
152 G. Da Prato and J. Zabczyk *Stochastic Equations in Infinite Dimensions (Second Edition)*
153 D. Hofmann, G. J. Seal and W. Tholen (eds.) *Monoidal Topology*
154 M. Cabrera García and Á. Rodríguez Palacios *Non-Associative Normed Algebras I: The Vidav–Palmer and Gelfand–Naimark Theorems*
155 C. F. Dunkl and Y. Xu *Orthogonal Polynomials of Several Variables (Second Edition)*
156 L. W. Beineke and R. J. Wilson (eds.) with B. Toft *Topics in Chromatic Graph Theory*
157 T. Mora *Solving Polynomial Equation Systems III: Algebraic Solving*
158 T. Mora *Solving Polynomial Equation Systems IV: Buchberger Theory and Beyond*
159 V. Berthé and M. Rigo (eds.) *Combinatorics, Words and Symbolic Dynamics*
160 B. Rubin *Introduction to Radon Transforms: With Elements of Fractional Calculus and Harmonic Analysis*
161 M. Ghergu and S. D. Taliaferro *Isolated Singularities in Partial Differential Inequalities*
162 G. Molica Bisci, V. D. Radulescu and R. Servadei *Variational Methods for Nonlocal Fractional Problems*
163 S. Wagon *The Banach–Tarski Paradox (Second Edition)*
164 K. Broughan *Equivalents of the Riemann Hypothesis I: Arithmetic Equivalents*
165 K. Broughan *Equivalents of the Riemann Hypothesis II: Analytic Equivalents*
166 M. Baake and U. Grimm (eds.) *Aperiodic Order II: Crystallography and Almost Periodicity*
167 M. Cabrera García and Á. Rodríguez Palacios *Non-Associative Normed Algebras II: Representation Theory and the Zel'manov Approach*
168 A. Yu. Khrennikov, S. V. Kozyrev and W. A. Zúñiga-Galindo *Ultrametric Pseudodifferential Equations and Applications*
169 S. R. Finch *Mathematical Constants II*
170 J. Krajíček *Proof Complexity*
171 D. Bulacu, S. Caenepeel, F. Panaite and F. Van Oystaeyen *Quasi-Hopf Algebras*
172 P. McMullen *Geometric Regular Polytopes*
173 M. Aguiar and S. Mahajan *Bimonoids for Hyperplane Arrangements*
174 M. Barski and J. Zabczyk *Mathematics of the Bond Market: A Lévy Processes Approach*
175 T. R. Bielecki, J. Jakubowski and M. Niewęgłowski *Structured Dependence between Stochastic Processes*
176 A. A. Borovkov, V. V. Ulyanov and Mikhail Zhitlukhin *Asymptotic Analysis of Random Walks: Light-Tailed Distributions*
177 Y.-K. Chan *Foundations of Constructive Probability Theory*
178 L. W. Beineke, M. C. Golumbic and R. J. Wilson (eds.) *Topics in Algorithmic Graph Theory*
179 H.-L. Gau and P. Y. Wu *Numerical Ranges of Hilbert Space Operators*

Encyclopedia of Mathematics and its Applications

Time-Domain Scattering

P. A. MARTIN
Colorado School of Mines

CAMBRIDGE
UNIVERSITY PRESS

CAMBRIDGE
UNIVERSITY PRESS

University Printing House, Cambridge CB2 8BS, United Kingdom

One Liberty Plaza, 20th Floor, New York, NY 10006, USA

477 Williamstown Road, Port Melbourne, VIC 3207, Australia

314–321, 3rd Floor, Plot 3, Splendor Forum, Jasola District Centre,
New Delhi – 110025, India

79 Anson Road, #06–04/06, Singapore 079906

Cambridge University Press is part of the University of Cambridge.

It furthers the University's mission by disseminating knowledge in the pursuit of education, learning, and research at the highest international levels of excellence.

www.cambridge.org
Information on this title: www.cambridge.org/9781108835596
DOI: 10.1017/9781108891066

First published 2021

A catalogue record for this publication is available from the British Library.

ISBN 978-1-108-83559-6 Hardback

To Thomas, Emma, Samantha and, last but not least, Ann

Contents

Preface

I have been studying scattering problems since I was a graduate student in the late 1970s. Specifically, I have studied time-harmonic problems where the solutions depend on time t through a factor $\exp(-\mathrm{i}\omega t)$ in which ω is the frequency; they are *frequency-domain problems*. More recently, I decided to study *time-domain problems*; the prototype problem is: solve the wave equation in the unbounded region exterior to a bounded obstacle (the scatterer) subject to appropriate boundary and initial conditions. One result is this book.

My frequency-domain work resulted in a book, *Multiple Scattering: Interaction of Time-Harmonic Waves with N Obstacles*, published in 2006. The present book is written in the same spirit. The problems considered come mainly from acoustics (governed by the three-dimensional scalar wave equation) with some discussion of problems from electromagnetics, elastodynamics and hydrodynamics. The emphasis is on exact methods, primarily separation of variables and boundary integral equations. As far as I know, there is no comparable book. For more information on the topics covered, see the Table of Contents and Section 1.7.

I am grateful to Gerhard Kristensson for his detailed comments on an early draft of the whole book. Thomas Anderson made many useful comments on a later draft. I also thank many others for comments and help with various parts of the book.

Preface

1
Acoustics and the Wave Equation

Acoustics is the study of small-amplitude disturbances in a compressible fluid: it is a branch of continuum mechanics. General texts include [711, 644, 696, 207, 227].[1] To simplify our analysis, we suppose that the fluid is inviscid. We derive the governing equations in Section 1.1. Our derivations cover many interesting situations, including inhomogeneous fluids (with properties that can vary spatially and temporally) and non-uniform background flows. However, (at the present time) few of these situations have been combined with scattering phenomena, such as when a sound wave interacts with an object immersed in the flow. Indeed, for most of the book, we shall restrict ourselves to homogeneous fluids. Then, for most (but not all) purposes, it is found that the governing equation is the scalar wave equation; the relevant equations are collected in Section 1.2. Section 1.3 is dedicated to waves on strings, where the motion is governed by the one-dimensional wave equation. For a general survey of the mathematics underlying the wave equation, see the paper by Leis [540]. For another survey, with more emphasis on inverse problems, try [79].

Formal properties of Laplace transforms are collected in Section 1.4, with more detailed discussions reserved for later. The vexed question of *causality* is discussed in Section 1.5. Finally, the governing equations for electromagnetic, elastodynamic and hydrodynamic problems are collected in Section 1.6.

1.1 Governing Equations

The exact equations for the motion of a compressible inviscid fluid are as follows [57, §3.6], [697, §I], [675, §2.1.1]. Conservation of mass gives the continuity equation,

$$\frac{\mathrm{D}\rho_{\mathrm{ex}}}{\mathrm{D}t} + \rho_{\mathrm{ex}} \operatorname{div} \boldsymbol{v}_{\mathrm{ex}} = 0, \tag{1.1}$$

[1] Citations such as this will be listed in chronological order.

where ρ_{ex} is the mass density, $\boldsymbol{v}_{\text{ex}}$ is the fluid velocity and t is time. (The subscript 'ex' denotes 'exact'.) The material derivative is defined by

$$\frac{\mathrm{D}f}{\mathrm{D}t} = \frac{\partial f}{\partial t} + \boldsymbol{v}_{\text{ex}} \cdot \operatorname{grad} f.$$

In the absence of body forces, conservation of linear momentum gives

$$\rho_{\text{ex}} \frac{\mathrm{D}\boldsymbol{v}_{\text{ex}}}{\mathrm{D}t} + \operatorname{grad} p_{\text{ex}} = \boldsymbol{0}, \tag{1.2}$$

where p_{ex} is the pressure. For isentropic flows [57, p. 156], [696, eqn (1-4.3)], the entropy per unit mass, E_{ex}, satisfies

$$\frac{\mathrm{D}E_{\text{ex}}}{\mathrm{D}t} = 0. \tag{1.3}$$

There is also an equation of state which we take as the statement that p_{ex} is a function of ρ_{ex} and E_{ex} [696, §1-4],

$$p_{\text{ex}} = p_{\text{ex}}(\rho_{\text{ex}}, E_{\text{ex}}). \tag{1.4}$$

Differentiating, we obtain

$$\operatorname{grad} p_{\text{ex}} = c_{\text{ex}}^2 \operatorname{grad} \rho_{\text{ex}} + h_{\text{ex}} \operatorname{grad} E_{\text{ex}} \tag{1.5}$$

and

$$\frac{\mathrm{D}p_{\text{ex}}}{\mathrm{D}t} = c_{\text{ex}}^2 \frac{\mathrm{D}\rho_{\text{ex}}}{\mathrm{D}t} + h_{\text{ex}} \frac{\mathrm{D}E_{\text{ex}}}{\mathrm{D}t} = -\rho_{\text{ex}} c_{\text{ex}}^2 \operatorname{div} \boldsymbol{v}_{\text{ex}}, \tag{1.6}$$

using (1.1) and (1.3), where

$$c_{\text{ex}}^2(\rho_{\text{ex}}, E_{\text{ex}}) = \frac{\partial p_{\text{ex}}}{\partial \rho_{\text{ex}}} \quad \text{and} \quad h_{\text{ex}}(\rho_{\text{ex}}, E_{\text{ex}}) = \frac{\partial p_{\text{ex}}}{\partial E_{\text{ex}}}. \tag{1.7}$$

Finally, the temperature T_{ex} satisfies [57, eqn (3.6.6)]

$$\frac{1}{T_{\text{ex}}} \frac{\mathrm{D}T_{\text{ex}}}{\mathrm{D}t} = \frac{\varkappa}{\rho_{\text{ex}}} \frac{\mathrm{D}p_{\text{ex}}}{\mathrm{D}t} = -\varkappa c_{\text{ex}}^2 \operatorname{div} \boldsymbol{v}_{\text{ex}}, \tag{1.8}$$

using (1.6), where $\varkappa$ is the ratio of the coefficient of thermal expansion to the specific heat at constant pressure ($\varkappa = \beta / c_p$ in Batchelor's notation [57]).

1.1.1 Linearisation: Ambient Flows

Consider an ambient flow in which $\boldsymbol{v}_{\text{ex}} = \boldsymbol{U}$, a constant velocity. (The case $\boldsymbol{U} = \boldsymbol{0}$ will be of most interest to us.) For such a flow, let $\rho_{\text{ex}} = \rho_0$, $p_{\text{ex}} = p_0$, $E_{\text{ex}} = E_0$, $T_{\text{ex}} = T_0$, $c_{\text{ex}}^2 = c_0^2$ and $h_{\text{ex}} = h_0$. We have $p_0 = p_{\text{ex}}(\rho_0, E_0)$, $c_0^2 = c_{\text{ex}}^2(\rho_0, E_0)$ and $h_0 = h_{\text{ex}}(\rho_0, E_0)$. Then (1.1), (1.2), (1.3), (1.6) and (1.8) give the following constraints on the ambient flow,

$$\frac{\mathscr{D}\rho_0}{\mathscr{D}t} = 0, \quad \operatorname{grad} p_0 = \boldsymbol{0}, \quad \frac{\mathscr{D}E_0}{\mathscr{D}t} = 0, \quad \frac{\mathscr{D}p_0}{\mathscr{D}t} = 0 \quad \text{and} \quad \frac{\mathscr{D}T_0}{\mathscr{D}t} = 0, \tag{1.9}$$

where $\mathscr{D}f/\mathscr{D}t = \partial f/\partial t + \boldsymbol{U}\cdot\operatorname{grad} f$. Combining $(1.9)_2$ and $(1.9)_4$ shows that p_0 is a constant, whereas (1.5) gives

$$\operatorname{grad} p_0 = c_0^2 \operatorname{grad} \rho_0 + h_0 \operatorname{grad} E_0 = \mathbf{0}. \tag{1.10}$$

The easiest way to satisfy $(1.9)_3$ and $(1.9)_5$ is to suppose that E_0 and T_0 are constants. Then $(1.9)_1$ and (1.10) imply that ρ_0 is constant. In this situation, we say the fluid is *homogeneous*: its properties do not vary with position (or time).

1.1.2 Linearisation: Acoustics

For linear acoustics, we consider small perturbations about the ambient state, and write

$$\begin{aligned} p_{\mathrm{ex}} &= p_0 + \varepsilon p_1 + \cdots, & \rho_{\mathrm{ex}} &= \rho_0 + \varepsilon \rho_1 + \cdots, & \boldsymbol{v}_{\mathrm{ex}} &= \boldsymbol{U} + \varepsilon \boldsymbol{v}_1 + \cdots, \\ E_{\mathrm{ex}} &= E_0 + \varepsilon E_1 + \cdots, & c_{\mathrm{ex}} &= c_0 + \varepsilon c_1 + \cdots, & h_{\mathrm{ex}} &= h_0 + \varepsilon h_1 + \cdots, \end{aligned}$$

where ε is a small parameter. Substitution in the equation of state (1.4) gives

$$\begin{aligned} p_{\mathrm{ex}}(\rho_{\mathrm{ex}}, E_{\mathrm{ex}}) &= p_{\mathrm{ex}}(\rho_0 + \varepsilon\rho_1 + \cdots, E_0 + \varepsilon E_1 + \cdots) \\ &= p_{\mathrm{ex}}(\rho_0, E_0) + \varepsilon \rho_1 \frac{\partial p_{\mathrm{ex}}}{\partial \rho_{\mathrm{ex}}}(\rho_0, E_0) + \varepsilon E_1 \frac{\partial p_{\mathrm{ex}}}{\partial E_{\mathrm{ex}}}(\rho_0, E_0) + \cdots, \end{aligned}$$

giving $p_0 = p_{\mathrm{ex}}(\rho_0, E_0)$ and

$$p_1 = c_0^2 \rho_1 + h_0 E_1 \quad \text{with} \quad c_0^2 = c_{\mathrm{ex}}^2(\rho_0, E_0) \quad \text{and} \quad h_0 = h_{\mathrm{ex}}(\rho_0, E_0).$$

Substitution in (1.1), (1.2), (1.3) and (1.8) gives, at first order in ε,

$$\frac{\mathscr{D}\rho_1}{\mathscr{D}t} + \operatorname{div}(\rho_0 \boldsymbol{v}_1) = 0, \quad \rho_0 \frac{\mathscr{D}\boldsymbol{v}_1}{\mathscr{D}t} + \operatorname{grad} p_1 = \mathbf{0}, \tag{1.11}$$

$$\frac{\mathscr{D}E_1}{\mathscr{D}t} + \boldsymbol{v}_1 \cdot \operatorname{grad} E_0 = 0, \quad \frac{\mathscr{D}T_1}{\mathscr{D}t} + \boldsymbol{v}_1 \cdot \operatorname{grad} T_0 = -\varkappa c_0^2 T_0 \operatorname{div} \boldsymbol{v}_1. \tag{1.12}$$

We are mainly interested in perturbations from the ambient state. Therefore we define the excess pressure p by $p_{\mathrm{ex}} = p_0 + p$, and we accept the linear approximation, giving $p = \varepsilon p_1$. We make similar definitions for other relevant quantities. Thus

$$\begin{aligned} p &= p_{\mathrm{ex}} - p_0 = \varepsilon p_1, & \boldsymbol{v} &= \boldsymbol{v}_{\mathrm{ex}} - \boldsymbol{U} = \varepsilon \boldsymbol{v}_1, \\ \tilde{\rho} &= \rho_{\mathrm{ex}} - \rho_0 = \varepsilon \rho_1, & \widetilde{E} &= E_{\mathrm{ex}} - E_0 = \varepsilon E_1, & \widetilde{T} &= T_{\mathrm{ex}} - T_0 = \varepsilon T_1. \end{aligned}$$

The equations relating these quantities are readily found, making use of (1.9). They are

$$p = c_0^2 \tilde{\rho} + h_0 \widetilde{E}, \quad \frac{\mathscr{D}\tilde{\rho}}{\mathscr{D}t} + \operatorname{div}(\rho_0 \boldsymbol{v}) = 0, \quad \rho_0 \frac{\mathscr{D}\boldsymbol{v}}{\mathscr{D}t} + \operatorname{grad} p = \mathbf{0}, \tag{1.13}$$

$$\frac{\mathscr{D}\widetilde{E}}{\mathscr{D}t} + \boldsymbol{v} \cdot \operatorname{grad} E_0 = 0, \quad \frac{\mathscr{D}\widetilde{T}}{\mathscr{D}t} + \boldsymbol{v} \cdot \operatorname{grad} T_0 = -\varkappa c_0^2 T_0 \operatorname{div} \boldsymbol{v}. \tag{1.14}$$

These are the basic equations for acoustic small-amplitude perturbations. We examine several special cases below.

1.1.3 Zero Ambient Velocity: Bergmann's Equation

When $\boldsymbol{U} = \boldsymbol{0}$, (1.9) implies that ρ_0, E_0 and T_0 do not depend on t, whereas p_0 is a constant. The constraint (1.10) permits us to have spatial variations in c_0^2 and ρ_0 within a stationary fluid (but not if E_0 is constant).

For the acoustic perturbation, (1.13) and (1.14) give

$$\frac{\partial \tilde{\rho}}{\partial t} + \operatorname{div}(\rho_0 \boldsymbol{v}) = 0, \quad \rho_0 \frac{\partial \boldsymbol{v}}{\partial t} + \operatorname{grad} p = \boldsymbol{0}, \quad \frac{\partial \widetilde{E}}{\partial t} + \boldsymbol{v} \cdot \operatorname{grad} E_0 = 0. \tag{1.15}$$

As $p = c_0^2 \tilde{\rho} + h_0 \widetilde{E}$ in which c_0^2 and h_0 do not depend on t, we can combine $(1.15)_1$ and $(1.15)_3$ to give

$$\frac{\partial p}{\partial t} + c_0^2 \operatorname{div}(\rho_0 \boldsymbol{v}) + h_0 \, \boldsymbol{v} \cdot \operatorname{grad} E_0 = 0. \tag{1.16}$$

Eliminating $h_0 \operatorname{grad} E_0$ using (1.10), we obtain

$$\frac{\partial p}{\partial t} + \rho_0 c_0^2 \operatorname{div} \boldsymbol{v} = 0. \tag{1.17}$$

Finally, eliminating $\boldsymbol{v}$, using the second of (1.15), gives

$$\rho_0 \operatorname{div}\left(\rho_0^{-1} \operatorname{grad} p\right) = \frac{1}{c_0^2} \frac{\partial^2 p}{\partial t^2}, \tag{1.18}$$

in which $\rho_0(\boldsymbol{r})$ and $c_0^2(\boldsymbol{r})$ can be functions of position $\boldsymbol{r} = (x, y, z)$ (but not of t). This is *Bergmann's equation* for the (excess) pressure [97, eqn (14)], [518, eqn (76.1)], [797, eqn (5.15)].

Suppose that the motion is known to be irrotational, meaning that the vorticity $\boldsymbol{\omega} = \operatorname{curl} \boldsymbol{v} = \boldsymbol{0}$. Then we can write $\boldsymbol{v} = \operatorname{grad} u$, where u is a velocity potential. (Note that some authors prefer to write $\boldsymbol{v} = -\operatorname{grad} u$; see, for example, [516, §285] and [644, §6.1].) It follows from $(1.15)_2$ that $p = -\rho_0(\boldsymbol{r}) \, \partial u / \partial t$ and then (1.17) yields

$$\nabla^2 u = \frac{1}{c_0^2(\boldsymbol{r})} \frac{\partial^2 u}{\partial t^2}. \tag{1.19}$$

1.1.4 Zero Ambient Velocity and Constant Ambient Density

When $\boldsymbol{U} = \boldsymbol{0}$ and ρ_0 is a constant, Bergmann's equation (1.18) reduces to

$$\nabla^2 p = \frac{1}{c_0^2(\boldsymbol{r})} \frac{\partial^2 p}{\partial t^2}. \tag{1.20}$$

As ρ_0 is constant, taking the curl of $(1.15)_2$ shows that the vorticity $\boldsymbol{\omega} = \operatorname{curl} \boldsymbol{v}$ does not depend on t. Therefore if the motion starts from a state in which $\boldsymbol{v}$ is constant, then $\boldsymbol{\omega} = \boldsymbol{0}$: the motion is irrotational, and we can write $\boldsymbol{v} = \operatorname{grad} u$. Then, as in Section 1.1.3, we have $p = -\rho_0 \, \partial u / \partial t$, whereas (1.20) shows that u satisfies the wave equation (1.19).

Note that irrotationality was *assumed* in Section 1.1.3 in order to obtain (1.19), whereas it can be *proved* when ρ_0 is constant.

Equation (1.19) often appears in the context of seismic inversion ('migration'); see, for example, [737], [797, eqn (5.9)]. It also appears in other imaging contexts [720, 521, 653], [326, eqn (2.1)]. Stochastic versions of (1.20), in which $c_0^2(\boldsymbol{r})$ is a random function of position, have also been studied and used; see, for example, [655], [786, eqn (3.17)], [326, eqn (12.1)] and [116].

1.1.5 Zero Ambient Velocity and Homogeneous Fluid

This is the textbook case, in which $\rho_0 \equiv \rho$ and $c_0 \equiv c$ are constants and $\boldsymbol{U} = \boldsymbol{0}$. In most of the book, we shall be concerned with this case.

The governing equations are the wave equation,

$$\nabla^2 u = \frac{1}{c^2}\frac{\partial^2 u}{\partial t^2}, \tag{1.21}$$

for the velocity potential u, together with $p = -\rho\,\partial u/\partial t$ and $\boldsymbol{v} = \operatorname{grad} u$. Evidently, p and any Cartesian component of $\boldsymbol{v}$ also solve the wave equation.

A simpler derivation of the governing equations can be given when the fluid is homogeneous, a derivation in which the entropy does not play a role. We take an equation of state which says that p_{ex} is a function of ρ_{ex}, $p_{\text{ex}} = p_{\text{ex}}(\rho_{\text{ex}})$. Let $\rho_{\text{ex}} = \rho$ and $p_{\text{ex}} = p_0$ when there is no motion, $\boldsymbol{v}_{\text{ex}} = \boldsymbol{0}$. Then (1.1) and (1.2) imply that p_0 is a constant and ρ does not depend on t. Then, in the notation of Section 1.1.2, we find that

$$p_1 = c^2\rho_1, \quad \text{where} \quad c^2 = p'_{\text{ex}}(\rho) \tag{1.22}$$

is the (constant) speed of sound. Also, from (1.1) and (1.2), we obtain

$$\frac{\partial \rho_1}{\partial t} + \rho \operatorname{div} \boldsymbol{v}_1 = 0 \quad \text{and} \quad \rho\frac{\partial \boldsymbol{v}_1}{\partial t} + \operatorname{grad} p_1 = \boldsymbol{0}. \tag{1.23}$$

Eliminating $\boldsymbol{v}_1$ gives

$$\nabla^2 p_1 = \frac{\partial^2 \rho_1}{\partial t^2} = \frac{1}{c^2}\frac{\partial^2 p_1}{\partial t^2}. \tag{1.24}$$

The rest of the derivation, leading to (1.21), proceeds as before. For more details on the derivation of the equations above, see, for example, [644, Chapter 6], [559, Chapter 1], [518, §64] or [696, Chapter 1].

1.1.6 Non-Zero Ambient Velocity and Homogeneous Fluid

In this case, the governing equations are (1.13) and (1.14), in which $\rho_0 \equiv \rho$, $c_0 \equiv c$, h_0 and E_0 are constants:

$$p = c^2\tilde{\rho} + h_0\widetilde{E}, \quad \frac{\mathscr{D}\tilde{\rho}}{\mathscr{D}t} + \rho \operatorname{div} \boldsymbol{v} = 0, \quad \frac{\mathscr{D}\widetilde{E}}{\mathscr{D}t} = 0, \quad \rho\frac{\mathscr{D}\boldsymbol{v}}{\mathscr{D}t} + \operatorname{grad} p = \boldsymbol{0}. \tag{1.25}$$

The first three of these give

$$\frac{\mathscr{D}p}{\mathscr{D}t} = -\rho c^2 \operatorname{div} \boldsymbol{v}$$

from which we can eliminate $\boldsymbol{v}$ using $(1.25)_4$ to obtain

$$\nabla^2 p = \frac{1}{c^2}\frac{\mathscr{D}^2 p}{\mathscr{D}t^2} = \frac{1}{c^2}\left(\frac{\partial}{\partial t} + \boldsymbol{U}\cdot\text{grad}\right)^2 p. \tag{1.26}$$

This is the *convected wave equation* [697, eqn (6)], [421, eqn (1.6.30)]. If the flow is irrotational, with $\boldsymbol{v} = \text{grad}\, u$, we find that the potential u also satisfies (1.26) with $p = -\rho(\partial u/\partial t + \boldsymbol{U}\cdot\text{grad}\, u)$. Applications of (1.26) will be mentioned in Section 7.5.

As the fluid is homogeneous and $\boldsymbol{U}$ is a constant vector, we can also obtain (1.26) using a Galilean transformation. Thus, if $(\boldsymbol{r},t)$ is a fixed frame in which the ambient velocity is $\boldsymbol{U}$, introduce a translating frame $(\boldsymbol{r}',t')$ with $\boldsymbol{r} = \boldsymbol{r}' + \boldsymbol{U}t'$ and $t = t'$. The chain rule gives, for example, $\partial u/\partial x' = \partial u/\partial x$, $\partial^2 u/\partial x'^2 = \partial^2 u/\partial x^2$ and $\partial u/\partial t' = \partial u/\partial t + \boldsymbol{U}\cdot\text{grad}\, u$. Hence, if u satisfies the wave equation with $\boldsymbol{r}' = (x',y',z')$ and t' as independent variables, then u satisfies the convected wave equation with independent variables $\boldsymbol{r} = (x,y,z)$ and t.

Equation (1.26) was used by Tatarski [814, eqn (5.1)] with $\boldsymbol{U}$ replaced by $\boldsymbol{U}(\boldsymbol{r})$, the local ambient velocity at position $\boldsymbol{r}$; see also [697, eqn (4)]. There are other versions of the convected wave equation that are intended for inhomogeneous fluids with a non-uniform ambient flow; see [697, 675, 146] and Section 1.1.7.

1.1.7 Non-Uniform Ambient Flows and Dynamic Materials

Sound transmission through a fluctuating ocean [302] exemplifies a problem in which the ambient flow is non-uniform in both space and time. We have already seen examples in which the background medium varies spatially (Section 1.1.3) but not in time. Media with temporal variations may be called *dynamic materials*. Such materials arise naturally (the oceans and the atmosphere are obvious examples) but there is also a growing interest in their creation. For some background and many applications, see [723, 582].

Perhaps the simplest model of dynamic materials is obtained by allowing c to be a function of time, giving [157]

$$\nabla^2 w = \frac{1}{c^2(t)}\frac{\partial^2 w}{\partial t^2}. \tag{1.27}$$

More generally, models of the form $\text{div}\{a(\boldsymbol{r},t)\,\text{grad}\, w\} = \partial^2 w/\partial t^2$ have been used [585]. In such models, including (1.27), no physical meaning is attributed to w. For some related one-dimensional studies, see [292, 755, 2, 891].

Pierce's Equation

For acoustic problems, Pierce [697] has derived a Bergmann-like wave equation, under certain assumptions about the dynamic medium: he assumes that it is 'slowly varying with position over distances comparable to a representative acoustic wavelength and that it is slowly varying with time over times comparable to a representative acoustic period' [697, p. 2293]. A stochastic version of Pierce's equation has been used recently [117].

For zero ambient velocity ($\boldsymbol{U} = \boldsymbol{0}$), Pierce's equation [697, eqn (23)] reduces to

$$\frac{1}{\rho_0(\boldsymbol{r})} \operatorname{div}\{\rho_0(\boldsymbol{r}) \operatorname{grad} u\} = \frac{\partial}{\partial t}\left(\frac{1}{c_0^2(\boldsymbol{r},t)} \frac{\partial u}{\partial t}\right), \tag{1.28}$$

where $u(\boldsymbol{r},t)$ is a velocity potential: $\boldsymbol{v} = \operatorname{grad} u$ and $p = -\rho_0\, \partial u/\partial t$. Equation (1.28) is W3 in the collection compiled by Campos [146]. Flatté [302, eqn (5.1.11) with eqn (6.1.1)] uses another equation for u,

$$\nabla^2 u = \frac{1}{c_0^2(\boldsymbol{r},t)} \frac{\partial^2 u}{\partial t^2}, \tag{1.29}$$

which reduces to Bergmann's equation (1.19) when c_0^2 does not depend on t. It has been remarked that the 'apparent simplicity of linearity [in (1.29)] is superseded by the complexity brought in by space-time inhomogeneity and [is] pregnant of exotic wave-like effects' [723, p. 928].

Note that we have written $\rho_0(\boldsymbol{r})$ in (1.28), not $\rho_0(\boldsymbol{r},t)$. This is because we showed in Section 1.1.3 that conservation of mass combined with $\boldsymbol{U} = \boldsymbol{0}$ implies that ρ_0 cannot depend on t. In other words, if we want to have $\rho_0(\boldsymbol{r},t)$, then we must have a moving ambient flow or we must abandon conservation of mass.

Note also that if c_0^2 does not depend on t, then (1.28) does not reduce to Bergmann's equation (1.19). Pierce [697, eqn (30)] attributes the discrepancy to a second-order effect that may be discarded.

In [39], the authors model a dynamic material by modifying Bergmann's equation (1.18), which we write as

$$\operatorname{div}\left(\frac{1}{\rho_0(\boldsymbol{r})} \operatorname{grad} p\right) = \kappa_0(\boldsymbol{r}) \frac{\partial^2 p}{\partial t^2}, \tag{1.30}$$

in which $\kappa_0 = (\rho_0 c_0^2)^{-1}$ is the (adiabatic) *compressibility* [696, p. 30]. In [39, eqn (4)], (1.30) is used but with $\rho_0(\boldsymbol{r},t)$ in place of $\rho_0(\boldsymbol{r})$. We have seen that such an equation is inconsistent with conservation of mass. This provides one motivation for relaxing the constraint of mass conservation. Another comes from continuum models of growing materials [349, Part IV], [260].

Exponential Growth

Let us abandon conservation of mass, replacing (1.1) by

$$\frac{\mathrm{D}\rho_{\mathrm{ex}}}{\mathrm{D}t} + \rho_{\mathrm{ex}} \operatorname{div} \boldsymbol{v}_{\mathrm{ex}} = \rho_{\mathrm{ex}} \gamma, \tag{1.31}$$

where $\gamma(\boldsymbol{r})$ is a given function of position, the *growth rate function*; for this model, see [349, eqn (13.5)]. We retain the other governing equations, namely (1.2), (1.3) and (1.4).

Linearising about an ambient state in which $\boldsymbol{U} = \boldsymbol{0}$ (and ignoring any temperature dependence), we find that p_0 is constant, E_0 does not depend on t and

$$\frac{\partial \rho_0}{\partial t} = \rho_0 \gamma \quad \text{whence} \quad \rho_0(\boldsymbol{r},t) = \rho_{00}(\boldsymbol{r})\, \mathrm{e}^{t\gamma(\boldsymbol{r})}, \tag{1.32}$$

where $\rho_{00}(\boldsymbol{r}) = \rho_0(\boldsymbol{r},0)$. As (1.10) also holds, we substitute ρ_0 and obtain

$$\mu\, \mathrm{e}^{\gamma t}\, (\operatorname{grad}\rho_{00} + t\rho_{00}\operatorname{grad}\gamma) + \operatorname{grad}E_0 = \mathbf{0}$$

where $\mu(\boldsymbol{r},t) = c_0^2/h_0$. To eliminate the term containing $t\rho_{00}$, we are forced to take $\operatorname{grad}\gamma(\boldsymbol{r}) = \mathbf{0}$: γ is a constant, γ_0, say. Then, we infer that $\mu\,\mathrm{e}^{\gamma_0 t}$ cannot depend on t, whence

$$\mu(\boldsymbol{r},t) = \mu_0(\boldsymbol{r})\,\mathrm{e}^{-\gamma_0 t} \tag{1.33}$$

and $\kappa_0^{-1} = \rho_0 c_0^2 = \rho_0\mu h_0 = \rho_{00}(\boldsymbol{r})\mu_0(\boldsymbol{r})h_{\mathrm{ex}}(\rho_0(\boldsymbol{r},t),E_0(\boldsymbol{r}))$, which depends on t, in general.

For the acoustic perturbation, we obtain a slightly modified form of (1.15):

$$\frac{\partial\tilde{\rho}}{\partial t} + \operatorname{div}(\rho_0\boldsymbol{v}) = \tilde{\rho}\gamma_0, \quad \rho_0\frac{\partial\boldsymbol{v}}{\partial t} + \operatorname{grad}p = \mathbf{0}, \tag{1.34}$$

$$p = c_0^2\tilde{\rho} + h_0\widetilde{E}, \quad \frac{\partial\widetilde{E}}{\partial t} + \boldsymbol{v}\cdot\operatorname{grad}E_0 = 0. \tag{1.35}$$

As E_0 does not depend on t, differentiating $(1.35)_2$ gives

$$\frac{\partial^2\widetilde{E}}{\partial t^2} = -\frac{\partial\boldsymbol{v}}{\partial t}\cdot\operatorname{grad}E_0 = \frac{1}{\rho_0}(\operatorname{grad}p)\cdot(\operatorname{grad}E_0)$$

after use of $(1.34)_2$. Eliminating $\widetilde{E}$ using $(1.35)_1$ and $\operatorname{grad}E_0$ using (1.10), we arrive at

$$\frac{h_0}{c_0^2}\frac{\partial^2}{\partial t^2}\left(\frac{p - c_0^2\tilde{\rho}}{h_0}\right) = -\frac{1}{\rho_0}(\operatorname{grad}p)\cdot(\operatorname{grad}\rho_0), \tag{1.36}$$

which is an equation relating p and $\tilde{\rho}$.

For a second equation, we start by integrating $(1.34)_2$. Let $\boldsymbol{g}(\boldsymbol{r},t) = \rho_0^{-1}\operatorname{grad}p$ and suppose that $\boldsymbol{v}(\boldsymbol{r},0) = \mathbf{0}$. Then

$$\boldsymbol{v}(\boldsymbol{r},t) = -\int_0^t \boldsymbol{g}(\boldsymbol{r},\tau)\,\mathrm{d}\tau. \tag{1.37}$$

We substitute this expression in $(1.34)_1$:

$$\frac{\partial\tilde{\rho}}{\partial t} - \tilde{\rho}\gamma_0 = \operatorname{div}\left(\rho_0(\boldsymbol{r},t)\int_0^t \boldsymbol{g}(\boldsymbol{r},\tau)\,\mathrm{d}\tau\right) = F(\boldsymbol{r},t), \tag{1.38}$$

say. Assuming that $\tilde{\rho}(\boldsymbol{r},0) = 0$, we can solve for $\tilde{\rho}$:

$$\tilde{\rho}(\boldsymbol{r},t) = \int_0^t \mathrm{e}^{\gamma_0(t-\tau')}F(\boldsymbol{r},\tau')\,\mathrm{d}\tau'. \tag{1.39}$$

Hence $\partial\tilde{\rho}/\partial t = F + \gamma_0\tilde{\rho}$ and $\partial^2\tilde{\rho}/\partial t^2 = \partial F/\partial t + \gamma_0 F + \gamma_0^2\tilde{\rho}$. Using these relations, we substitute (1.39) in (1.36), recalling that $\mu(\boldsymbol{r},t) = c_0^2/h_0$ is given by (1.33):

$$\frac{1}{\mu}\frac{\partial^2(\mu\tilde{\rho})}{\partial t^2} = \frac{\partial^2\tilde{\rho}}{\partial t^2} + \frac{2}{\mu}\frac{\partial\mu}{\partial t}\frac{\partial\tilde{\rho}}{\partial t} + \frac{\tilde{\rho}}{\mu}\frac{\partial^2\mu}{\partial t^2} = \frac{\partial F}{\partial t} - \gamma_0 F. \tag{1.40}$$

Next, let us evaluate F, defined by (1.38). Making use of $(1.32)_2$,

$$F(\boldsymbol{r},t) = \operatorname{div} \int_0^t e^{\gamma_0(t-\tau)} \operatorname{grad} p(\boldsymbol{r},\tau)\, d\tau = \int_0^t e^{\gamma_0(t-\tau)} \nabla^2 p(\boldsymbol{r},\tau)\, d\tau.$$

Hence

$$\frac{\partial F}{\partial t} = \nabla^2 p + \gamma_0 \int_0^t e^{\gamma_0(t-\tau)} \nabla^2 p\, d\tau = \nabla^2 p + \gamma_0 F. \tag{1.41}$$

Using (1.40) and (1.41), (1.36) becomes

$$\frac{h_0}{c_0^2}\frac{\partial^2}{\partial t^2}\left(\frac{p}{h_0}\right) = \frac{1}{\mu}\frac{\partial^2(\mu\tilde{\rho})}{\partial t^2} - \frac{1}{\rho_0}(\operatorname{grad} p)\cdot(\operatorname{grad}\rho_0) = \rho_0 \operatorname{div}\left(\frac{\operatorname{grad} p}{\rho_0}\right). \tag{1.42}$$

Evidently, this is a generalisation of Bergmann's equation (1.18).

More General Growth Models

We have seen that if we start from (1.31) with growth rate function $\gamma(\boldsymbol{r})$, then we are forced to take $\gamma = \gamma_0$, a constant, so that spatial variation of γ is lost. For a more general model, we could replace (1.31) with

$$\frac{D\rho_{ex}}{Dt} + \rho_{ex} \operatorname{div} \boldsymbol{v}_{ex} = \rho_{ex}\frac{\partial\eta}{\partial t}, \tag{1.43}$$

where $\eta(\boldsymbol{r},t)$ is specified. Proceeding as with the model (1.31), it turns out that $p(\boldsymbol{r},t)$ satisfies a complicated integrodifferential equation; see [604] for details.

Further growth models could be developed. Notice that the model (1.43) is simple (and linear in ρ_{ex}), so there is plenty of scope for alternative models.

No Growth Model at All: Specify the Background Density

Instead of replacing conservation of mass by a growth model, such as (1.31) or (1.43), let us simply specify $\rho_0(\boldsymbol{r},t)$, assuming that this specification is contrived by some external means. This is a plausible approach if we wish to create dynamic materials. As before, we take $\boldsymbol{U} = \boldsymbol{0}$, and we find that p_0 is constant and $\partial E_0/\partial t = 0$. Then, from (1.6), we obtain

$$0 = \frac{\partial p_0}{\partial t} = c_0^2\frac{\partial \rho_0}{\partial t} + h_0\frac{\partial E_0}{\partial t},$$

which reduces to $\partial\rho_0/\partial t = 0$. In other words, if we want $\partial\rho_0/\partial t \neq 0$, then we must modify (1.3), $DE_{ex}/Dt = 0$. This could be done, perhaps by retaining temperature effects [57, eqn (3.6.3)], [696, eqn (1-4.6)]. However, as far as we know, this option has not been contemplated.

Final Comments

The discussion in this section is essentially exact, within the limits of perturbation theory. We have not introduced additional approximations, such as those arising from relevant time scales. For example, the time scale associated with acoustic disturbances is much shorter than those associated with biological growth [349, §13.1].

Nevertheless, we should keep in mind that technological progress may lead to dynamic materials that can change rapidly, thereby making material and acoustic time scales comparable.

1.1.8 Nonlinear Acoustics

The linear equations derived above are sufficient for most of what follows. However, occasionally, an exact formulation is needed. If we restrict to flows that are irrotational ($\operatorname{curl}\boldsymbol{v}_{\text{ex}} = \boldsymbol{0}$) and homentropic ($E_{\text{ex}}$ is constant), an exact equation for the exact velocity potential u_{ex} can be derived,

$$\mathscr{C}^2\nabla^2 u_{\text{ex}} - \frac{\partial^2 u_{\text{ex}}}{\partial t^2} = \frac{\partial v_{\text{ex}}^2}{\partial t} + \frac{1}{2}\boldsymbol{v}_{\text{ex}} \cdot \operatorname{grad} v_{\text{ex}}^2, \tag{1.44}$$

where $\boldsymbol{v}_{\text{ex}} = \operatorname{grad} u_{\text{ex}}$ and $v_{\text{ex}} = |\boldsymbol{v}_{\text{ex}}|$. The quantity $\mathscr{C}^2$ depends on the fluid and the flow. For a polytropic gas with ratio of specific heats γ, we have

$$\mathscr{C}^2 = c_{\text{ex}}^2 - (\gamma - 1)\left(\frac{\partial u_{\text{ex}}}{\partial t} + \frac{1}{2}v_{\text{ex}}^2\right), \tag{1.45}$$

where c_{ex}^2 is the usual speed of sound, (1.7). Substituting (1.45) and $\boldsymbol{v}_{\text{ex}} = \operatorname{grad} u_{\text{ex}}$ in (1.44) gives a complicated nonlinear partial differential equation for the velocity potential u_{ex}. For a derivation of (1.44) and (1.45), see [384, §3.2] or [161, §4.4].

Equations (1.44) and (1.45) were written down in a paper by Longhorn [569, §6]. An equation similar to (1.44) can be found in [108, eqn (1.85)]. Equations (1.44) and (1.45) provide a firm foundation for quantifying nonlinear effects arising from inviscid, irrotational, compressible flows generated by the motions of spherical objects, for example; see [569, 317, 544]. For direct numerical simulation of such flows, see [683]. For nonlinear acoustics generally, see [383, 257], for example.

1.2 Acoustic Scattering

We have seen (Section 1.1.5) that linear acoustics in a homogeneous inviscid compressible fluid is governed by the wave equation. In three dimensions, this equation is

$$\nabla^2 u \equiv \frac{\partial^2 u}{\partial x^2} + \frac{\partial^2 u}{\partial y^2} + \frac{\partial^2 u}{\partial z^2} = \frac{1}{c^2}\frac{\partial^2 u}{\partial t^2}, \tag{1.46}$$

where x, y and z are Cartesian coordinates, t is time and c is the (positive) constant speed of sound. We always consider u to be a velocity potential, so that the velocity and (excess) pressure in the fluid are given by

$$\boldsymbol{v} = \operatorname{grad} u \quad \text{and} \quad p = -\rho\frac{\partial u}{\partial t}, \tag{1.47}$$

respectively, where ρ is the constant ambient density of the fluid. Solutions of (1.46) are called *wavefunctions*.

We also write the wave equation concisely as $\Box^2 u = 0$, where

$$\Box^2 u = \nabla^2 u - \frac{1}{c^2}\frac{\partial^2 u}{\partial t^2} \tag{1.48}$$

defines the *d'Alembertian* (or *Dalembertian*) of u. The notation and terminology varies. In his 1901 book, Poincaré [701, p. 456] writes 'nous allons introduire le symbole suivant' $\Box u$ for our $\Box^2 u$. Bateman [62, §4] writes that Dalembertian 'is the name suggested by Lorentz [571, p. 17]. Many writers use Cauchy's symbol $\Box$ to denote the Dalembertian [whereas] Wilson and Lewis [890, eqn (58)] use the symbol $\Diamond^2 u$'. Bateman writes Ωu 'because the form suggests a wave'. Our definition of $\Box^2 u$ agrees with that used in [643, eqn (1.7.6)]. Barton [56, eqn (10.1.1)] uses $\Box^2$ for our $-\Box^2$, whereas John [446, p. 126] uses $\Box$ for our $-c^2\Box^2$.

One branch of scattering theory concerns time-harmonic wavefunctions. These are of the form $u(x,y,z,t) = \mathrm{Re}\{U(x,y,z)\,\mathrm{e}^{-\mathrm{i}\omega t}\}$, where U is complex valued and ω is the circular frequency: solutions are constructed in the *frequency domain*. From (1.46), U satisfies the Helmholtz equation,

$$\nabla^2 U + (\omega/c)^2 U = 0.$$

We are mainly interested in retaining the unspecified dependence on time: we work in the *time domain*. Of course, the two domains are connected via Fourier or Laplace transforms:

> *The connection between time-dependent and time-harmonic waves is often presented as a direct consequence of the use of a Fourier transform with respect to the time variable. But such a presentation is formal, for this Fourier transform hides a basic notion: the direction of the passing of time.* (Ben Amar and Hazard [89, p. 942])

We shall return to 'the direction of the passing of time' later.

In the time domain, it is usually most convenient if we can formulate an initial-boundary value problem, in which we specify one boundary condition and two initial conditions. To fix ideas, consider a bounded obstacle B with smooth boundary S. Denote the unbounded region exterior to S by B_e. The problem is to find a wavefunction $u(P,t)$, where $P(x,y,z)$ is a typical point in B_e, subject to a boundary condition when $P \in S$ and initial conditions at $t = 0$. All these will be discussed and defined in Chapter 4.

By way of comparison, in the frequency domain, we usually require that $U(P)$ satisfies a boundary condition when $P \in S$ and a 'condition at infinity' as P recedes away from B. The latter is usually taken as the Sommerfeld radiation condition; it ensures that waves generated or scattered by B travel away from B.

1.3 Waves on a String

It will be helpful to recall some properties of waves on strings. This is the prototypical context for understanding one-dimensional wave motion. The string is along

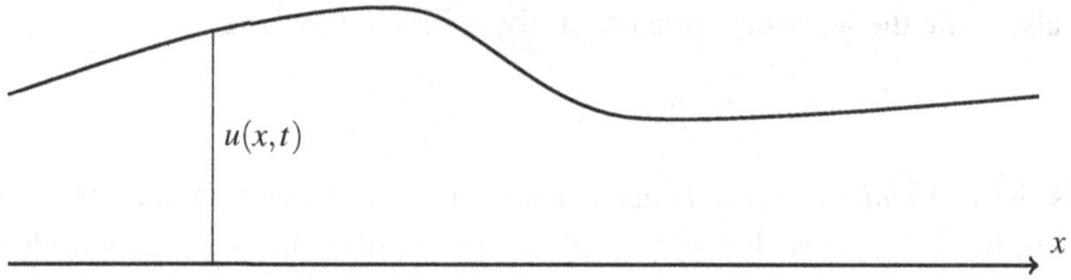

Figure 1.1 A snapshot of the displaced string at time t.

the x-axis, its density per unit length is ρ, and it is under uniform tension T. The displacement of the string from its equilibrium position is $u(x,t)$; see Fig. 1.1. Assuming the slope is small (so that the exact governing equation can be linearised) leads to the *one-dimensional wave equation* for u,

$$\frac{\partial^2 u}{\partial x^2} = \frac{1}{c^2}\frac{\partial^2 u}{\partial t^2}, \tag{1.49}$$

where $c = \sqrt{T/\rho}$ is a constant. For textbook derivations of (1.49), see, for example, [643, §2.1], [871, §12], [517, §22], [104] and [644, §4.1]. Antman [25] is critical of most textbook derivations and then shows that an 'honest' derivation can be given.

There are studies on the wave equation (1.49) in which c is replaced by $c(t)$ or $c(x,t)$; for example, see [292] and [755], respectively.

The general solution of (1.49) is

$$u(x,t) = f(x-ct) + g(x+ct), \tag{1.50}$$

where f and g are arbitrary (piecewise smooth) functions. This is *d'Alembert's solution*. (Actually, d'Alembert derived (1.49) for the vibrations of a stretched string and gave (1.50) with $c = 1$ in 1747, whereas (1.50) was given in 1749 by Euler; see [801, §V.16] and [834, 879].)

Notice that if $u(x,t)$ solves (1.49), then so do $u(-x,t)$ and $u(x,-t)$; the latter result is known as 'time-reversibility'. Time-reversibility has been exploited in many publications by Fink and his collaborators, especially in the context of inverse problems; see, for example, [297, 298, 52, 18].

Consider one piece of (1.50) (set $g \equiv 0$),

$$u(x,t) = f(x-ct) = u(x+c\tau, t+\tau).$$

This shows that there is a disturbance that propagates to the right (in the direction of x increasing), unchanged, with speed c, as t increases. Similarly, the term $g(x+ct)$ in (1.50) represents a disturbance propagating to the left.

1.3.1 An Initial-Value Problem

If we want to determine f and g in (1.50), we have to give more information. Perhaps the simplest problem arises when we specify two initial conditions,

$$u(x,0) = u_0(x), \quad (\partial u/\partial t)(x,0) = u_1(x), \quad -\infty < x < \infty, \tag{1.51}$$

where $u_0(x)$ and $u_1(x)$ are given functions. The problem is to find $u(x,t)$ for $t > 0$: we have an initial-value problem known as the *Cauchy problem*. Applying the initial conditions (1.51) to (1.50) gives

$$f(x) + g(x) = u_0(x), \quad -cf'(x) + cg'(x) = u_1(x). \tag{1.52}$$

Note that these two equations do not determine f and g uniquely: adding an arbitrary constant to f and subtracting the same constant from g does not violate (1.52).

Eliminating g from (1.52) gives $2cf' = cu_0' - u_1$ and then integration gives

$$f(x) = \frac{1}{2}u_0(x) - \frac{1}{2c}\int_{x_0}^{x} u_1(\xi)\,\mathrm{d}\xi,$$

where x_0 is an arbitrary constant (introduced so as to take account of the constant of integration). As $g = u_0 - f$,

$$g(x) = \frac{1}{2}u_0(x) + \frac{1}{2c}\int_{x_0}^{x} u_1(\xi)\,\mathrm{d}\xi.$$

Hence, from (1.50),

$$u(x,t) = \frac{1}{2}\{u_0(x-ct) + u_0(x+ct)\} + \frac{1}{2c}\int_{x-ct}^{x+ct} u_1(\xi)\,\mathrm{d}\xi, \quad t > 0; \tag{1.53}$$

the dependence on x_0 has cancelled. This is a classical result 'attributable to d'Alembert (1747)' [546, eqn (1.19)] and often referred to as 'd'Alembert's formula' [322, eqn (4.17)], [248, eqn (2.2.8)], [247, p. 29], [105, eqn (4.1.16)], [56, eqn (12.1.5)], [718, eqn (2.33)], [275, §2.4.1, eqn (8)]. In fact, (1.53) was given later by Euler [274, p. 16]; see [879, eqn (4)] and [540, p. 340].

For the one-dimensional wave equation, we have the luxury of knowing the general solution, (1.50), and this can be used to solve a variety of initial-value and boundary-value problems for $u(x,t)$. We shall do this later, and there are more examples in books such as [40, Chapter 2], [85, §2.4], [718, §2.7] and [173, Chapter 3]; for another example, see [115]. But, once we move away from (1.49), we shall require other methods. For example, (1.53) can be derived by using a Fourier transform with respect to x [785, §16.2]. Also, as (1.53) solves an initial-value problem, it is natural to try using a Laplace transform with respect to t or, equivalently, a one-sided Fourier transform with respect to t; we shall return to these methods later.

1.3.2 A Bead on an Infinite String: Solution for All t

Suppose there is a bead of mass m located at $x = 0$. A wave is incident from the left; it is partially reflected and partially transmitted, and the bead has its own equation of

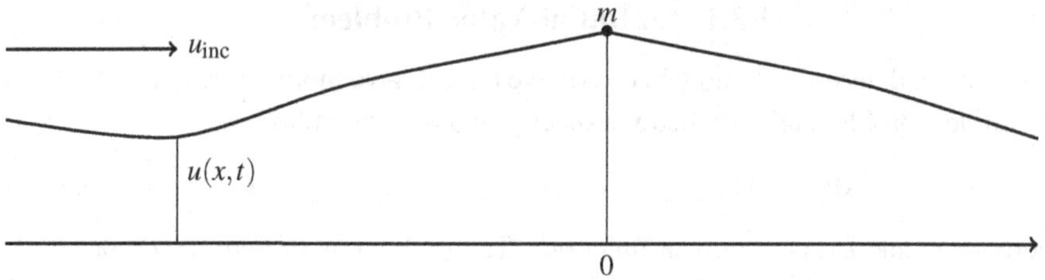

Figure 1.2 A snapshot of the bead on the displaced string at time t.

motion. See Fig. 1.2. For a bead of negligible width, $u(x,t)$ must satisfy the following conditions,

$$u(0+,t) = u(0-,t) \quad \text{and} \quad m\frac{\partial^2 u}{\partial t^2}(0,t) = T\frac{\partial u}{\partial x}(0+,t) - T\frac{\partial u}{\partial x}(0-,t). \tag{1.54}$$

Here we have indicated right-hand and left-hand limits by $f(0\pm) = \lim_{\varepsilon\to 0} f(\pm\varepsilon^2)$. The first of (1.54) imposes continuity of u at the bead (see Fig. 1.2) whereas the second is the equation of motion of the bead.

Problems of this type have a long history. Lamb [514] supposed that the bead is attached to springs, giving an additional term proportional to $u(0,t)$ on the left-hand side of (1.54). Damping terms, proportional to $(\partial u/\partial t)(0,t)$, can also be included.

Returning to (1.54), as these conditions hold at $x = 0$, it is more convenient when using (1.50) to write

$$u(x,t) = \begin{cases} u_{inc}(t - x/c) + g(t + x/c), & x < 0, \\ f(t - x/c), & x > 0. \end{cases} \tag{1.55}$$

Here, u_{inc} is the given incident wave, f is the unknown transmitted wave and g is the unknown reflected wave. Note that, for a given time t, we have to find $f(\tau)$ and $g(\tau)$ for all $\tau < t$.

Notice that, in (1.55), we did not include a function of $t + x/c$ when $x > 0$: Why not? A physical reason is that we want the bead to induce waves that propagate outwards, away from the bead: we can say that we seek a *causal solution.* (We shall return to causality in Section 1.5.) We also know that in the absence of the bead (put $m = 0$) $u(x,t) = u_{inc}(t - x/c)$ for all x and for all t; indeed, this is what we mean when we say that u_{inc} is the *incident* wave.

Application of (1.54) to (1.55) gives

$$u_{inc}(t) + g(t) = f(t) \quad \text{and} \quad mf''(t) = (T/c)\{-f'(t) + u'_{inc}(t) - g'(t)\}.$$

Eliminating g gives $mf'' = (2T/c)(u'_{inc} - f')$, a differential equation for f. The general solution is

$$f(t) = A + B\mathrm{e}^{-\alpha t} + \alpha\,\mathrm{e}^{-\alpha t}\int_{-\infty}^{t} \mathrm{e}^{\alpha\tau} u_{inc}(\tau)\,\mathrm{d}\tau \quad \text{with} \quad \alpha = \frac{2T}{mc}, \tag{1.56}$$

where A and B are arbitrary constants. An application of l'Hôpital's rule shows that the term containing the integral approaches $u_{\text{inc}}(t)$ as $t \to \pm\infty$. As we do not want exponential growth as $t \to -\infty$, we take $B = 0$. We can also take $A = 0$ without loss of generality: $u(x,t) = A$ is a wavefunction that satisfies (1.54) so that A cannot be determined without providing additional information (such as the position of the bead at some value of t). Finally, having determined f, $g = f - u_{\text{inc}}$.

As a check on (1.56) (with $A = B = 0$), let us examine what happens as $m \to 0$, that is, as $\alpha \to \infty$. The dominant contribution to the integral in (1.56) comes from near $\tau = t$. A standard argument (using Laplace's method [107, §5.1] or Watson's lemma [107, §4.1]) shows that $f(t) \to u_{\text{inc}}(t)$ as $\alpha \to \infty$, which is the desired result.

Variants of the bead problem have been discussed in [104, pp. 81–85], [546, §16] and [40, §2.6], but our treatment is different. We consider a general incident wave whereas, in the cited books, it is assumed that the incident wave has a front: $u_{\text{inc}}(t) = 0$ for $t < 0$ whence $u_{\text{inc}}(t - x/c) = 0$ for $x > ct$, implying that the wavefront reaches the bead at $t = 0$. Consequently, such problems can be reduced to an initial-value problem: we do this next.

1.3.3 A Bead on an Infinite String: Initial-Value Problem

For an incident wave with a front at $x = ct$, we can write

$$u(x,t) = u_{\text{inc}}(t - x/c) + v(x,t), \quad 0 < |x| < \infty, \quad t > 0,$$

with $v(x,0) = (\partial v/\partial t)(x,0) = 0$, $0 < |x| < \infty$. The symmetry of the problem implies that $v(x,t)$ is an even function of x. (Assuming that $v(x,t)$ is an odd function of x leads to the conclusion $v \equiv 0$.) Thus, we can write

$$v(x,t) = f(t - x/c) + h(t + x/c), \quad x > 0, \quad t > 0, \tag{1.57}$$

where the functions f and h are to be found. (We may anticipate that h can be omitted because it represents a wave travelling backwards, towards the bead.) Applying the initial conditions on $v(x,t)$ gives $f(-\xi) + h(\xi) = 0$ and $f'(-\xi) + h'(\xi) = 0$ for $\xi > 0$. Differentiating the first of these equations with respect to ξ gives $-f'(-\xi) + h'(\xi) = 0$, and then the second equation gives $f'(-\xi) = h'(\xi) = 0$. Hence

$$f(-\xi) = f_0, \quad h(\xi) = -f_0, \quad \xi > 0, \tag{1.58}$$

where f_0 is an arbitrary constant. Thus, from (1.57),

$$v(x,t) = 0, \quad x > ct, \quad t > 0:$$

ahead of the wavefront at $x = ct$, there is no disturbance. To determine what is happening behind the wavefront ($0 < x < ct$, $t > 0$), we use the conditions at the bead (1.54). The first of these is satisfied automatically (because $v(x,t)$ is an even function of x) and the second gives

$$m\{u''_{\text{inc}}(t) + f''(t)\} = -(2T/c)f'(t), \quad t > 0,$$

where we have used $h'(t) = 0$ for $t > 0$. Solving this equation gives

$$f(t) = A + B\mathrm{e}^{-\alpha t} - \mathrm{e}^{-\alpha t}\int_0^t \mathrm{e}^{\alpha\tau} u'_{\mathrm{inc}}(\tau)\,\mathrm{d}\tau, \quad t > 0, \quad \text{with} \quad \alpha = \frac{2T}{mc}, \tag{1.59}$$

where A and B are arbitrary constants.

We have $f(0+) = A + B$ and $f'(0+) = -\alpha B - u'_{\mathrm{inc}}(0)$. From (1.58), we also have $f(0-) = f_0$ and $f'(0-) = 0$. If we insist on continuity of $f(t)$ and $f'(t)$ at $t = 0$, we obtain $B = -\alpha^{-1} u'_{\mathrm{inc}}(0)$ and $A = f_0 - B$. Hence

$$f(t) = f_0 + \frac{1}{\alpha}\left(1 - \mathrm{e}^{-\alpha t}\right) u'_{\mathrm{inc}}(0) - \mathrm{e}^{-\alpha t}\int_0^t \mathrm{e}^{\alpha\tau} u'_{\mathrm{inc}}(\tau)\,\mathrm{d}\tau \tag{1.60}$$

and $v(x,t) = f(t - x/c) - f_0$ for $0 < x < ct$, $t > 0$; as in Section 1.3.1, the arbitrary constant f_0 cancels. One can check that imposing continuity conditions on $f(t)$ at $t = 0$ implies that $v(x,t)$ will be continuously differentiable across the wavefront at $x = ct$. One can also check that $v \to 0$ as $\alpha \to \infty$, which is the correct zero-mass limit.

1.3.4 A Simple Initial-Boundary Value Problem

The complications in the bead problem stem from the conditions at the bead, (1.54). Let us consider a simpler problem, one that is actually closer to the scattering problems encountered later. Thus let us seek a wavefunction $u(x,t)$ for $x > 0$ and $t > 0$, with

$$u(0,t) = d(t) \text{ for } t > 0, \quad u(x,0) = u_0(x) \text{ and } (\partial u/\partial t)(x,0) = 0 \text{ for } x > 0, \tag{1.61}$$

where d and u_0 are specified functions; this problem is posed in one of Hadamard's books [374, §24]. When the conditions (1.61) are imposed on (1.50), we obtain

$$u(x,t) = \begin{cases} \dfrac{1}{2}\{u_0(x+ct) + u_0(x-ct)\}, & 0 < ct < x, \\ d(t - x/c) + \dfrac{1}{2}\{u_0(x+ct) - u_0(ct-x)\}, & 0 < x < ct. \end{cases} \tag{1.62}$$

We notice several things about this result. First, we did not impose any condition on $u(x,t)$ as $x \to \infty$: we solved in a quadrant with two initial conditions and one boundary condition, see (1.61).

Second, we see that the effect of the boundary condition is not felt at location x until $t > x/c$: there is a wavefront at $x = ct$ moving away from $x = 0$ at speed c. As the wavefront passes, $u(x,t)$ jumps by an amount $d(0) - u_0(0)$; if we want $u(x,t)$ to be continuous, we must enforce a compatibility condition on the data, $d(0) = u_0(0)$. Furthermore, if we want u to be twice differentiable, then we must impose $d'(0) = 0$ and $d''(0) = c^2 u_0''(0)$; these compatibility conditions are discussed in [445, pp. 7–9] and in [734, pp. 11–12], where (1.62) and the generalisation to non-zero $(\partial u/\partial t)(x,0)$ can be found. For more on compatibility conditions, see [546, §15], [204, §2.6] and Section 4.4.

The special case with a zero boundary condition ($d \equiv 0$) can be treated by reflection in the boundary at $x = 0$; see, for example, [275, p. 69], [709, §1.6] and [204, §2.6].

For another special case, consider zero initial conditions ($u_0 \equiv 0$), whence

$$u(x,t) = \begin{cases} 0, & 0 < ct < x, \\ d(t - x/c), & 0 < x < ct. \end{cases} \tag{1.63}$$

This elementary solution can be found in [546, eqn (2.4)], for example. It is representative of a scattering problem, where u is the scattered wave and d is determined by the incident wave. We see that $u(x,t) = 0$ ahead of the wavefront. This, and the sole occurrence of the function of $x - ct$ (through $d(t - x/c)$), is sometimes said to define a causal solution: the wavefront propagates *away* from the boundary at $x = 0$, reaching a distance $x = ct$ after time t. But notice that this causal nature was deduced, not enforced.

For an alternative derivation of (1.62), using a Laplace transform of (1.49) with respect to t, see [869, example E4.2.1]. A similar method is used to derive (1.63) in [546, §10] and in [216, §4.2].

1.3.5 A Bead on an Infinite String: Use of Laplace Transforms

Let us solve the problem for $v(x,t)$ formulated in Section 1.3.3, using Laplace transforms. (See Section 1.4 for a brief summary of basic properties.) Thus define

$$V(x,s) = \mathscr{L}\{v(x,t)\} = \int_0^\infty v(x,t)\,\mathrm{e}^{-st}\,\mathrm{d}t. \tag{1.64}$$

Proceeding formally, we Laplace-transform the wave equation, making use of the zero initial conditions, and obtain

$$\frac{\partial^2 V}{\partial x^2} - \frac{s^2}{c^2}V = 0.$$

Discarding a solution that grows exponentially with x, we obtain

$$V(x,s) = F(s)\,\mathrm{e}^{-sx/c} = \mathscr{L}\{f(t - x/c)\,\mathrm{H}(t - x/c)\},$$

where $\mathrm{H}(t)$ is the Heaviside unit function ($\mathrm{H}(t) = 1$ for $t > 0$ and $\mathrm{H}(t) = 0$ for $t < 0$) and $F(s) = \mathscr{L}\{f\}$. To find f, we use the bead's equation of motion,

$$m\left\{u''_{\mathrm{inc}}(t) + \frac{\partial^2 v}{\partial t^2}(0,t)\right\} = 2T\,\frac{\partial v}{\partial x}(0,t), \quad t > 0.$$

The Laplace transform of this equation gives

$$\mathscr{L}\{u''_{\mathrm{inc}}\} + s^2 F(s) = -\alpha s F(s),$$

where we have used $V(x,s) = F(s)\,\mathrm{e}^{-sx/c}$ and α is defined by (1.59). Hence

$$F(s) = -\frac{\mathscr{L}\{u''_{\mathrm{inc}}\}}{s(s+\alpha)} = \frac{1}{\alpha}\mathscr{L}\{\mathrm{e}^{-\alpha t} - 1\}\mathscr{L}\{u''_{\mathrm{inc}}\},$$

a Laplace convolution (see (1.76)). Inverting

$$f(t) = \frac{1}{\alpha} \int_0^t \left(e^{-\alpha(t-\tau)} - 1 \right) u''_{\rm inc}(\tau)\, d\tau.$$

An integration by parts reduces this expression to (1.60) (apart from f_0 in (1.60), which was removed in Section 1.3.3 by the addition of $h = -f_0$).

The solution given here can be found in [148, §67]. As noted below (1.64), the process used is formal because it assumes that v is sufficiently smooth: in particular, it assumes that $v(x,t)$ is smooth across the wavefront at $x = ct$; this is why we obtained agreement with the solution obtained in Section 1.3.3 after continuity conditions had been imposed there so as to obtain (1.60) from (1.59).

1.3.6 Discussion and Summary

In the calculations above, we have mentioned 'causal solutions'. It is useful to review what we did. In Section 1.3.1, we solved an initial-value problem for $u(x,t)$, $t > 0$, starting from d'Alembert's general solution of the wave equation. Doing this, we implicitly specified a direction of time, forward from $t = 0$, thus breaking the time-reversibility of the wave equation. This remark also applies to the initial-value problems solved in Sections 1.3.3–1.3.5. On the other hand, for the unrestricted bead problem solved in Section 1.3.2, we had to specify the direction of time by requiring that the bead causes waves that propagate away from the bead as time advances.

Extensions to problems involving N beads on a string, or composite strings with N sections joined together (equivalently, waves in layered media), can be solved by patching together appropriate solutions of d'Alembert type (1.50). This is simple in principle, but it becomes complicated as N increases. For details and applications, see, for example, [646, 149, 337] and [303, §3.5].

1.3.7 Damping, Dissipation, Absorption and Losses

Ideal models, such as those based on the wave equation (1.46), do not account for energy dissipation. In real materials, wave energy can be converted into other forms, so we may wish to build models that account for this conversion. Historically, the earliest lossy wave equation is due to Stokes [796, p. 302],

$$\frac{1}{c^2}\frac{\partial^2 u}{\partial t^2} = \frac{\partial^2 u}{\partial x^2} + \frac{4\mu}{3\rho c^2}\frac{\partial^3 u}{\partial x^2 \partial t}, \tag{1.65}$$

intended to model viscous losses; μ is the viscosity coefficient. See, for example, [711, §346], [561, eqn (31)] and [559, p. 79, eqn (205)]. In three dimensions, (1.65) becomes

$$\frac{1}{c^2}\frac{\partial^2 u}{\partial t^2} = \nabla^2 u + \tau \frac{\partial}{\partial t}\nabla^2 u, \tag{1.66}$$

where τ is a time constant [417, eqn (2.36)]. If it can be argued that the dominant effects are well represented by the basic wave equation (1.46), then this equation may be used to eliminate $\nabla^2 u$ from the last term in (1.66), resulting in

$$\frac{1}{c^2}\frac{\partial^2 u}{\partial t^2} = \nabla^2 u + \frac{\tau}{c^2}\frac{\partial^3 u}{\partial t^3}. \tag{1.67}$$

The idea leading to (1.67) is attributed to Blackstock [103]; see [417, eqn (2.48)]. Combinations of (1.66) and (1.67) have also been investigated [417, eqn (2.52)].

For a model without third derivatives, we have the *telegrapher's equation*,

$$\frac{1}{c^2}\frac{\partial^2 u}{\partial t^2} = \frac{\partial^2 u}{\partial x^2} - \frac{\alpha+\beta}{c^2}\frac{\partial u}{\partial t} - \frac{\alpha\beta}{c^2}u = 0, \tag{1.68}$$

where α and β are constants [203, p. 192, eqn (19)]; see also [67, p. 74] and [799, p. 550, eqn (27)].

The case $\alpha = \beta$ is special; its general solution is [203, p. 193]

$$u(x,t) = \mathrm{e}^{-\alpha t}\left\{f(x-ct) + g(x+ct)\right\},$$

where f and g are arbitrary functions. For an application, see [448, eqn (15)].

When $\beta = 0$, (1.68) reduces to

$$\frac{1}{c^2}\frac{\partial^2 u}{\partial t^2} = \frac{\partial^2 u}{\partial x^2} - \gamma\frac{\partial u}{\partial t} = 0, \tag{1.69}$$

where γ is a constant. This is usually known as the *damped wave equation*. It appears in a paper from 1876 by Heaviside [397, eqn (3)] on the propagation of signals along cables, a problem of telegraphy. For some discussion of (1.69), see [643, pp. 865–869], [644, p. 127], [197, §4.10] and [275, §4.3, example 5].

One feature of the models above is that they are *dispersive*. Thus, for one-dimensional equations such as (1.65) and (1.69), if we seek solutions in the form $u(x,t) = \mathrm{e}^{\mathrm{i}(kx-\omega t)}$, we obtain an equation relating ω and k, the *dispersion relation*. This shows that, in general, the phase speed $c_{\mathrm{p}} = \omega/k$ depends on the wavelength $2\pi/k$, as does the group speed $c_{\mathrm{g}} = \mathrm{d}\omega/\mathrm{d}k$, with $c_{\mathrm{p}} \neq c_{\mathrm{g}} \neq c$.

Another feature of the models above is that they predict attenuation: wave motions die out. However, it turns out that the predictions do not always match experimental data. Consequently, there have been efforts to employ modified models involving fractional time derivatives or fractional Laplacians [565]. See, for example, [830, 418, 417]. The guiding principle is to build models that obey conservation of mass and momentum (see Section 1.1) together with empirical constitutive relations; fractional derivatives may be introduced into the constitutive relations. For more details and references, see Holm's book [417]. At present, these fractional models have not been used in the context of scattering problems, but that may change in the future.

1.3.8 The Forced Wave Equation

The wave equation is homogeneous. The inhomogeneous (or forced) version is also of interest. Thus consider solving

$$\frac{\partial^2 u}{\partial x^2} - \frac{1}{c^2}\frac{\partial^2 u}{\partial t^2} = -q(x,t) \tag{1.70}$$

where q is given. Linearity implies that it is sufficient to solve (1.70) for $t > 0$ with zero initial conditions; non-zero initial conditions can be accommodated using (1.53). We claim that a solution of this problem is given by

$$u(x,t) = \frac{c}{2}\int_0^t Q(x,t;\tau)\,\mathrm{d}\tau \tag{1.71}$$

for $t > 0$ with

$$Q(x,t;\tau) = \int_{x_-}^{x_+} q(\xi,\tau)\,\mathrm{d}\xi \quad \text{and} \quad x_\pm(x,t;\tau) = x \pm c(t-\tau).$$

Let us verify (1.71). We have

$$\frac{\partial Q}{\partial x}(x,t;\tau) = q(x_+,\tau) - q(x_-,\tau), \quad \frac{\partial^2 Q}{\partial x^2}(x,t;\tau) = q_x(x_+,\tau) - q_x(x_-,\tau),$$

where $q_x(x,\tau) \equiv (\partial q/\partial x)(x,\tau)$. Also

$$\frac{\partial Q}{\partial t}(x,t;\tau) = cq(x_+,\tau) + cq(x_-,\tau), \quad \frac{\partial^2 Q}{\partial t^2}(x,t;\tau) = c^2 q_x(x_+,\tau) - c^2 q_x(x_-,\tau),$$

whence $Q(x,t;\tau)$ is seen to solve the (homogeneous) wave equation (1.49). Next, we differentiate (1.71) giving

$$\begin{aligned}
\frac{\partial^2 u}{\partial x^2} &= \frac{c}{2}\int_0^t \frac{\partial^2 Q}{\partial x^2}(x,t;\tau)\,\mathrm{d}\tau,\\
\frac{\partial u}{\partial t} &= \frac{c}{2}Q(x,t;t) + \frac{c}{2}\int_0^t \frac{\partial Q}{\partial t}(x,t;\tau)\,\mathrm{d}\tau,\\
\frac{\partial^2 u}{\partial t^2} &= \frac{c}{2}\frac{\partial Q}{\partial t}(x,t;t) + \frac{c}{2}\int_0^t \frac{\partial^2 Q}{\partial t^2}(x,t;\tau)\,\mathrm{d}\tau,
\end{aligned}$$

having noted that $Q(x,t;t) = 0$. As $(\partial Q/\partial t)(x,t;t) = 2cq(x,t)$, we find that $u(x,t)$ solves (1.70). We also confirm that $u(x,0) = 0$ and $(\partial u/\partial t)(x,0) = 0$.

For derivations of (1.71), see, for example, [105, eqn (4.1.16)] or [204, §2.5]. The fact that solutions of the forced wave equation can be written in terms of solutions of the (unforced) wave equation (in this case, $Q(x,t;\tau)$) is an example of *Duhamel's principle*. For more examples, see [203, pp. 202–204], [446, Chapter 5, §1 (c)] and [275, §2.4.2]. Kingston [479] discusses how to choose q so that waves travel in one direction only. For the three-dimensional forced wave equation, see Section 4.7.

1.4 Laplace Transforms: Formal Properties

Having just used Laplace transforms in Section 1.3.5, it is appropriate to recall some basic properties. For more information, see, for example, [216, 869] and Chapter 5.

Suppose that $f(t)$ is a smooth function for $t > 0$, and is such that

$$F(s) = \mathscr{L}\{f(t)\} = \int_0^\infty f(t)\,\mathrm{e}^{-st}\,\mathrm{d}t \tag{1.72}$$

exists for $\mathrm{Re}(s) > A$, where s is the (complex) transform variable and A is a positive constant. Typically, $F(s)$ is an analytic function of s in the half-plane $\mathrm{Re}(s) > A$. We also assume that

$$f(t) = 0 \quad \text{for } t < 0; \tag{1.73}$$

evidently, this is not required for the definition (1.72) but it will be natural in our applications later.

Integration by parts gives

$$\mathscr{L}\{f'\} = sF(s) - f(0+), \quad \mathscr{L}\{f''\} = s^2F(s) - sf(0+) - f'(0+),$$

provided f' and f'' have Laplace transforms.

Functions with *retarded arguments*, such as $f(t - x/c)$, arise frequently; we have

$$\mathscr{L}\{f(t - x/c)\} = \int_{x/c}^\infty f(t - x/c)\,\mathrm{e}^{-st}\,\mathrm{d}t = \mathrm{e}^{-sx/c}F(s), \tag{1.74}$$

where $x \geq 0$ and we have used (1.73). We also have

$$\mathscr{L}\{f(\alpha t)\} = \alpha^{-1}F(s/\alpha), \quad \alpha > 0. \tag{1.75}$$

The *convolution theorem* is

$$\mathscr{L}\left\{\int_0^t f(\tau)g(t-\tau)\,\mathrm{d}\tau\right\} = F(s)G(s). \tag{1.76}$$

In particular, as $\mathscr{L}\{1\} = s^{-1}$ and $\mathscr{L}\{t\} = s^{-2}$, we have

$$\mathscr{L}\left\{\int_0^t f(\tau)\,\mathrm{d}\tau\right\} = \frac{F(s)}{s}, \quad \mathscr{L}\left\{\int_0^t \tau f(\tau)\,\mathrm{d}\tau\right\} = \frac{F(s)}{s^2}. \tag{1.77}$$

Laplace transforms can be inverted,

$$f(t) = \frac{1}{2\pi\mathrm{i}}\int_{\mathrm{Br}} F(s)\,\mathrm{e}^{st}\,\mathrm{d}s, \tag{1.78}$$

for $t > 0$, where Br is the Bromwich contour in the complex s-plane, defined by $s = C + \mathrm{i}\eta$, $-\infty < \eta < \infty$, with $C > A$.

Laplace transforms will arise in many places throughout the book. Further properties will be developed as needed.

1.5 Causality

'Causality' is a philosopher's plaything. You can spend 16 pages on definitions before getting to what it was you wanted to ask. (Schulman [749, p. 149])

The notion of 'causality' has exercised philosophers for centuries. For example, in 1912, Bertrand Russell famously wrote: 'The law of causality, I believe, like much that passes muster among philosophers, is a relic of a bygone age, surviving, like the monarchy, only because it is erroneously supposed to do no harm' [726, p. 180]. A century later, Norton could ask, 'Is there an independent principle of causality in physics?' [667]. In the context of scattering theory, the following question arises. Is causality something that has to be *enforced* when solving a problem, or is it something that we should *deduce* from the solution after it has been obtained?

In his book, Nussenzveig [670] gives several different causality conditions, starting with an intuitive statement [670, p. 5]:

C1. Primitive causality: the effect cannot precede the cause.

Toll [828] calls C1 'strict causality': 'no output can occur before the input'. He goes on to assert that, 'in the case of a scattering system, [C1] becomes' [828, p. 1760]

C2. No scattered wave can appear until the primary wave has reached some part of the scatterer.

Implicitly, then, Toll is considering a scattering problem with an incident pulse; Nussenzveig [670, p. 59] does the same: 'The outgoing wave cannot appear before the incoming wave has reached the scatterer.' In other words, the incident pulse generates an outgoing wave, thus specifying what Schulman calls the 'radiative arrow of time' [749, p. 81]:

[Earlier] *I described the radiative arrow as 'the fact that one uses outgoing wave boundary conditions for electromagnetic radiation, that retarded Green's functions should be used for ordinary calculations, that radiation reaction has a certain sign, that more radiation escapes to the cosmos than comes in'. ... It does not take a cynic to realize four vague definitions are used because there isn't a single firm one.* (Schulman [749, p. 91])

Fortunately, perhaps, we shall be interested in making 'ordinary calculations'.

Jones [454, p. 537] characterises causality as follows:

C3. Causality insists that there shall be no disturbance until the source is switched on. Thereafter the field propagates behind wavefronts which travel with speeds and in directions characteristic of the medium. No energy flow can be detected by an observer before such a wavefront has passed over him.

The first sentence is primitive causality (C1) but the second introduces wave speeds and wavefronts. Kranyš [497, p. 306] connects these notions with the hyperbolic nature of the governing PDEs:

C4. The characteristic surfaces in hyperbolic theories always propagate at a finite velocity (implying the existence of wave fronts), which is in agreement with the classical causality principle. This principle says that the transport or propagation processes in nature are due to (causal) chains of interactions (cause–effect) between a source of perturbation (emission of signal) and the response (reception) and this can be realized (due to the inertia of the interacting bodies) only after a certain time delay (relaxation), so that the transport occurs at a finite velocity.

In the electromagnetic context, Nussenzveig [670, p. 5] gives the following condition:

C5. Relativistic (or macroscopic) causality: no signal can propagate with velocity greater than the speed of light in vacuum.

Here is another definition with a relativistic flavour [533, p. 335]:

C6. There exist events that are consequences of other events that occurred earlier in some inertial reference frame. For these events, the temporal order must be identical in all reference frames: the cause must precede the effect in all referential frames.

Later in his book, Nussenzveig [670, p. 141] gives yet another formulation:

C7. Causality condition: for an incident wave with a plane wavefront, the scattered wave cannot arrive at any point in space before the incident wave arrives there.

This condition can be violated in acoustic problems when a pulse is scattered by a penetrable obstacle through which the waves can travel faster than the waves outside the scatterer [787, 664].

Let us try to quantify some of the statements above [670, §1.3]. We consider a physical system for which an input $f(t)$ produces an output $u(t)$. Assume that u depends linearly on f so that we can write

$$u(t) = \int_{-\infty}^{\infty} \kappa(t,\tau) f(\tau)\, \mathrm{d}\tau \tag{1.79}$$

for some function κ (which is called the *impulse response function*). Next, assume that the system is time-invariant: delaying the input by t_0 delays the output by t_0. This implies that $\kappa(t,\tau) = k(t-\tau)$ where $k(t)$ is a function of one variable. Hence (1.79) becomes

$$u(t) = \int_{-\infty}^{\infty} k(t-\tau) f(\tau)\, \mathrm{d}\tau = \int_{-\infty}^{\infty} k(\tau) f(t-\tau)\, \mathrm{d}\tau. \tag{1.80}$$

Finally, assume that we have primitive causality (C1 above), meaning that the output $u(t)$ cannot precede the input $f(t)$, whence (1.80) reduces to

$$u(t) = \int_{-\infty}^{t} k(t-\tau) f(\tau)\, \mathrm{d}\tau = \int_{0}^{\infty} k(\tau) f(t-\tau)\, \mathrm{d}\tau. \tag{1.81}$$

Equivalently, primitive causality implies that $k(t) = 0$ for all $t < 0$.

In the discussion above (which is standard in books on linear systems theory, such as [461, Chapter 1]), all dependence on spatial variations has been suppressed.

Several of the other statements on causality (C2, C3, C4 and C7) assume incident waves with a wavefront; ahead of the front, there is no disturbance. Therefore, calculating the disturbance due to scattering reduces to solving an initial-value problem: we assume that

$$u(t) = 0 \quad \text{for all } t < 0, \tag{1.82}$$

and then the problem is to find $u(t)$ for $t > 0$. Equivalently, to achieve (1.82), we see that the input $f(t)$ must vanish for all $t < 0$, reducing (1.81) to

$$u(t) = \int_0^t k(t-\tau) f(\tau)\,\mathrm{d}\tau = \int_0^t k(\tau) f(t-\tau)\,\mathrm{d}\tau, \quad t > 0, \tag{1.83}$$

which we recognise as a Laplace convolution (1.76).

Functions satisfying (1.82) are often called *causal functions* [742, p. 1], *causal signals* [839, §14.3] or simply *causal* [858, 377]. We shall often seek causal solutions. Note that we invoked (1.82) as (1.73).

Returning to primitive causality, we can say that the 'use of *initial* conditions implies this form of causality, *by definition*' [749, p. 149]. For problems that do not have a specified (forward) direction of 'time's arrow', additional conditions will be needed; for a prototypical example, see Section 1.3.2.

For another example, in abstract scattering theory [635, Chapter 1, §4], [718, Chapter 6], time's arrow is implicit, going from $t = -\infty$ to $t = +\infty$. One compares two solutions, one solution coming from a problem without a scatterer (the incident field) and one when there is a scatterer. The comparisons are done far away as $t \to \pm\infty$.

1.6 Electromagnetics, Elastodynamics and Hydrodynamics

The main focus of this book is on acoustics and the scalar wave equation. However, we shall make a few references to related work on electromagnetic waves, elastodynamic waves and water waves (hydrodynamics). The relevant governing equations are collected in this section.

1.6.1 Electromagnetics

Electromagnetic waves are governed by Maxwell's equations [799, 451, 440, 506]. In their simplest form, they are

$$\operatorname{curl}\boldsymbol{E} + \mu\frac{\partial \boldsymbol{H}}{\partial t} = \boldsymbol{0}, \quad \operatorname{curl}\boldsymbol{H} - \varepsilon\frac{\partial \boldsymbol{E}}{\partial t} = \boldsymbol{0}, \quad \operatorname{div}\boldsymbol{E} = 0 \quad \text{and} \quad \operatorname{div}\boldsymbol{H} = 0, \tag{1.84}$$

where $\boldsymbol{E}(\boldsymbol{r},t)$ is the electric field and $\boldsymbol{H}(\boldsymbol{r},t)$ is the magnetic field. We have assumed that we have a homogeneous, isotropic, dielectric medium, so that the electric permittivity ε and the magnetic permeability μ are constants. Eliminating $\boldsymbol{H}$ from the first two of (1.84) gives

$$\operatorname{curl}\operatorname{curl}\boldsymbol{E} + \frac{1}{c^2}\frac{\partial^2 \boldsymbol{E}}{\partial t^2} = \boldsymbol{0} \quad \text{with} \quad c^2 = \frac{1}{\mu\varepsilon}. \tag{1.85}$$

As $\operatorname{div}\boldsymbol{E} = 0$, use of the vector identity

$$\operatorname{curl}\operatorname{curl}\boldsymbol{E} = \operatorname{grad}\operatorname{div}\boldsymbol{E} - \nabla^2\boldsymbol{E} \tag{1.86}$$

in (1.85) shows that each Cartesian component of $\boldsymbol{E}$ satisfies the scalar wave equation (1.46). Also, from the divergence of $(1.84)_2$, we deduce that $\operatorname{div}\boldsymbol{E}$ does not depend on t. Therefore, if $\boldsymbol{E}(\boldsymbol{r},t) = \boldsymbol{0}$ at one value of t (such as $t = 0$, as in certain initial-value problems), then $(1.84)_3$ is satisfied for all t. Exactly the same results can be deduced for $\boldsymbol{H}$. For integral representations of $\boldsymbol{E}$ and $\boldsymbol{H}$, see Section 10.5.

Equations (1.84) are Maxwell's equations as distilled by Heaviside. Maxwell himself used potential functions, a scalar potential $\varphi(\boldsymbol{r},t)$ and a vector potential $\boldsymbol{A}(\boldsymbol{r},t)$. They are related to $\boldsymbol{E}$ and $\boldsymbol{H}$ by

$$\mu\boldsymbol{H} = \operatorname{curl}\boldsymbol{A} \quad \text{and} \quad \boldsymbol{E} = -\frac{\partial \boldsymbol{A}}{\partial t} - \operatorname{grad}\varphi. \tag{1.87}$$

These ensure that $(1.84)_1$ and $(1.84)_4$ are satisfied automatically. The other two of (1.84) reduce to

$$\nabla^2\varphi + \frac{\partial}{\partial t}\operatorname{div}\boldsymbol{A} = 0 \quad \text{and} \quad \operatorname{curl}\operatorname{curl}\boldsymbol{A} + \frac{1}{c^2}\frac{\partial^2 \boldsymbol{A}}{\partial t^2} + \frac{1}{c^2}\operatorname{grad}\frac{\partial\varphi}{\partial t} = \boldsymbol{0}. \tag{1.88}$$

As (1.87) introduces an extra degree of freedom, we need a constraint; this is usually taken as the *Lorenz gauge condition*, $c^2\operatorname{div}\boldsymbol{A} + \partial\varphi/\partial t = 0$. When used in (1.88), together with (1.86), we find that φ and each Cartesian component of $\boldsymbol{A}$ satisfies the wave equation (1.46). For more information on electromagnetic potentials, see [799, §1.9], [451, §1.7] and [440, §6.4]. For extensions to inhomogeneous and anisotropic media, see [660, 175].

1.6.2 Elastodynamics

Consider a homogeneous isotropic elastic solid. Waves in such a medium are governed by

$$\frac{\partial \tau_{ij}}{\partial x_j} = \rho\frac{\partial^2 u_i}{\partial t^2}, \quad i = 1,2,3, \tag{1.89}$$

where u_i are the components of the displacement vector $\boldsymbol{u}(\boldsymbol{r},t)$, $\boldsymbol{r} = (x_1,x_2,x_3)$, ρ is the density, the summation convention is used and we have ignored body forces. The stresses τ_{ij} are related to u_i by Hooke's law,

$$\tau_{ij} = \lambda\,\delta_{ij}\frac{\partial u_k}{\partial x_k} + \mu\left(\frac{\partial u_i}{\partial x_j} + \frac{\partial u_j}{\partial x_i}\right), \tag{1.90}$$

where λ and μ are the Lamé moduli and δ_{ij} is the Kronecker delta ($\delta_{ij} = 1$ when $i = j$ and $\delta_{ij} = 0$ when $i \neq j$). Eliminating the stresses from (1.89) gives

$$(\lambda + 2\mu)\,\mathrm{grad}\,\mathrm{div}\,\boldsymbol{u} - \mu\,\mathrm{curl}\,\mathrm{curl}\,\boldsymbol{u} = \rho\,\frac{\partial^2 \boldsymbol{u}}{\partial t^2}. \tag{1.91}$$

Taking the divergence of (1.91) shows that $\mathrm{div}\,\boldsymbol{u}$ satisfies the wave equation (1.46) with speed c_p, where $c_p^2 = (\lambda + 2\mu)/\rho$; c_p is the speed of compressional waves in the solid. Similarly, taking the curl of (1.91) followed by use of (1.86) shows that each Cartesian component of $\mathrm{curl}\,\boldsymbol{u}$ satisfies the wave equation with speed $c_s = \sqrt{\mu/\rho}$; c_s is the speed of transverse waves in the solid.

In the other direction, it is known [5, §2.10] that we can write $\boldsymbol{u} = \mathrm{grad}\,\varphi + \mathrm{curl}\,\boldsymbol{A}$ with $\mathrm{div}\,\boldsymbol{A} = 0$ provided $c_p^2 \nabla^2 \varphi = \partial^2 \varphi/\partial t^2$ and $c_s^2 \nabla^2 A_i = \partial^2 A_i/\partial t^2$, $i = 1, 2, 3$. This decomposition is convenient if we want to construct valid displacements from solutions of uncoupled scalar wave equations. However, this approach is less convenient if we have to impose boundary conditions because these inevitably couple the potentials φ and $\boldsymbol{A}$. For integral representations of $\boldsymbol{u}$, see Section 10.6.

1.6.3 Hydrodynamics

Consider an infinite ocean occupying the three-dimensional half-space $z < 0$. The flat boundary at $z = 0$ is the undisturbed free surface, and gravity acts in the negative z-direction. The water is assumed to be inviscid and incompressible. For irrotational motion, there is a velocity potential $u(x, y, z, t)$, and it is harmonic,

$$\nabla^2 u = 0, \quad z < 0. \tag{1.92}$$

For small-amplitude motions, u satisfies a linear boundary condition,

$$\frac{\partial^2 u}{\partial t^2} + g\,\frac{\partial u}{\partial z} = 0 \quad \text{at } z = 0, \tag{1.93}$$

where g is the acceleration due to gravity. The free surface is at $z = \zeta(x, y, t)$ where

$$\zeta(x, y, t) = -\frac{1}{g}\frac{\partial u}{\partial t} \quad \text{evaluated at } z = 0.$$

There is no motion at great depth: $|\mathrm{grad}\,u| \to 0$ as $z \to -\infty$.

Volterra [859] emphasised the distinction between water waves (where time enters through the boundary condition (1.93) and not via Laplace's equation (1.92)) and sound waves (where time enters into the governing wave equation (1.46)).

Another distinction is that water waves are *dispersive*. To see this, consider the harmonic function

$$u(x, y, z, t) = A\,\mathrm{e}^{kz} \sin\{k(x - ct)\},$$

where k and c are constants. This potential satisfies the free-surface condition (1.93) provided k and c are related by

$$c^2 = g/k.$$

This is the dispersion relation. As the wavelength is $2\pi/k$, we see that long waves travel faster than short waves. This dispersive character of water waves makes them quite different from acoustic, electromagnetic and elastic waves; for example, acoustic waves of all wavelengths travel at the same speed c.

For more information on the linear theory of water waves, see [516, Chapter IX], [795, Part I] and [444, 873]. For integral representations of u, see Section 10.7.

1.7 Overview of the Book

We consider scattering problems in the time domain. The prototypical problem is: solve the three-dimensional wave equation in the region exterior to a bounded obstacle B with smooth boundary S, with appropriate boundary and initial conditions. (To limit the scope, we do not consider two-dimensional problems in detail.)

We start in Chapter 2 by constructing a wide variety of solutions to the three-dimensional wave equation. Some of these are well known (for example, plane waves, solutions with spherical symmetry, and solutions built by separation of variables in spherical polar coordinates) but others are less familiar (such as spheroidal wavefunctions or solutions built using similarity variables). We also consider solutions that represent moving singularities.

The wave equation is a hyperbolic partial differential equation (PDE). The theory of such equations is well developed. We describe the salient features in Chapter 3: characteristics, discontinuities and weak solutions. For applications to acoustics, we connect the mathematical theory with constraints placed on the physical solution arising from the underlying continuum mechanics.

Most scattering problems can be formulated as initial-boundary value problems, and this is done in Chapter 4. We go beyond the prototypical problem and consider problems in which waves can pass through S into the interior B (transmission problems) and problems where the motion of S is taken into account (such as with locally reacting surfaces or thin elastic shells). We also give some discussion of the forced (inhomogeneous) wave equation.

As noted in Section 1.4, a natural tool for solving problems with initial conditions is the Laplace transform, $\mathscr{L}$. Properties of $\mathscr{L}$ are developed in Chapter 5. The non-standard aspects concern the action of $\mathscr{L}$ on discontinuous functions: we know that hyperbolic PDEs admit discontinuous solutions, so we must ensure that we take account of such discontinuities when using $\mathscr{L}$. We also describe the action of the Fourier transform.

Chapter 6 is dedicated to problems with spherical symmetry. These problems are essentially one-dimensional, which means they can usually be solved explicitly; the discussion of waves on strings in Section 1.3 will be relevant here.

Much of the literature on time-domain scattering concerns the problem of scattering by a sphere. In fact, there are many different but related problems, and their solutions are often used as benchmarks for testing methods intended to treat non-spherical scatterers. We solve sphere-scattering problems in Chapter 7. We also discuss problems involving moving spheres and scattering by spheroids.

Inverting a Laplace transform motivates a study of singularities in an appropriate complex plane. This idea has led to distinct research areas, ranging from the very theoretical (Lax–Phillips scattering theory) to the very formal (Baum's Singularity Expansion Method). These areas are reviewed and related in Chapter 8. The singularities themselves are known as scattering frequencies, natural frequencies, or complex resonances.

The remainder of the book is concerned with time-domain boundary integral equations. These are especially useful for non-spherical scatterers. The starting point is a suitable integral representation. Thus, in Chapter 9, we derive Kirchhoff's formula and we introduce single-layer and double-layer potentials. Integral-equation methods are then developed in Chapter 10. Unlike in the frequency domain, the integral equations are usually solved numerically by marching forward in time: there is a temporal discretisation and a spatial discretisation. We discuss a variety of numerical algorithms, including the popular convolution quadrature method. We also discuss electromagnetic problems, elastodynamic problems, hydrodynamic problems, the method of fundamental solutions (often known as the equivalent source method in the context of time-domain problems), and problems involving thin scatterers such as screens or cracks.

Finally, the book includes a substantial bibliography, from Abboud et al. [1] to Zworski [910]. All references listed have been cited at least once in the book, as indicated. Refer to the Citation Index for an alphabetical list of authors cited.

2
Wavefunctions

Many up-to-date expositions fail ... to show the link with the older results, and give the erroneous impression that the modern theories have no roots and are cut off from a rich past. The truth, of course, is that progress comes not only from pushing further and further into new territory but also from frequent returns to the familiar grounds, from seeking an ever-deeper understanding of their nature, and finding there new inspiration and guidance.

(Treves [833, p. ix])

The three-dimensional wave equation (1.46), which we repeat here,

$$\Box^2 u \equiv \frac{\partial^2 u}{\partial x^2} + \frac{\partial^2 u}{\partial y^2} + \frac{\partial^2 u}{\partial z^2} - \frac{1}{c^2}\frac{\partial^2 u}{\partial t^2} = 0, \tag{2.1}$$

has many different solutions. These solutions are called *wavefunctions*. In this chapter, we describe a wide variety of wavefunctions.

We start in Section 2.1 with plane waves and wavefunctions with spherical symmetry. The construction of more general (and complicated) solutions starts in Section 2.2; the time-domain far-field pattern is defined. Wavefunctions proportional to spherical harmonics are described in Section 2.3. Some of these are familiar (such as those involving Bessel functions) but others involve similarity variables or integral representations. Section 2.4 surveys Bateman's work on constructing wavefunctions. He found many new solutions of the wave equation and interconnections between wavefunctions. In Section 2.5, we construct wavefunctions using spheroidal coordinates. Finally, in Section 2.6, we construct wavefunctions that represent moving singularities.

2.1 Simple Solutions

2.1.1 Plane Waves

The simplest solutions represent plane waves. Thus

$$u(x,y,z,t) = f(z-ct) + g(z+ct) \tag{2.2}$$

and, more generally,

$$u(x,y,z,t) = f(\widehat{\boldsymbol{e}}\cdot\boldsymbol{r} - ct) + g(\widehat{\boldsymbol{e}}\cdot\boldsymbol{r} + ct), \tag{2.3}$$

are wavefunctions, for arbitrary twice-differentiable functions f and g; in (2.3), $\widehat{\boldsymbol{e}}$ is a constant unit vector and $\boldsymbol{r} = (x,y,z)$.

Reflection of plane waves by plane boundaries has been discussed [300], [125, §9], [126, Chapter 5]; most textbooks give frequency-domain analyses.

Evidently, (2.2) is d'Alembert's general solution of the one-dimensional wave equation, (1.50). An obvious extension of (2.2) gives the wavefunction

$$u(x,y,z,t) = V_2(x,y)\, f(z - ct),$$

where V_2 is an arbitrary solution of Laplace's equation in two dimensions. More generally, one can seek wavefunctions in the form

$$u(x,y,z,t) = p(x,y,z)\, f\{q(x,y,z) - ct\}. \tag{2.4}$$

This search was conducted by Friedlander [309]; see also [804]. A familar example of (2.4) occurs with spherical symmetry; see (2.10) below.

2.1.2 Spherical Polar Coordinates

For three-dimensional scattering problems, we shall often use wavefunctions constructed using spherical polar coordinates, (r,θ,ϕ), where

$$x = r\sin\theta\cos\phi, \quad y = r\sin\theta\sin\phi, \quad z = r\cos\theta, \tag{2.5}$$

with $r \geq 0$, $0 \leq \theta \leq \pi$ and $-\pi < \phi \leq \pi$. In terms of these coordinates, the wave equation (2.1) becomes

$$\frac{1}{r^2}\frac{\partial}{\partial r}\left(r^2\frac{\partial u}{\partial r}\right) + \frac{1}{r^2\sin\theta}\frac{\partial}{\partial\theta}\left(\sin\theta\frac{\partial u}{\partial\theta}\right) + \frac{1}{r^2\sin^2\theta}\frac{\partial^2 u}{\partial\phi^2} = \frac{1}{c^2}\frac{\partial^2 u}{\partial t^2}. \tag{2.6}$$

The first term on the left-hand side can be written as $r^{-1}(\partial^2/\partial r^2)(ru)$, whence

$$\frac{\partial^2(ru)}{\partial r^2} + \frac{1}{r}\Lambda u = \frac{1}{c^2}\frac{\partial^2(ru)}{\partial t^2}, \tag{2.7}$$

where Λ is the *Laplace–Beltrami operator*, defined by

$$\Lambda u = \frac{1}{\sin\theta}\frac{\partial}{\partial\theta}\left(\sin\theta\frac{\partial u}{\partial\theta}\right) + \frac{1}{\sin^2\theta}\frac{\partial^2 u}{\partial\phi^2}. \tag{2.8}$$

2.1.3 Spherically Symmetric Wavefunctions

From (2.7), wavefunctions with spherical symmetry (no dependence on θ and ϕ), $u(r,t)$, satisfy

$$\frac{\partial^2(ru)}{\partial r^2} = \frac{1}{c^2}\frac{\partial^2(ru)}{\partial t^2}, \tag{2.9}$$

the one-dimensional wave equation for the quantity ru. Thus, d'Alembert's general solution gives

$$u(r,t) = r^{-1}\{f_1(r-ct) + f_2(r+ct)\}, \tag{2.10}$$

where f_1 and f_2 are arbitrary smooth functions. See, for example, [711, §279], [516, §285], [881, §7.3], [696, §1-12].

A special case of (2.10) is the *simple source*, which we define by

$$s_0(r,t) = r^{-1}f(t-r/c), \tag{2.11}$$

where f is arbitrary. The wavefunction s_0 represents a spherical wave propagating away from the origin ($r=0$) as t increases: it is an example of an *outgoing* wavefunction, as is $r^{-1}f_1(r-ct)$ in (2.10). The other term in (2.10), $r^{-1}f_2(r+ct)$, is an *incoming* wavefunction.

Simple sources (2.11) in which f is replaced by the Dirac delta δ have been used as a fundamental solution for the three-dimensional wave equation; see (4.39).

Cartesian derivatives of a simple source (2.11) give more wavefunctions (but these are no longer spherically symmetric). With $x_1 = x$, $x_2 = y$ and $x_3 = z$,

$$\frac{\partial s_0}{\partial x_i} = -\left(\frac{1}{cr}f'(t-r/c) + \frac{1}{r^2}f(t-r/c)\right)\frac{\partial r}{\partial x_i}$$

represents a dipole at the origin, directed along the x_i-axis, $i = 1,2,3$. Here, $f'(t) = \mathrm{d}f/\mathrm{d}t$.

We can also translate the simple source. Let $R = |\boldsymbol{r} - \boldsymbol{r}'|$ where $\boldsymbol{r} = (x_1,x_2,x_3) = (x,y,z)$ and $\boldsymbol{r}' = (x_1',x_2',x_3') = (x',y',z')$. Then

$$s(R,t) = R^{-1}f(t-R/c) \tag{2.12}$$

is an outgoing wavefunction when regarded as a function of x, y, z and t for fixed $\boldsymbol{r}'$. Evidently, s is singular at $\boldsymbol{r} = \boldsymbol{r}'$.

Translated simple sources (2.12) will be used to define single-layer potentials in Section 9.4.1. They are used to define *retarded volume potentials*,

$$u(\boldsymbol{r},t) = \int_B q(\boldsymbol{r}',t-R/c)\,\frac{\mathrm{d}\boldsymbol{r}'}{4\pi R}, \tag{2.13}$$

where B is a bounded volume and q is specified; see (4.34). They have been used within time-domain versions of the *method of fundamental solutions* (also called the *equivalent source method*); see Section 10.8. Simple sources and dipoles have also been used within the *linear sampling method* for certain time-domain inverse problems [367, 375, 705].

Another wavefunction (which will be used in Section 9.4) is

$$\frac{\partial s}{\partial x_i'} = -\left(\frac{1}{cR}f'(t-R/c) + \frac{1}{R^2}f(t-R/c)\right)\frac{\partial R}{\partial x_i'}. \tag{2.14}$$

It is interesting to note that solutions of the form (2.10), depending on $r = |\boldsymbol{r}|$ and

t only, are special in the sense that such solutions do not exist in two dimensions or in more than three dimensions; for a proof, see [641]. More generally,

one may still ask what distinguishes our three-dimensional world from, say, a possible seven-dimensional world in which signals are propagated according to the seven-dimensional wave equation? Indeed, there is a decisive distinction for the three-dimensional wave equation; it is the only one that permits as solutions relatively undistorted spherical waves of arbitrary shape,

$$u(x_1, x_2, \ldots, x_n, t) = h(r) f(r - ct), \tag{2.15}$$

where $r^2 = x_1^2 + x_2^2 + \cdots + x_n^2$*. The function f is called the* wave form *or* wave shape, *while the function h is known as the* attenuating factor. *To verify our statement, we substitute the expression* (2.15) *into the n-dimensional wave equation and note that, as the function f was supposed to be arbitrary, the coefficients of f, f', and f'' in the resulting expression must be equal to zero, imposing incompatible conditions on the function h unless $n = 3$, in which case setting $h(r) = 1/r$ does make all these coefficients zero. Now signals, which are usually radiated from a central source, have the form of spherical waves; therefore such signals can remain undistorted in transmission only in our three-dimensional world.*

(Courant [201, p. 101])

2.2 The Time-Domain Far-Field Pattern

One generalisation of the simple source (2.11) arises by seeking solutions of the wave equation (2.7) in the form

$$u(r, \theta, \phi, t) = \sum_{n=0}^{\infty} r^{-n-1} f_n(\theta, \phi, t + t_0 - r/c), \tag{2.16}$$

where t_0 is a constant and the functions f_n are to be found. Substituting in (2.7) and balancing like powers of r (as in the method of Frobenius for ordinary differential equations) shows that $f_n(\theta, \phi, \tau)$ satisfies the recurrence relation

$$\frac{2n}{c}\frac{\partial f_n}{\partial \tau} + \{n(n-1) + \Lambda\} f_{n-1} = 0, \quad n = 1, 2, \ldots. \tag{2.17}$$

If we give initial conditions, such as $f_n(\theta, \phi, 0) = 0$ for $n = 1, 2, \ldots$, then (2.17) determines $f_n(\theta, \phi, \tau)$ in terms of $f_0(\theta, \phi, \tau)$. This process has been studied by Friedlander [311]. Then, in a series of papers [311, 312, 313, 314], he investigated the inverse problem of reconstructing u from f_0. See Sections 2.3.2 and 4.7.4.

We will see later that we can often arrange that $u = 0$ for $r \geq a + ct$ and some constant a; wavefunctions with this property are known as *expanding waves* [314, p. 551]. Then the choice $t_0 = a/c$ will lead to $f_0(\theta, \phi, 0) = 0$.

The expansion (2.16) was the starting point for the development by Bayliss & Turkel [75] of approximate boundary conditions designed so that computational domains can be truncated without introducing spurious reflections; see also [366, eqn (27)]. For reviews on this topic, see [343, 379, 380].

Let us examine the behaviour of $u(\boldsymbol{r}, t)$ for large r. This will be of most interest when t is also large, so that we take into account the finite speed of propagation. Consequently, we may ignore the time shift t_0 in (2.16), which then gives

$$f_0(\theta,\phi,t) = \lim_{r\to\infty} \{ru(r,\theta,\phi,t+r/c)\}. \tag{2.18}$$

Thus $ru \sim f_0(\theta,\phi,t-r/c)$ for large r. An observer situated at a great distance from the sources would therefore record a disturbance proportional to $f_0(\theta,\phi,\tau)$ in terms of a local time τ, the origin of which is related to the onset of the disturbance. Hence $f_0(\theta,\phi,\tau)$ may be called the radiation field *of the pulse u.* (Friedlander [311, p. 54])

A better name for f_0 is (time-domain) *far-field pattern* [387, p. 185] by analogy with frequency-domain usage. In [387, eqn (5.43)], a formula similar to (2.18) is used to define the far-field pressure. In [143, eqn (2.2)], $u_\infty(\boldsymbol{r}/r,t) = f_0(\theta,\phi,t)$ is called the *far field*. Knowledge of u_∞ is used to determine the shape of a sound-soft (Dirichlet boundary condition) scatterer. This is an example of a time-domain inverse problem. For more examples, see [580, 167, 367, 144]. See also Section 4.7.4 for inverse source problems.

2.3 Use of Spherical Harmonics

Returning to the wave equation (2.7), let us look for solutions in the form

$$u(r,\theta,\phi,t) = w(r,t)Y(\theta,\phi).$$

Substituting and separating gives

$$\frac{r}{w}\left(\frac{1}{c^2}\frac{\partial^2(rw)}{\partial t^2} - \frac{\partial^2(rw)}{\partial r^2}\right) = \frac{1}{Y}\Lambda Y,$$

where Λ is the Laplace–Beltrami operator (2.8). As usual, we choose the separation constant as $-n(n+1)$, where n is an integer ($n \geq 0$); then the solutions Y are the spherical harmonics, Y_n^m [599, eqn (3.8)]. As w depends on n, we henceforth write $w = u_n$ giving $u = u_n Y_n^m$ and then, by superposition,

$$u(r,\theta,\phi,t) = \sum_{n,m} c_n^m u_n(r,t) Y_n^m(\theta,\phi) \tag{2.19}$$

with coefficients c_n^m and the shorthand notation

$$\sum_{n,m} \equiv \sum_{n=0}^{\infty}\sum_{m=-n}^{n}. \tag{2.20}$$

The equation for u_n becomes

$$\frac{1}{c^2}\frac{\partial^2(ru_n)}{\partial t^2} - \frac{\partial^2(ru_n)}{\partial r^2} + n(n+1)\frac{u_n}{r} = 0. \tag{2.21}$$

This is a hyperbolic PDE for $u_n(r,t)$.

If we substitute the expansion (2.19) in (2.18), we find that the far-field pattern has the expansion

$$f_0(\theta,\phi,t) = \sum_{n,m} c_n^m g_n(t) Y_n^m(\theta,\phi) \tag{2.22}$$

with

$$g_n(t) = \lim_{r\to\infty} \{r u_n(r,t+r/c)\}. \tag{2.23}$$

The substitution $u_n = r^n W$ in (2.21) shows that W satisfies

$$\frac{\partial^2 W}{\partial r^2} + \frac{2}{r}(n+1)\frac{\partial W}{\partial r} = \frac{1}{c^2}\frac{\partial^2 W}{\partial t^2} \tag{2.24}$$

whereas the substitution $u_n = r^{-n-1}W$ gives

$$\frac{\partial^2 W}{\partial r^2} - \frac{2n}{r}\frac{\partial W}{\partial r} = \frac{1}{c^2}\frac{\partial^2 W}{\partial t^2}. \tag{2.25}$$

The substitution $u_n = r^{-1}W$ will be investigated in Section 2.3.2.

Equations (2.24) and (2.25) are examples of what is sometimes known as the Euler–Darboux–Poisson equation [197, §6.6], [275, p. 70]. For (2.24), see Lamb [516, §301, eqn (3)]. Direct substitution shows that one solution of (2.25) is $W(r,t) = (r^2 - c^2t^2)^n$, which gives

$$u_n(r,t) = r^{-n-1}\left(r^2 - c^2t^2\right)^n = r^{n-1}\left(1 - c^2t^2/r^2\right)^n. \tag{2.26}$$

This solution was found and used by Boström [120, eqn (2.11)].

When $n = 0$, (2.21) reduces to (2.9), u is spherically symmetric and we recover (2.10). More generally, there are many ways to construct solutions of (2.21). We discuss some of these below.

2.3.1 Separated Solutions

The standard procedure is to look for separated solutions of (2.21), $u_n(r,t) = R(r)\,\mathrm{e}^{st}$, where s is a parameter. We use the letter s because we often use Laplace transforms with s as the transform variable. (For application to scattering by a sphere, see Section 7.2.) When $s = -\mathrm{i}\omega$, we are in the realm of time-harmonic scattering theory (ω is the real circular frequency).

The function $R(r)$ satisfies

$$r^2R'' + 2rR' - \{(sr/c)^2 + n(n+1)\}R = 0,$$

which is the differential equation satisfied by modified spherical Bessel functions [661, eqn 10.47.2]. Thus separated solutions of (2.21) are

$$u_n(r,t) = \{A\, i_n(sr/c) + B k_n(sr/c)\}\,\mathrm{e}^{st}, \tag{2.27}$$

where i_n and k_n are modified spherical Bessel functions and A and B are arbitrary constants.

Properties of i_n and k_n can be found in [661, Chapter 10] (where our i_n is denoted by $\mathrm{i}_n^{(1)}$). In particular

$$i_0(x) = \frac{1}{x}\sinh x, \quad k_0(x) = \frac{\pi}{2x}\,\mathrm{e}^{-x}. \tag{2.28}$$

More generally, $i_n(x)$ is bounded at $x = 0$ but exponentially large as $x \to \infty$ whereas $k_n(x)$ is unbounded at $x = 0$ but exponentially small as $x \to \infty$. In detail,

$$i_n(x) \sim x^n/(2n+1)!!, \quad k_n(x) \sim (2n-1)!!/x^{n+1}, \quad \text{as } x \to 0.$$

From [661, eqn 10.49.12], we have

$$\mathrm{e}^x\, k_n(x) = \frac{\pi}{2}\sum_{j=0}^{n}\frac{a_{jn}}{x^{j+1}} \quad \text{with} \quad a_{jn} = \frac{(n+j)!}{2^j\, j!\,(n-j)!}. \tag{2.29}$$

Thus $x\mathrm{e}^x\, k_n(x)$ is a polynomial in $1/x$ of degree n. Equivalently,

$$(2/\pi)x^{n+1}\,\mathrm{e}^x\, k_n(x) = \theta_n(x), \tag{2.30}$$

a polynomial in x of degree n known as a *reverse Bessel polynomial.*

Similarly, from [661, eqn 10.49.8],

$$i_n(x) = \frac{\mathrm{e}^x}{2}\sum_{j=0}^{n}(-1)^j\frac{a_{jn}}{x^{j+1}} + (-1)^{n+1}\frac{\mathrm{e}^{-x}}{2}\sum_{j=0}^{n}\frac{a_{jn}}{x^{j+1}},$$

whence

$$2(-x)^{n+1} i_n(x) = \mathrm{e}^{-x}\,\theta_n(x) - \mathrm{e}^x\,\theta_n(-x). \tag{2.31}$$

Equivalently, using (2.30), we can write (2.31) as

$$i_n(x) = -\pi^{-1}\left\{k_n(-x) + (-1)^n k_n(x)\right\}. \tag{2.32}$$

Separated solutions will be used in Chapter 7 for the problem of scattering by a sphere. The scattered field will be represented using (2.27) with $A = 0$. From (2.29), $x\mathrm{e}^x\, k_n(x) \to \frac{1}{2}\pi$ as $x \to \infty$, and so the corresponding far-field pattern is given by (2.23),

$$g_n(t) = \lim_{r\to\infty}\left\{r k_n(sr/c)\,\mathrm{e}^{s(t+r/c)}\right\} = \frac{\pi c}{2s}\,\mathrm{e}^{st}.$$

2.3.2 Similarity Solutions

Making the substitution $u_n(r,t) = r^{-1}V(r,t)$ in (2.21) gives

$$\frac{1}{c^2}\frac{\partial^2 V}{\partial t^2} - \frac{\partial^2 V}{\partial r^2} + \frac{n(n+1)}{r^2}V = 0. \tag{2.33}$$

Next, introduce a dimensionless similarity variable, $\zeta(r,t) = ct/r$, and look for a solution of (2.33) in the form $V(r,t) = v(\zeta)$. We have

$$\frac{\partial V}{\partial r} = -\frac{ct}{r^2}v', \quad \frac{\partial^2 V}{\partial r^2} = \frac{2ct}{r^3}v' + \frac{c^2t^2}{r^4}v'',$$
$$\frac{\partial V}{\partial t} = \frac{c}{r}v', \quad \frac{\partial^2 V}{\partial t^2} = \frac{c^2}{r^2}v''.$$

Then, substitution in (2.33) gives

$$(1-\zeta^2)v''(\zeta) - 2\zeta v'(\zeta) + n(n+1)v(\zeta) = 0, \tag{2.34}$$

which is Legendre's equation for $v(\zeta)$ [661, §18.8]. Solutions are $P_n(\zeta)$ and $Q_n(\zeta)$, where P_n is a Legendre polynomial and Q_n is a Legendre function. The solution

$$u_n(r,t) = r^{-1}P_n(ct/r) \tag{2.35}$$

was found by Bateman in 1938 [68]; for other occurrences, see [311, p. 63], [140, §IV], [773, eqn (6)] and [32, eqn (8)].

Evidently, we can replace ζ by $c(t-\tau)/r$, where τ is a parameter, and then construct more solutions by integrating with respect to τ. For example,

$$u_n(r,t) = \frac{1}{r}\int_{t_1}^{t_2} f(\tau)P_n(c[t+t_0-\tau]/r)\,\mathrm{d}\tau$$

is a wavefunction, where f is an arbitrary function and t_0, t_1 and t_2 are constants.

Solutions can also be constructed with variable limits of integration,

$$u_n(r,t) = \frac{1}{r}\int_{t_1}^{t+t_0-r/c} f(\tau)P_n(c[t+t_0-\tau]/r)\,\mathrm{d}\tau. \tag{2.36}$$

This can be verified by direct calculation; see below. (The upper integration limit can be replaced by $t+t_0+r/c$, but wavefunctions with the retarded argument $t-r/c$ are more useful.) We shall use the similarity representation (2.36) in Section 7.4.

It is interesting to note that if we replace P_n in (2.36) by the Legendre function Q_n, then we do not obtain a wavefunction; see below (2.40).

Friedlander [311, eqn (3.10)] used the representation (2.36) with $t_0 = t_1 = 0$ and $f(\tau) = g_n'(\tau)$,

$$u_n(r,t) = \frac{1}{r}\int_0^{t-r/c} g_n'(\xi)P_n(c[t-\xi]/r)\,\mathrm{d}\xi. \tag{2.37}$$

Hence, motivated by (2.23),

$$r\,u_n(r,t+r/c) = \int_0^t g_n'(\xi)P_n(1+c[t-\xi]/r)\,\mathrm{d}\xi.$$

Letting $r\to\infty$ (and using $P_n(1) = 1$), we recover (2.23) (assuming that $g_n(0) = 0$). Thus (2.37) provides a mechanism for reconstructing a wavefunction u from its far-field pattern f_0; see (2.19) and (2.22).

Proof that (2.36) Defines a Wavefunction

Write (2.36) as $u_n = r^{-1}V$, where

$$V(r,t) = \int_{t_1}^{T} f(\tau)P_n(\eta)\,\mathrm{d}\tau, \tag{2.38}$$

$$T(r,t) = t + t_0 - r/c \quad \text{and} \quad \eta(\tau;r,t) = (t + t_0 - \tau)c/r. \tag{2.39}$$

We have to check that V satisfies (2.33). We start with

$$\begin{aligned}\frac{\partial V}{\partial t} &= f(T)P_n(\eta(T;r,t))\frac{\partial T}{\partial t} + \int_{t_1}^{T} f(\tau)P_n'(\eta)\frac{\partial \eta}{\partial t}\,\mathrm{d}\tau \\ &= f(T) + \frac{c}{r}\int_{t_1}^{T} f(\tau)P_n'(\eta)\,\mathrm{d}\tau\end{aligned}$$

using $\eta(T;r,t) = 1$ and $P_n(1) = 1$. Similarly

$$\begin{aligned}\frac{\partial^2 V}{\partial t^2} &= f'(T) + \frac{c}{r}f(T)P_n'(1) + \frac{c^2}{r^2}\int_{t_1}^{T} f(\tau)P_n''(\eta)\,\mathrm{d}\tau, \\ \frac{\partial V}{\partial r} &= -\frac{1}{c}f(T) - \int_{t_1}^{T} f(\tau)\frac{c}{r^2}(t + t_0 - \tau)P_n'(\eta)\,\mathrm{d}\tau, \\ \frac{\partial^2 V}{\partial r^2} &= \frac{1}{c^2}f'(T) + \frac{1}{cr}f(T)P_n'(1) + \int_{t_1}^{T} f(\tau)\left\{\frac{\eta^2}{r^2}P_n''(\eta) + \frac{2\eta}{r^2}P_n'(\eta)\right\}\mathrm{d}\tau.\end{aligned} \tag{2.40}$$

Substituting in (2.33), making use of (2.34), gives the desired result.

The proof above makes essential use of $P_n(1) = 1$. This explains why we do not obtain a wavefunction when P_n is replaced by the Legendre function Q_n: $Q_n(x)$ is singular at $x = \pm 1$.

2.3.3 Smirnov's Representation

The representation (2.36) has some similarities with one given by Smirnov in 1937 [782] (see also [350, eqn (1.129)]),

$$u_n(r,t) = \int_{0}^{t-r/c} f(\tau)\mathscr{P}_n(c[t-\tau]/r)\,\mathrm{d}\tau, \tag{2.41}$$

where

$$\mathscr{P}_n(\eta) = \int_{1}^{\eta} P_n(\xi)\,\mathrm{d}\xi. \tag{2.42}$$

Proof that (2.41) Defines a Wavefunction

Generalise Smirnov's representation (2.41) slightly to

$$u_n(r,t) = \int_{t_1}^{T} f(\tau)\mathscr{P}_n(\eta)\,\mathrm{d}\tau, \tag{2.43}$$

where T and η are defined by (2.39). Thus $\mathscr{P}_n'(\eta) = P_n(\eta)$, $\eta(T;r,t) = 1$ and $\mathscr{P}_n(1) = 0$. Differentiating,

$$\frac{\partial u_n}{\partial t} = \frac{c}{r}\int_{t_1}^{T} f(\tau)P_n(\eta)\,\mathrm{d}\tau, \quad \frac{\partial u_n}{\partial r} = -\frac{1}{r}\int_{t_1}^{T} f(\tau)\eta P_n(\eta)\,\mathrm{d}\tau,$$

$$\frac{\partial^2(ru_n)}{\partial t^2} = cf(T) + \frac{c^2}{r}\int_{t_1}^{T} f(\tau)P_n'(\eta)\,\mathrm{d}\tau,$$

$$\frac{\partial^2(ru_n)}{\partial r^2} = \frac{1}{c}f(T) + \frac{1}{r}\int_{t_1}^{T} f(\tau)\eta^2 P_n'(\eta)\,\mathrm{d}\tau.$$

Substituting in (2.21), its left-hand side becomes

$$\frac{1}{r}\int_{t_1}^{T} f(\tau)\{(1-\eta^2)P_n'(\eta) + n(n+1)\mathscr{P}_n(\eta)\}\,\mathrm{d}\tau,$$

and this vanishes identically, as we see after putting $v(\zeta) = P_n(\zeta)$ in (2.34) and then integrating from $\zeta = 1$ to $\zeta = \eta$.

2.3.4 A Bateman-Like Wavefunction

There is a useful generalisation of the similarity solution (2.35). Instead of $u_n = r^{-1}P_n(ct/r)$, look for a solution of (2.21) in the form

$$u_n(r,t) = r^{-1}P_n(\varphi/r), \tag{2.44}$$

where $\varphi(r,t)$ is to be found. Proceeding as in Section 2.3.2, substitute $V = P_n(\zeta)$ in (2.33) and compare with the differential equation satisfied by $P_n(\zeta)$, (2.34). Doing this gives

$$\frac{r^2}{c^2}\left(\frac{\partial \zeta}{\partial t}\right)^2 - r^2\left(\frac{\partial \zeta}{\partial r}\right)^2 = 1 - \zeta^2,$$

$$\frac{r^2}{c^2}\frac{\partial^2 \zeta}{\partial t^2} - r^2\frac{\partial^2 \zeta}{\partial r^2} = -2\zeta.$$

The substitution $\zeta = \varphi/r$ then gives

$$\frac{1}{c^2}\left(\frac{\partial \varphi}{\partial t}\right)^2 - \left(\frac{\partial \varphi}{\partial r}\right)^2 + \frac{2\varphi}{r}\frac{\partial \varphi}{\partial r} - 1 = 0, \tag{2.45}$$

$$\frac{1}{c^2}\frac{\partial^2 \varphi}{\partial t^2} - \frac{\partial^2 \varphi}{\partial r^2} + \frac{2}{r}\frac{\partial \varphi}{\partial r} = 0. \tag{2.46}$$

We seek $\varphi(r,t)$ satisfying both of these PDEs. To do this, we look for solutions of the linear homogeneous PDE (2.46) and then see if any of these satisfy (2.45).

A typical separated solution, $\varphi(r,t) = R(r)T(t)$, is $\varphi(r,t) = r^2 j_1(\gamma r)\cos(\gamma ct)$, where j_1 is a spherical Bessel function and γ is a constant; we have not found a linear combination of such solutions satisfying (2.45). Exceptionally ($\gamma = 0$), we obtain $\varphi(r,t) = (A+Bt)(C+Dr^3)$ but then substitution in (2.45) leads back to a time-shifted version of (2.35).

More interestingly, writing $\varphi(r,t) = R(r) + T(t)$ in (2.46) gives $\varphi = A(r^2 - c^2t^2) + Bct + C + Dr^3$. This function also solves (2.45) provided $B^2 + 4AC = 1$ and $D = 0$. Thus (2.44) gives a wavefunction when

$$\varphi(r,t) = A(r^2 - c^2t^2) + Bct + C \quad \text{with} \quad B^2 + 4AC = 1. \tag{2.47}$$

Dimensionally, because φ/r is dimensionless, the constant A is an inverse length, B is dimensionless and C is a length.

The choices $A = 0$ and $B = 1$ give a time-shifted version of (2.35). The choices $A = \frac{1}{2}r_0^{-1}$, $B = 0$ and $C = \frac{1}{2}r_0$, where r_0 is a constant (length), give the solution

$$u_n(r,t) = \frac{1}{r} P_n\left(\frac{r^2 + r_0^2 - c^2t^2}{2rr_0}\right). \tag{2.48}$$

This solution can be found in Copson's 1958 survey of known *Riemann functions* [196, p. 341]; for some later occurrences, see [145, eqn (7)], [119, eqn (12)], [140, eqn (8)], [549, eqn (14)] and [551, eqn (23)].

2.3.5 Multipole Representation

Another representation, reminiscent of (2.16), can be obtained by substituting

$$u_n(r,t) = \sum_{j=0}^{\infty} r^{-j-1} f_j(t + t_0 - r/c) \tag{2.49}$$

in (2.21), resulting in

$$2(j+1)f'_{j+1} = c(n-j)(n+j+1)f_j, \quad j = 0,1,2,\dots.$$

The factor $(n - j)$ implies that we can put $f_j \equiv 0$ for $j > n$, reducing (2.49) to

$$u_n(r,t) = \sum_{j=0}^{n} r^{-j-1} f_j(t + t_0 - r/c). \tag{2.50}$$

(This formula has been used in [430, eqn (15)].) Next, put $f_0 = f^{(n)}$ for some sufficiently smooth function $f(\tau)$, where $f^{(n)}(\tau) = \mathrm{d}^n f/\mathrm{d}\tau^n$. We can then express f_j in terms of $f^{(n-j)}$; we find that

$$u_n(r,t) = \frac{1}{r} \sum_{j=0}^{n} \frac{(n+j)!}{j!\,(n-j)!} \left(\frac{c}{2r}\right)^j f^{(n-j)}(t + t_0 - r/c). \tag{2.51}$$

Representations of this kind have a long history. Equation (2.51) can be found in a paper by Bromwich [131, eqn (3.492)]:

> *The solution* (2.51) *was originally worked out in 1899; and has been found also by Prof. H. M. Macdonald. It was published as a question in Part II of the Mathematical Tripos, 1910; and has formed the basis of a paper on the scattering of plane waves by spheres, communicated to the Royal Society in 1916* [132]. (Bromwich [131, p. 144])

Most subsequent derivations start from frequency-domain expansions. See [354], [355, eqn (1)], [215], [145, eqn (19)], [405, eqn (3.6)] and [147, eqn (4)]. In 1979, Auphan & Matthys [31, eqn (3)] wrote down (2.51) without references to earlier work (apart from stating that it is 'the classical general solution'), and then they used (2.51) to solve the problem of scattering of a sound pulse by a rigid sphere (Neumann boundary condition). This problem will be discussed further in Chapter 7.

2.3.6 Lamb, Grote and Keller

Let $u_0(r,t)$ be any spherically symmetric wavefunction, given in general by (2.10). Then, as already noted in Section 2.1.3, any Cartesian derivative of u_0 is another wavefunction. Hobson's theorem (specifically, use Corollary 3.5 and (3.23) from [599]) then shows that

$$u_n(r,t) = r^n \left(\frac{1}{r} \frac{\partial}{\partial r} \right)^n u_0(r,t), \tag{2.52}$$

a formula given by Bromwich [131, eqn (3.46)] and by Lamb [516, §301, eqn (5)]. For an application, see [139, eqn (2.15)].

We know that $ru_0(r,t)$ satisfies the one-dimensional wave equation (2.9). Therefore, if we invert (2.52), we should be able to build solutions of (2.9) in terms of u_n; this idea was pursued by Grote & Keller [365].

Assume that all u_n are expanding waves, so that $u_n(r,t) = 0$ for sufficiently large r. Then integrating (2.52) when $n = 1$ gives

$$ru_0(r,t) = -r \int_r^\infty u_1(s,t)\,\mathrm{d}s. \tag{2.53}$$

More generally, we have [365, eqn (3.11)]

$$ru_0(r,t) = -r \int_r^\infty \frac{(r^2 - s^2)^{n-1}}{(2s)^{n-1}(n-1)!} u_n(s,t)\,\mathrm{d}s, \quad n = 1,2,\ldots. \tag{2.54}$$

We give a proof below. Grote & Keller [365] gave a different proof, and then used their result to construct approximate non-reflecting boundary conditions on spherical surfaces; see also [379, 380].

Proof of (2.54)

We prove (2.54) by induction. From (2.52), we have

$$u_{n+1}(r,t) = r^n \frac{\partial}{\partial r} \left(\frac{u_n(r,t)}{r^n} \right)$$

whence

$$\frac{u_n(s,t)}{s^n} = -\int_s^\infty u_{n+1}(\sigma,t) \frac{\mathrm{d}\sigma}{\sigma^n}.$$

Substituting in (2.54) gives

$$
\begin{aligned}
r u_0(r,t) &= r \int_r^\infty \frac{s(r^2-s^2)^{n-1}}{2^{n-1}(n-1)!} \int_s^\infty u_{n+1}(\sigma,t)\,\frac{\mathrm{d}\sigma}{\sigma^n}\,\mathrm{d}s \\
&= r \int_r^\infty \frac{u_{n+1}(\sigma,t)}{2^{n-1}\sigma^n(n-1)!} \int_r^\sigma s(r^2-s^2)^{n-1}\,\mathrm{d}s\,\mathrm{d}\sigma.
\end{aligned}
$$

The inner integral evaluates to $-(r^2-\sigma^2)^n/(2n)$, and so we obtain (2.54) but with n increased to $n+1$. As (2.54) reduces to (2.53) when $n=1$, the inductive proof is complete.

2.3.7 Changing the Independent Variables

For wavefunctions $u_n(r,t)Y_n^m(\theta,\phi)$, u_n satisfies (2.21) whereas $V = r u_n$ satisfies (2.33), which we repeat here:

$$
\frac{1}{c^2}\frac{\partial^2 V}{\partial t^2} - \frac{\partial^2 V}{\partial r^2} + \frac{n(n+1)}{r^2} V = 0. \tag{2.55}
$$

Let us change the independent variables from r and t to ρ and τ; application of the chain rule gives

$$
\begin{aligned}
&\frac{\partial^2 V}{\partial \tau^2}\left\{\frac{1}{c^2}\left(\frac{\partial \tau}{\partial t}\right)^2 - \left(\frac{\partial \tau}{\partial r}\right)^2\right\} + \frac{\partial^2 V}{\partial \rho^2}\left\{\frac{1}{c^2}\left(\frac{\partial \rho}{\partial t}\right)^2 - \left(\frac{\partial \rho}{\partial r}\right)^2\right\} \\
&\quad + 2\frac{\partial^2 V}{\partial \rho\,\partial \tau}\left\{\frac{1}{c^2}\frac{\partial \rho}{\partial t}\frac{\partial \tau}{\partial t} - \frac{\partial \rho}{\partial r}\frac{\partial \tau}{\partial r}\right\} + \frac{\partial V}{\partial \rho}\left\{\frac{1}{c^2}\frac{\partial^2 \rho}{\partial t^2} - \frac{\partial^2 \rho}{\partial r^2}\right\} \\
&\quad + \frac{\partial V}{\partial \tau}\left\{\frac{1}{c^2}\frac{\partial^2 \tau}{\partial t^2} - \frac{\partial^2 \tau}{\partial r^2}\right\} + \frac{n(n+1)}{r^2} V = 0.
\end{aligned} \tag{2.56}
$$

Before discussing specific choices for ρ and τ, we pause to give a brief discussion of characteristic curves.

Characteristic Curves

Equations (2.55) and (2.56) are second-order hyperbolic PDEs. Their understanding is aided by introducing characteristic curves. Thus, in general, consider

$$
A\frac{\partial^2 u}{\partial x^2} + B\frac{\partial^2 u}{\partial x\,\partial y} + C\frac{\partial^2 u}{\partial y^2} + D = 0, \tag{2.57}
$$

where A, B and C are functions of x and y, and D can depend on x, y, u, $\partial u/\partial x$ and $\partial u/\partial y$. The two families of characteristic curves in the xy-plane are obtained by solving

$$
\frac{\mathrm{d}y}{\mathrm{d}x} = \frac{B \pm \sqrt{B^2-4AC}}{2A}; \tag{2.58}
$$

see, for example, [322, eqn (3.93)] or [446, p. 35, eqn (1.11)]. When applied to (2.55), with $x=r$, $y=t$, $A=-1$, $B=0$ and $C=c^{-2}$, we obtain $\mathrm{d}t/\mathrm{d}r = \pm c^{-1}$. Thus, one family is given by $r - ct = \text{constant}$ (we call these the *outgoing characteristics*) and the other is given by $r + ct = \text{constant}$ (*incoming characteristics*).

For characteristics in the context of the full three-dimensional wave equation, see Section 3.1.

Some Applications: Domain Truncation and Perfectly Matched Layers

Various changes of variables have been used for various purposes. Most of these have $\rho = \rho(r)$ (with no dependence on t) and $\tau = kct$, a dimensionless version of t, where k is a constant (an inverse length). These choices reduce (2.56) to

$$\frac{\partial^2 V}{\partial \tau^2} - \left(\frac{\rho'(r)}{k}\right)^2 \frac{\partial^2 V}{\partial \rho^2} - \frac{\rho''(r)}{k^2}\frac{\partial V}{\partial \rho} + \frac{n(n+1)}{(kr)^2}V = 0.$$

Specific choices of $\rho(r)$ have been made in the context of *perfectly matched layers*, as used to truncate computational domains. The main idea [176] is to make a complex-valued choice for $\rho(r)$ in the frequency domain [818, 188, 695] or in the Laplace-transform domain [380, §3]. For a review, see [448].

Other choices for $\rho(r)$ have been made so as to map an exterior region $r \geq a$ to a finite region $0 \leq \rho < 1$. One possibility is the algebraic mapping,

$$\rho(r) = \frac{r-a}{r+b}, \quad r(\rho) = \frac{a + b\rho}{1-\rho}, \tag{2.59}$$

where b is a constant (a length) and ρ is dimensionless. For alternatives and discussion, see [363, 122], [343, Chapter 6] and [123, §16.4].

Zenginoğlu's Truncation Method: Hyperboloidal Compactification

Zenginoğlu [903] has proposed combining the spatial mapping (2.59) with a temporal mapping $\tau(r,t)$ defined in terms of $\rho(r)$ by

$$\rho(r) - \tau(r,t) = k(r - ct). \tag{2.60}$$

He refers to this process as *hyperboloidal compactification.* Differentiating (2.60) gives $\partial\tau/\partial r = \rho' - k$, $\partial^2\tau/\partial r^2 = \rho''$, $\partial\tau/\partial t = kc$ and $\partial^2\tau/\partial t^2 = 0$, and then (2.56) becomes

$$\rho'(2k-\rho')\frac{\partial^2 V}{\partial \tau^2} - \rho'^2\frac{\partial^2 V}{\partial \rho^2} + 2\rho'(k-\rho')\frac{\partial^2 V}{\partial \rho\, \partial \tau} - \rho''\left(\frac{\partial V}{\partial \rho} + \frac{\partial V}{\partial \tau}\right) + \frac{n(n+1)}{r^2}V = 0. \tag{2.61}$$

Use of the spatial mapping (2.59) gives

$$\rho'(r) = \frac{\ell}{(r+b)^2} = \frac{1}{\ell}(1-\rho)^2 \quad \text{and} \quad \rho''(r) = -\frac{2}{\ell^2}(1-\rho)^3 \quad \text{with} \quad \ell = a+b.$$

Substitution in (2.61), using (2.59), we obtain

$$A\frac{\partial^2 V}{\partial \rho^2} + B\frac{\partial^2 V}{\partial \rho\, \partial \tau} + C\frac{\partial^2 V}{\partial \tau^2} + 2(1-\rho)\left(\frac{\partial V}{\partial \rho} + \frac{\partial V}{\partial \tau}\right) + \frac{n(n+1)\ell^2}{(a+b\rho)^2}V = 0, \tag{2.62}$$

where $A = -(1-\rho)^2$, $B = 2k\ell + 2A$ and $C = 2k\ell + A$.

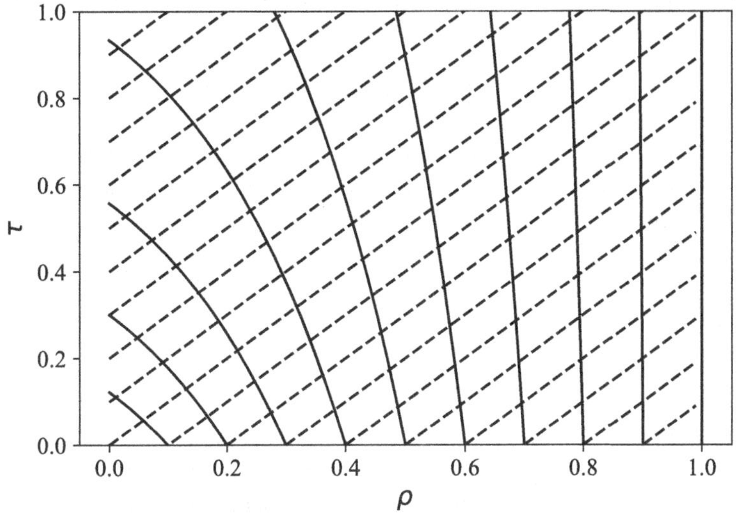

Figure 2.1 The characteristics for (2.62) in the $\rho\tau$-plane, with $k\ell = 1$. The straight dashed lines are the outgoing characteristics (2.64). The solid curved lines are the characteristics defined by (2.65); they do not meet the line $\rho = 1$ ($r = \infty$), and so no condition is needed there to eliminate incoming characteristics.

Let us determine the characteristics for (2.62) in the $\rho\tau$-plane. Comparing with (2.57), (2.58) gives

$$\frac{\mathrm{d}\tau}{\mathrm{d}\rho} = 1 \quad \text{and} \quad \frac{\mathrm{d}\tau}{\mathrm{d}\rho} = 1 - \frac{2k\ell}{(1-\rho)^2}. \tag{2.63}$$

The first of these gives

$$\rho - \tau = \text{constant}, \tag{2.64}$$

which is equivalent to the outgoing characteristics; see (2.60).

Next, consider the second of (2.63). Assume that $2k\ell > 1$ so that $\mathrm{d}\tau/\mathrm{d}\rho < 0$ for $0 \le \rho < 1$, with $\mathrm{d}\tau/\mathrm{d}\rho \to -\infty$ as $\rho \to 1$. Integrating,

$$\tau = \rho - \frac{2k\ell}{1-\rho} + \text{constant} = (\rho - \rho_0)\left\{1 - \frac{2k\ell}{(1-\rho)(1-\rho_0)}\right\}, \tag{2.65}$$

having specified that $\tau = 0$ when $\rho = \rho_0$ with $0 \le \rho_0 < 1$.

The characteristics are plotted in Fig. 2.1.[1] There are no incoming characteristics: the outgoing nature of the solution is incorporated automatically without requiring a perfectly matched layer or other non-reflecting boundary condition.

A complication with the mapped problem is that $t = 0$ is mapped to $\tau = \rho - kr(\rho) = \tau_0(\rho)$, say, with $r(\rho)$ defined by (2.59). Thus, for an initial-value problem,

[1] I am grateful to Anıl Zenginoğlu for making this figure.

we have to solve (2.62) for $\tau > \tau_0(\rho)$, $0 < \rho < 1$. For zero initial conditions, we can assume that $V \equiv 0$ for $\tau < \tau_0(\rho)$, $0 < \rho < 1$. These considerations suggest examining the Fourier transform with respect to τ. We know that (2.55) has solutions

$$V = \kappa r \mathrm{e}^{-\mathrm{i}\kappa ct} \left\{ \mathscr{A} h_n^{(1)}(\kappa r) + \mathscr{B} h_n^{(2)}(\kappa r) \right\}, \tag{2.66}$$

where $\mathscr{A}$, $\mathscr{B}$ and κ are arbitrary constants, and $h_n^{(1)}$ and $h_n^{(2)}$ are spherical Hankel functions. Use of the mappings (2.59) and (2.60) in (2.66) will generate exact solutions of (2.62). Using [661, eqn 10.52.4], the far-field approximation of (2.66) is

$$V \sim (-\mathrm{i})^{n+1} \mathscr{A}\, \mathrm{e}^{\mathrm{i}\kappa(r-ct)} + \mathrm{i}^{n+1} \mathscr{B} \mathrm{e}^{-\mathrm{i}\kappa(r+ct)} \quad \text{as } r \to \infty.$$

Then, using (2.59) and (2.60), we find that

$$V \sim \mathrm{e}^{\mathrm{i}(\kappa/k)(\rho-\tau)} \left\{ (-\mathrm{i})^{n+1} \mathscr{A} + \mathrm{i}^{n+1} \mathscr{B} \mathrm{e}^{-2\mathrm{i}\kappa\ell/(1-\rho)} \right\} \quad \text{as } \rho \to 1.$$

In this limit, the term involving $\mathscr{B}$ will vanish if we require that $\operatorname{Im} \kappa < 0$, leaving a purely outgoing solution. This provides another perspective on properties of Zenginoğlu's truncation method. His paper [903] contains some numerical examples; we may expect further developments and applications.

2.4 Harry Bateman and the Wave Equation

Harry Bateman (1882–1946) is justly famous for his work on the transformation of partial differential equations, and on the construction of 'general' solutions involving arbitrary functions. This work started while he was a student at Cambridge and carried on throughout his entire career.

We have already mentioned one of Bateman's solutions, (2.35). Here, we review some of his other work on constructing wavefunctions. Apart from its intrinsic interest, this also illustrates the great variety of solutions available.

Before describing subsequent work by Bateman and others, let us recall some simple solutions of the three-dimensional wave equation, $\Box^2 w = 0$, starting with (2.10),

$$w = r^{-1}\{f(r-ct) + g(r+ct)\}, \tag{2.67}$$

where f and g are arbitrary smooth functions. The choice $f(\xi) = g(\xi) = (2\xi)^{-1}$ yields

$$w = (r^2 - c^2t^2)^{-1}. \tag{2.68}$$

Generalising (2.67),

$$w = r^{-1} F_2(u, v)$$

is a wavefunction, where $u = r - ct$ and $v = (x \pm \mathrm{i}y)/(r+z)$; see [62, p. 42], [64, p. 184] and [67, p. 109]. Here, and throughout this section, F_n is an arbitrary function of n variables.

More generally, $w = F_2(u,v)$ is a wavefunction with the choices

$$u = x\cos\alpha + y\sin\alpha + \mathrm{i}z, \qquad v = x\sin\alpha - y\cos\alpha - ct, \tag{2.69}$$

where α is a parameter [62, p. 12], [67, p. 99]. For other linear choices of u and v, see [67, p. 105, eqn (J)]. Further solutions can be obtained by integrating or differentiating with respect to parameters, as we shall see.

2.4.1 Integral Representations

One of Bateman's teachers at Cambridge was E.T. Whittaker. In 1903, Whittaker [882] showed that the general solution of Laplace's equation in three dimensions, $\nabla^2 V = 0$, could be written as

$$V(x,y,z) = \int_0^{2\pi} F_2(z + \mathrm{i}x\cos\alpha + \mathrm{i}y\sin\alpha,\, \alpha)\,\mathrm{d}\alpha.$$

For the three-dimensional wave equation, $\square^2 w = 0$, Whittaker [882] also obtained

$$w(x,y,z,t) = \int_0^{2\pi}\int_0^{\pi} F_3(u,\alpha,\beta)\,\mathrm{d}\alpha\,\mathrm{d}\beta, \tag{2.70}$$

where $u = x\sin\alpha\cos\beta + y\sin\alpha\sin\beta + z\cos\alpha - ct$. We recognise (2.70) as a plane-wave representation for w. For further discussion and generalisations to the forced wave equation $\square^2 w = -q$ (Section 4.7), see [230] and [839, §3.4].

If we take $F_3 = \mathrm{e}^{\mathrm{i}ku} F_2(\alpha,\beta)$ in (2.70) we obtain

$$w(x,y,z,t) = \mathrm{e}^{-\mathrm{i}\omega t}\int_0^{2\pi}\int_0^{\pi} \mathrm{e}^{\mathrm{i}k(x\sin\alpha\cos\beta + y\sin\alpha\sin\beta + z\cos\alpha)} F_2(\alpha,\beta)\,\mathrm{d}\alpha\,\mathrm{d}\beta, \tag{2.71}$$

where $\omega = kc$. This defines a time-harmonic *Herglotz wavefunction* [190, §3.3].

In one of his early papers, Bateman [58, p. 457] gave the formula

$$w(x,y,z,t) = \int_0^{2\pi} F_3(u,v,\alpha)\,\mathrm{d}\alpha, \tag{2.72}$$

with u and v defined by (2.69). (For the special case $F_3(u,v,\alpha) = F_2(v/u,\alpha)$, see [62, p. 113].) The representation as a one-dimensional integral, (2.72), should be compared with Whittaker's two-dimensional integral, (2.70).

In (2.72), we can use any range of integration, $\alpha_1 < \alpha < \alpha_2$, say. However, if we choose an interval of length 2π and replace $F_3(u,v,\alpha)$ by $F_2(u,v)$, we obtain an axisymmetric wavefunction. Thus, with $x = \rho\cos\phi$ and $y = \rho\sin\phi$ (so that ρ, ϕ and z are cylindrical polar coordinates), we have

$$w(\rho,\phi,z,t) = \int_0^{2\pi} F_2(u,v)\,\mathrm{d}\alpha,$$

with $u = \rho\cos(\alpha - \phi) + \mathrm{i}z$ and $v = \rho\sin(\alpha - \phi) - ct$. For axisymmetry, we require $w(\rho,\phi+\delta,z,t) = w(\rho,\phi,z,t)$ for any δ. This is seen to be true, using $\alpha - (\phi+\delta) =$

$(\alpha-\delta)-\phi$ and the fact that the integrand is 2π-periodic in α. Thus, setting $\phi=0$, we see that

$$w(\rho,z,t)=\int_0^{2\pi} F_2(\rho\cos\alpha+\mathrm{i}z,\,\rho\sin\alpha-ct)\,\mathrm{d}\alpha \tag{2.73}$$

defines an axisymmetric wavefunction [58, p. 458]. For a direct proof, see [542, eqn (19.52)].

For a representation that is equivalent to (2.72), interchange x and z in (2.69), and then define

$$u_1=u-\mathrm{i}v=\mathrm{i}(ct+x)+(\mathrm{i}y+z)\lambda,$$
$$v_1=-\lambda(u+\mathrm{i}v)=\mathrm{i}(ct-x)\lambda+\mathrm{i}y-z,$$

where $\lambda=\mathrm{e}^{-\mathrm{i}\alpha}$. As $2u=u_1-v_1/\lambda$ and $2v=\mathrm{i}(u_1+v_1/\lambda)$, we can rewrite (2.72) as

$$w(x,y,z,t)=\int_{|\lambda|=1} F_3(u_1,v_1,\lambda)\,\mathrm{d}\lambda, \tag{2.74}$$

where the integration is around the unit circle in the complex λ-plane. The formula (2.74) was rediscovered by Penrose [690, eqn. (1.2)] in his development of 'twistor theory'; for historical remarks, see [691, 252, 30].

For an integral representation of a different kind, Bateman [67, p. 476] considered

$$w(x,y,z,t)=\int_{\alpha_1}^{\alpha_2} W(X,Y,Z,T)\tan\alpha\,\mathrm{d}\alpha, \tag{2.75}$$

where α_1 and α_2 are constants,

$$X=x\tan^2\alpha,\quad Y=y\tan^2\alpha,\quad Z=z\tan^2\alpha,\quad T=t-(r/c)\sec^2\alpha$$

and W is a wavefunction. Direct calculation shows that

$$\Box^2 w=\int_{\alpha_1}^{\alpha_2}\Omega\,\mathrm{d}\alpha,$$

where

$$\Omega=2\tan\alpha\sec^2\alpha\left\{\frac{\tan\alpha}{c^2}\frac{\partial^2 W}{\partial T^2}-\frac{1}{rc}\frac{\partial W}{\partial T}-\frac{\tan^2\alpha}{rc}\left(x\frac{\partial}{\partial X}+y\frac{\partial}{\partial Y}+z\frac{\partial}{\partial Z}\right)\frac{\partial W}{\partial T}\right\}$$
$$=-\frac{1}{rc}\frac{\partial}{\partial\alpha}\left(\frac{\partial W}{\partial T}\tan^2\alpha\right).$$

Hence

$$\Box^2 w=-\frac{1}{rc}\left[\frac{\partial W}{\partial T}\tan^2\alpha\right]_{\alpha=\alpha_1}^{\alpha_2}, \tag{2.76}$$

so that w will be a wavefunction if the right-hand side of (2.76) vanishes. The representation (2.75) has been used by Hillion in several papers; see [411], for example.

Bateman was also interested in wavefunctions representing moving sources. We shall discuss those in Section 2.6.

2.4.2 Transformations Similar to Kelvin Inversion

Kelvin inversion transforms one solution of Laplace's equation into another [64, p. 199], [474, p. 231], [67, p. 160]. Thus, if $V(x,y,z)$ solves $\nabla^2 V = 0$, another solution is

$$r^{-1}V(x/r^2, y/r^2, z/r^2).$$

There are similar transformations for the wave equation. Thus, if $W(x,y,z,t)$ is a wavefunction, then so are

$$s^{-2}W(x/s^2, y/s^2, z/s^2, t/s^2) \tag{2.77}$$

and

$$\frac{1}{z-ct}W\left(\frac{x}{z-ct}, \frac{y}{z-ct}, \frac{s^2-1}{2(z-ct)}, \frac{s^2+1}{2c(z-ct)}\right), \tag{2.78}$$

where $s^2 = r^2 - c^2t^2$; see [60], [62, p. 31] and [67, p. 161]. The transformation (2.77) was used by Morawetz in her analysis of energy decay outside star-shaped obstacles; see [633], [527, Appendix 3], [634], [532, Appendix 3], [524, Appendix E] and Section 4.2.3.

At first sight, the formula (2.78) appears to be dimensionally incorrect, because of the presence of $s^2 \pm 1$. However, this is easily rectified [61, p. 276]. Let a be a constant with dimensions of length. Then, we can replace (2.78) by

$$\frac{1}{z-ct}W\left(\frac{x}{z-ct}, \frac{y}{z-ct}, \frac{s^2-a^2}{2a(z-ct)}, \frac{s^2+a^2}{2ac(z-ct)}\right).$$

Bateman obtained the transformations (2.77) and (2.78) by first rewriting the wave equation as Laplace's equation in four dimensions: put $x_1 = x$, $x_2 = y$, $x_3 = z$ and $x_4 = \mathrm{i}ct$, so that

$$\Box^2 w = \sum_{j=1}^{4} \frac{\partial^2 w}{\partial x_j^2}.$$

Then, he introduced hexaspherical coordinates, defined by

$$\alpha_j = x_j, \quad \alpha_5 = \tfrac{1}{2}(s^2-1), \quad \alpha_6 = -\tfrac{1}{2}\mathrm{i}(s^2+1),$$

where $j = 1,2,3,4$, $s^2 = r^2 + x_4^2$ and $r^2 = x_1^2 + x_2^2 + x_3^2$. We note that $\sum_{j=1}^{6} \alpha_j^2 = 0$. Applications of coordinates satisfying this constraint have been made [462] so as to derive so-called R-separable solutions of $\Box^2 w = 0$. There are many such solutions but we are not aware of any substantial applications; for an exception, see [804].

Returning to Bateman, if we put

$$w(x,y,z,t) = f(\alpha_1, \alpha_2, \alpha_3, \alpha_4, \alpha_5, \alpha_6), \tag{2.79}$$

we obtain

$$\begin{aligned}\Box^2 w = \sum_{j=1}^{4} \frac{\partial^2 f}{\partial \alpha_j^2} + 2\sum_{j=1}^{4} \alpha_j \left(\frac{\partial^2 f}{\partial \alpha_j \partial \alpha_5} - \mathrm{i} \frac{\partial^2 f}{\partial \alpha_j \partial \alpha_6} \right) \\ + 4\left(\frac{\partial f}{\partial \alpha_5} - \mathrm{i} \frac{\partial f}{\partial \alpha_6} \right) + s^2 \left(\frac{\partial^2 f}{\partial \alpha_5^2} - 2\mathrm{i} \frac{\partial^2 f}{\partial \alpha_5 \partial \alpha_6} - \frac{\partial^2 f}{\partial \alpha_6^2} \right).\end{aligned} \tag{2.80}$$

Next, suppose that f is a homogeneous function of order -1, so that

$$\lambda f(\lambda \alpha_1, \lambda \alpha_2, \dots, \lambda \alpha_6) = f(\alpha_1, \alpha_2, \dots, \alpha_6)$$

for any $\lambda \neq 0$. Differentiating this relation with respect to λ gives

$$f(\alpha_1, \alpha_2, \dots, \alpha_6) + \sum_{j=1}^{6} \alpha_j \frac{\partial}{\partial \alpha_j} f(\alpha_1, \alpha_2, \dots, \alpha_6) = 0.$$

Then, differentiating with respect to α_i gives

$$\sum_{j=1}^{6} \alpha_j \frac{\partial^2 f}{\partial \alpha_i \partial \alpha_j} = -2 \frac{\partial f}{\partial \alpha_i}, \quad i = 1, 2, \dots, 6.$$

Use of this formula (with $i = 5$ and $i = 6$) in (2.80) gives

$$\Box^2 w = \sum_{j=1}^{6} \frac{\partial^2 f}{\partial \alpha_j^2}.$$

Thus, $\Box^2 w = 0$ becomes the six-dimensional form of Laplace's equation in α-space. This equation continues to hold if the coordinate system is rotated in any manner; such rotations will transform one solution into another.

Now, following Bateman, introduce six homogeneous coordinates, ℓ, m, n, ℓ', m' and n', defined by

$$\begin{aligned} \ell &= \alpha_1 + \mathrm{i}\alpha_2 = x + \mathrm{i}y, & \ell' &= \alpha_1 - \mathrm{i}\alpha_2 = x - \mathrm{i}y, \\ m &= \alpha_3 + \mathrm{i}\alpha_4 = z - ct, & m' &= \alpha_3 - \mathrm{i}\alpha_4 = z + ct, \\ n &= \alpha_5 + \mathrm{i}\alpha_6 = s^2, & n' &= \alpha_5 - \mathrm{i}\alpha_6 = -1. \end{aligned}$$

Then, as

$$w(x, y, z, t) = -\frac{1}{n'} w\left(\frac{\ell + \ell'}{-2n'}, \frac{\ell - \ell'}{-2\mathrm{i}n'}, \frac{m + m'}{-2n'}, \frac{m - m'}{2cn'} \right),$$

we see that we can always write w as (2.79), where f is a homogeneous function of order -1.

For a simple example of a transformation, interchange n' with $-n$; this generates the solution (2.77). For another example, interchange m with n and m' with n'; this generates the solution (2.78) (if we also change the signs of z, c and t). For a third example, consider a rotation that takes α_3 and α_4 to

$$\alpha_3' = \alpha_3 \cos\theta + \alpha_4 \sin\theta \quad \text{and} \quad \alpha_4' = \alpha_4 \cos\theta - \alpha_3 \sin\theta,$$

respectively, but leaves α_1, α_2, α_5 and α_6 unchanged. Choose an imaginary angle of rotation, $\theta = \mathrm{i}\chi$, so that $\tan\theta = \mathrm{i}\tanh\chi = \mathrm{i}v/c$, say, where $|v/c| < 1$. Then, we find that a wavefunction $W(x,y,z,t)$ is transformed into $W(x,y,\beta[z-vt],\beta[t-vz/c^2])$, where $\beta = (1-v^2/c^2)^{-1/2}$. This is the famous Lorentz transformation [680, §16-3], [848, §1.5].

For a simple application of the Bateman transformation (2.78), choose a one-dimensional wavefunction, $W(x,y,z,t) = F_1(z+ct+A)$, where F_1 is an arbitrary function of one variable and A is an arbitrary constant; this generates the wavefunction

$$(z-ct)^{-1}F_1(A+s^2/(z-ct)).$$

If we translate the time variable (replace t by $t-A/c$), we obtain the wavefunction

$$w(x,y,z,t) = gF_1(\theta), \tag{2.81}$$

with

$$g(x,y,z,t) = \frac{1}{z-ct+A} \quad \text{and} \quad \theta(x,y,z,t) = z+ct+\frac{x^2+y^2}{z-ct+A}.$$

Solutions of the form (2.81) are called ‘relatively undistorted’ by Courant & Hilbert [203, p. 762]; see also (2.15). In general, (2.81) will be a wavefunction, for arbitrary F_1, provided g and θ satisfy

$$\mathscr{H}\theta \equiv \left(\frac{\partial\theta}{\partial x}\right)^2 + \left(\frac{\partial\theta}{\partial y}\right)^2 + \left(\frac{\partial\theta}{\partial z}\right)^2 - \frac{1}{c^2}\left(\frac{\partial\theta}{\partial t}\right)^2 = 0, \tag{2.82}$$

$$\frac{\partial g}{\partial x}\frac{\partial\theta}{\partial x} + \frac{\partial g}{\partial y}\frac{\partial\theta}{\partial y} + \frac{\partial g}{\partial z}\frac{\partial\theta}{\partial z} - \frac{1}{c^2}\frac{\partial g}{\partial t}\frac{\partial\theta}{\partial t} + \frac{g}{2}(\Box^2\theta) = 0 \tag{2.83}$$

and $\Box^2 g = 0$. (The equation $\mathscr{H}\theta = 0$ is known as the *eikonal equation*; see (3.13).) For examples, generalisations and applications to ‘focus wave modes’ and ‘acoustic bullets’, see [410, 486, 809] and references therein. The causal nature of these solutions has caused some controversy [404, 412, 407, 756].

2.4.3 Some Two-Dimensional Wavefunctions

In his first book [62, p. 138], Bateman states the following result for two-dimensional wavefunctions. Suppose that $W_2(x,y,t)$ solves the two-dimensional wave equation,

$$\frac{\partial^2 W_2}{\partial x^2} + \frac{\partial^2 W_2}{\partial y^2} = \frac{1}{c^2}\frac{\partial^2 W_2}{\partial t^2}, \tag{2.84}$$

and that W_2 is a homogeneous function of degree $\frac{1}{2}$,

$$W_2(\lambda x,\lambda y,\lambda t) = \lambda^{1/2}W_2(x,y,t).$$

Suppose that $\tau(x,y,t)$ is defined by

$$c^2(t-\tau)^2 - [x-\xi(\tau)]^2 - [y-\eta(\tau)]^2 = 0,$$

where ξ and η are smooth functions. (We shall see a similar three-dimensional construction in Section 2.6; see (2.108).) Then

$$W_2(x-\xi, y-\eta, t-\tau)$$

is another two-dimensional wavefunction. As a special case, suppose that $V_2(x,y)$ solves Laplace's equation in two dimensions and that $V_2(\lambda x, \lambda y) = \lambda^{1/2} V_2(x,y)$. Then

$$W_2(x,y,t) = V_2(x-\xi, y)$$

is a two-dimensional wavefunction. This solution was used by Eshelby [270] in his much-cited study of cracks propagating under anti-plane loadings.

Another two-dimensional wavefunction, given in [62, p. 86] and [67, p. 486], is

$$W_2(x,y,t) = r^{-1/2} e^{i\theta/2} \{f(r-ct) + g(r+ct)\}, \tag{2.85}$$

where $x = r\cos\theta$, $y = r\sin\theta$, and f and g are arbitrary smooth functions; compare with (2.67) and with the simple solution

$$W_2(x,y,t) = \left|c^2t^2 - r^2\right|^{-1/2}. \tag{2.86}$$

The solution (2.85) can be found in one of Lamb's papers [515, eqn (6)]. Bateman refers to (2.85) as 'Poisson's wave-function' [62, p. 86], and he cites an article by Poisson from 1823.

2.4.4 Further Transformations

As well as transforming wavefunctions into wavefunctions, Bateman was interested in other transformations. For example, suppose that $H(x,y,t)$ solves the two-dimensional heat equation,

$$\frac{\partial^2 H}{\partial x^2} + \frac{\partial^2 H}{\partial y^2} = \frac{\partial H}{\partial t}.$$

Then, $w(x,y,z,t) = H(x,y,(z-ct)/4)e^{-(z+ct)}$ is a wavefunction [62, §13]. For another example, suppose that $W_2(x,y,t)$ solves (2.84). Then

$$w(x,y,z,t) = gW_2(X,Y,T)$$

is a three-dimensional wavefunction when

$$g = (x+iy)^{-1/2}, \quad X = \sqrt{x^2+y^2}, \quad Y = z \quad \text{and} \quad T = t;$$

see [67, p. 476]. Hillion [410, p. 2750] gave a related solution with

$$g = (z-ct)^{-1/2}, \quad X = x, \quad Y = \sqrt{z^2-c^2t^2} \quad \text{and} \quad T = iy/c.$$

A simpler transformation is

$$w(x,y,z,t) = W_2(x\sin\alpha, y\sin\alpha, t \pm (z/c)\cos\alpha),$$

where α is a parameter. When applied to the elementary solution $W_2(x,y,t) = (x^2 + y^2 - c^2t^2)^{-1/2}$ (see (2.86)), it yields

$$w(x,y,z,t) = \mathscr{M}^{-1/2} \quad \text{with} \quad \mathscr{M} = (x^2+y^2)\sin^2\alpha - (z\cos\alpha - ct)^2,$$

which is the axisymmetric 'X-wave' of Lu & Greenleaf [576, eqn (17)]. More generally, define $\theta(x,y,z,t)$ by

$$(x+\mathrm{i}y)\theta = \sqrt{\mathscr{M}} - \mathrm{i}(z\cos\alpha - ct).$$

Then, routine calculations show that $\mathscr{H}\theta = 0$ and $\Box^2\theta = 0$. In addition, if $g = \mathscr{M}^{-1/2}$, we find that (2.83) is satisfied, so that $\mathscr{M}^{-1/2}F_1(\theta)$ is a wavefunction. This generalises the solution of Lu & Greenleaf [576, eqn (14)]; they have $F_1(\theta) = \theta^{-n}$, where n is a non-negative integer. For other generalisations, see [736].

Bateman also obtained transformations between harmonic functions and wavefunctions. Thus, suppose that $V(x,y,z)$ solves $\nabla^2 V = 0$. Then, trivially, $\Box^2 V = 0$, so that the transformations (2.77) and (2.78) can be used, with W replaced by V. For example,

$$w(x,y,z,t) = s^{-2}V(x/s^2, y/s^2, z/s^2)$$

is a wavefunction; as usual, $s^2 = x^2+y^2+z^2-c^2t^2$. In particular, as $r^\nu P_\nu(\cos\theta)$ is an axisymmetric harmonic function (with $r^2 = x^2+y^2+z^2$ and $z = r\cos\theta$), where ν is arbitrary and P_ν is a Legendre function,

$$w(x,y,z,t) = r^\nu s^{-2-2\nu} P_\nu(\cos\theta) \tag{2.87}$$

is a wavefunction. If we put $\nu = -\frac{1}{2}$ in (2.87), we obtain

$$w(x,y,z,t) = \frac{K(\sin[\theta/2])}{\sqrt{r}\sqrt{r^2-c^2t^2}}, \tag{2.88}$$

a solution that was pointed out to the author by C.J. Chapman in 2008. In this formula, K is the complete elliptic integral of the first kind [661, eqn 19.2.8].

For another transformation, Bateman [61] showed that

$$w(x,y,z,t) = \frac{1}{\sqrt{s+\mathrm{i}ct}}V\left(\frac{x}{s+\mathrm{i}ct}, \frac{y}{s+\mathrm{i}ct}, \frac{z}{s+\mathrm{i}ct}\right)$$

is a wavefunction; another [59, p. 112], [67, §6.54] is

$$w(x,y,z,t) = \frac{1}{\sqrt{x+\mathrm{i}y}}V\left(\sqrt{x^2+y^2}, z, \mathrm{i}ct\right).$$

2.5 Spheroidal Wavefunctions

For scattering problems involving prolate spheroids (described below), we can use *prolate spheroidal coordinates* ξ, η and ϕ, defined by

$$x=\ell\sqrt{(\xi^2-1)(1-\eta^2)}\cos\phi,\quad y=\ell\sqrt{(\xi^2-1)(1-\eta^2)}\sin\phi,\quad z=\ell\xi\eta, \tag{2.89}$$

where ℓ is a positive constant. The foci are at $(x,y,z)=(0,0,\pm\ell)$. The surface $\xi=\xi_0>1$ is a prolate spheroid with semi-major axis of length $a=\ell\xi_0$ and semi-minor axis of length $b=\ell\sqrt{\xi_0^2-1}<a$. Note that $a^2-b^2=\ell^2$. The exterior of the spheroid corresponds to $\xi>\xi_0$, $-1\le\eta\le 1$ and $-\pi\le\phi<\pi$. In the far field, we have $r\sim\ell\xi$ as $\xi\to\infty$.

We can think of ξ as being a radial-like variable and η as a polar-angle-like variable; ϕ is the same azimuthal angle as in spherical polar coordinates (2.5).

Spheroids are axisymmetric surfaces, sometimes called axisymmetric ellipsoids. Prolate spheroids are cigar-like surfaces, with spheres and needles as limiting cases; the major axis is along the z-axis. For oblate spheroids, the minor axis is along the z-axis; spheres and discs are limiting cases. One can introduce oblate spheroidal coordinates, similar to (2.89), but we do not give details; see [301], [3, §21.3], [753] or [661, §30.14].

In terms of prolate spheroidal coordinates, the wave equation (2.1) becomes

$$\frac{1}{\ell^2(\xi^2-\eta^2)}\left\{\frac{\partial}{\partial\xi}\left((\xi^2-1)\frac{\partial u}{\partial\xi}\right)+\frac{\partial}{\partial\eta}\left((1-\eta^2)\frac{\partial u}{\partial\eta}\right)+\frac{\xi^2-\eta^2}{(\xi^2-1)(1-\eta^2)}\frac{\partial^2 u}{\partial\phi^2}\right\}$$
$$=\frac{1}{c^2}\frac{\partial^2 u}{\partial t^2};$$

see [3, eqn 21.5.1] or [661, eqn 30.13.6]. Looking for solutions in the form

$$u(\xi,\eta,\phi,t)=U(\xi,\eta,\phi,s)\,\mathrm{e}^{st}, \tag{2.90}$$

where s is a parameter (later, in Section 7.6, s will be the Laplace-transform variable), we obtain

$$\frac{\partial}{\partial\xi}\left((\xi^2-1)\frac{\partial U}{\partial\xi}\right)+\frac{\partial}{\partial\eta}\left((1-\eta^2)\frac{\partial U}{\partial\eta}\right)+\frac{\xi^2-\eta^2}{(\xi^2-1)(1-\eta^2)}\frac{\partial^2 U}{\partial\phi^2}$$
$$=(\xi^2-\eta^2)p^2U, \tag{2.91}$$

where p is a dimensionless parameter,

$$p=s\ell/c. \tag{2.92}$$

Going further, look for solutions of (2.91) in the separated form

$$U(\xi,\eta,\phi,s)=R(\xi)\,S(\eta)\,\mathrm{e}^{\pm\mathrm{i}m\phi}, \tag{2.93}$$

where m is a non-negative integer. Substitution and rearrangement gives

$$\frac{1}{R}\frac{\mathrm{d}}{\mathrm{d}\xi}\left((\xi^2-1)\frac{\mathrm{d}R}{\mathrm{d}\xi}\right)-p^2\xi^2-\frac{m^2}{\xi^2-1}=-\frac{1}{S}\frac{\mathrm{d}}{\mathrm{d}\eta}\left((1-\eta^2)\frac{\mathrm{d}S}{\mathrm{d}\eta}\right)-p^2\eta^2+\frac{m^2}{1-\eta^2}.$$

Denoting the separation constant by $\mu_m(p)$, we obtain

$$\frac{\mathrm{d}}{\mathrm{d}\xi}\left((\xi^2-1)\frac{\mathrm{d}R}{\mathrm{d}\xi}\right)-\left(\mu_m(p)+p^2\xi^2+\frac{m^2}{\xi^2-1}\right)R=0, \quad \xi>\xi_0>1, \tag{2.94}$$

$$\frac{\mathrm{d}}{\mathrm{d}\eta}\left((1-\eta^2)\frac{\mathrm{d}S}{\mathrm{d}\eta}\right)+\left(\mu_m(p)+p^2\eta^2-\frac{m^2}{1-\eta^2}\right)S=0, \quad -1\le\eta\le 1; \tag{2.95}$$

recall that we want to solve the wave equation outside the spheroid defined by $\xi=\xi_0$.

2.5.1 Spheroidal Wave Equation

Evidently, (2.94) and (2.95) are the same ordinary differential equation (ODE) but with different ranges for the independent variable. The ODE, which is called the *spheroidal wave equation*, can be written in various ways, such as [262, eqn 16.9 (1)] and [661, eqn 30.2.1]. We choose the latter and consider

$$(1-x^2)\frac{\mathrm{d}^2y}{\mathrm{d}x^2}-2x\frac{\mathrm{d}y}{\mathrm{d}x}+\left(\lambda+\gamma^2(1-x^2)-\frac{m^2}{1-x^2}\right)y=0, \tag{2.96}$$

where we assume that x is real, m is a non-negative integer and γ is a complex parameter. The standard special cases of (2.96) are: associated Legendre equation, $\gamma=0$; axisymmetric, $m=0$; prolate, γ is real and positive; and oblate, $\gamma=\mathrm{i}\nu$, ν is real and positive. Comparison with (2.94) and (2.95) shows that we want $\gamma^2=-p^2$; we take

$$\gamma=\mathrm{i}p=\mathrm{i}s\ell/c. \tag{2.97}$$

(Although much of the literature uses c instead of γ, we reserve c for the speed of sound, as in (2.92) and (2.97).)

The first step is to determine eigenvalues $\lambda=\lambda_n^m(\gamma^2)$ so that $y(x)$ is a bounded solution of (2.96) for $-1\le x\le 1$. The integer n is a counter; we take it to satisfy $n\ge m$. Numerical methods for computing $\lambda_n^m(\gamma^2)$ are available; for surveys, see [554, 55, 754]. We are especially interested in methods that are effective when γ is an arbitrary complex number (because s in (2.97) will be the Laplace-transform variable in Section 7.6). One such method involves finding the eigenvalues of a tri-diagonal matrix [416, 254], but there are other options.

Analytically, it is known that [661, eqn 30.3.8]

$$\lambda_n^m(\gamma^2)=n(n+1)+\sum_{k=1}^{\infty}\ell_{2k}\gamma^{2k}, \quad |\gamma^2|<r_n^m, \tag{2.98}$$

where the coefficients ℓ_{2k} can be computed and estimates for the radii of convergence r_n^m have been given [613, §3.2]. In particular, $\lambda_n^m(0)=n(n+1)$. It is also known that there are branch points in the complex γ-plane; these were first noted and their locations computed by Oguchi [672]. See [435, 55, 780, 754] for further studies and references.

Some asymptotic approximations for $\lambda_n^m(\gamma^2)$ are available. Put $\gamma=|\gamma|\mathrm{e}^{\mathrm{i}\chi}$ and suppose that $|\gamma|$ is large. Then we have [661, eqn 30.9.1]

$$\lambda_n^m(\gamma^2) \sim -\gamma^2 + (2n - 2m + 1)\gamma \quad \text{when } \chi = 0 \text{ (prolate case)} \tag{2.99}$$

but [661, eqn 30.9.4]

$$\lambda_n^m(\gamma^2) \sim 2q|\gamma| \quad \text{when } \chi = \pi/2 \text{ (oblate case);} \tag{2.100}$$

here, $q = n+1$ when $n-m$ is even and $q = n$ when $n-m$ is odd. For other values of $\chi = \arg\gamma$, it appears that one obtains either the prolate approximation or the oblate approximation, depending on the value of χ and the locations of the branch points (which depend on n and m): 'At these branch points, two spheroidal eigenvalues merge and become analytic continuations of each other' [55, §3.3].

The complications for large γ and fixed n contrast strongly with the situation for fixed γ and large n. Then we have [661, eqn 30.3.2]

$$\lambda_n^m(\gamma^2) = n(n+1) - \gamma^2/2 + O(n^{-2}) \quad \text{as } n \to \infty. \tag{2.101}$$

This simple estimate (which comes from (2.98)) holds for arbitrary fixed γ.

2.5.2 Angular Spheroidal Wavefunctions

Once the eigenvalues $\lambda_n^m(\gamma)$ have been determined, the corresponding eigenfunctions can be constructed. These are solutions $y(x)$ of (2.96), bounded for $-1 \le x \le 1$. They are denoted by $\mathsf{Ps}_n^m(x, \gamma^2)$, $n = m, m+1, m+2, \ldots$, and they are called *angular prolate spheroidal wavefunctions* (PSWFs). This notation is used in [661, §30.4] (see also [262, §16.11] and [27, p. 170]) but other notations are common; see (2.106). The angular PSWFs are orthogonal and normalised so that [661, eqn 30.4.6]

$$\int_{-1}^{1} \mathsf{Ps}_n^m(x,\gamma^2)\, \mathsf{Ps}_{n'}^m(x,\gamma^2)\, \mathrm{d}x = \frac{2}{2n+1}\frac{(n+m)!}{(n-m)!}\,\delta_{nn'}. \tag{2.102}$$

The axisymmetric case ($m = 0$) has been studied extensively; for example, there is a book [674] dedicated to properties of $\mathsf{Ps}_n^0(x, \gamma^2)$ when γ^2 is real and positive.

2.5.3 Radial Spheroidal Wavefunctions

Returning to the spheroidal wave equation (2.96) with $\lambda = \lambda_n^m(\gamma^2)$, we can seek solutions $y(x)$ for $x > 1$. These are called *radial* PSWFs. Two independent solutions of (2.96) are denoted by $S_n^{m(1)}(x, \gamma)$ and $S_n^{m(2)}(x, \gamma)$ [661, eqn 30.11.3]. These can be written as infinite series of spherical Bessel functions, $j_l(\gamma x)$ and $y_l(\gamma x)$, respectively; see [262, §16.9], [3, eqn 21.9.1] and [661, eqn 30.11.3]. For scattering problems, we are also interested in the combination

$$S_n^{m(3)}(x,\gamma) = S_n^{m(1)}(x,\gamma) + \mathrm{i}\, S_n^{m(2)}(x,\gamma).$$

These functions are identified because of their behaviour as $x \to \infty$. Thus [661, eqn 30.11.6]

$$S_n^{m(3)}(x,\gamma) = h_n^{(1)}(\gamma x)\left(1 + O(x^{-1})\right) \quad \text{as } x \to \infty, \tag{2.103}$$

for fixed γ, where $h_n^{(1)}$ is a spherical Hankel function. Then, using [661, eqn 10.52.4], we obtain

$$S_n^{(3)}(x, \mathrm{i}p) \sim -\mathrm{i}^{-n}\frac{\mathrm{e}^{-px}}{px} \quad \text{as } x \to \infty. \tag{2.104}$$

Despite appearances, this estimate does not hold as $px \to \infty$: it is a far-field approximation, valid as $x \to \infty$ for fixed p. This distinction was emphasised by Silbiger [778] in a comment on a paper by Chertock [174], who in turn noted that the same error was made by Morse & Feshbach [643, eqns (11.3.91) and (11.3.93)]. (It can also be found in [3, eqns 21.9.4 and 21.9.5].)

2.5.4 Summary

Combining (2.90), (2.93), (2.94) and (2.95), we have solutions

$$u(\xi,\eta,\phi,t) = S_n^{m(3)}(\xi,\mathrm{i}p)\,\mathrm{Ps}_n^m(\eta,-p^2)\,\mathrm{e}^{\pm\mathrm{i}\phi}\,\mathrm{e}^{st}, \tag{2.105}$$

where $m \geq 0$, $n \geq m$ and $p = s\ell/c$.

In the literature on scattering problems, one often sees other notation; the following are common,

$$S_n^{m(3)}(\xi,\mathrm{i}p) = R_{mn}^{(3)}(\mathrm{i}p,\xi) \quad \text{and} \quad \mathrm{Ps}_n^m(\eta,-p^2) = S_{mn}(\mathrm{i}p,\eta), \tag{2.106}$$

and we shall use this notation in Section 7.6.

As general references on spheroidal wavefunctions, see [643, pp. 1502–1513], [612], [262, pp. 134–158], [301], [27, Chapter VIII], [3, Chapter 21] and [661, Chapter 30]. For computation of PWSFs, see [55, 480, 481, 7, 754, 849] and references therein. For applications of PWSFs to scattering problems, see Section 7.6.

2.6 Moving Sources

Electromagnetic radiation by a moving point source is a classical problem, mainly because of its relation to the motion of electrons. Its solution, for arbitrary prescribed motion of the source, was given by Liénard [558] in 1898 and by Wiechert [883] in 1900.

We give a method of solution due to Conway [191]; see also [62, §41], [192] and [67, §1.93]. Thus, define a function $\mathscr{S}$ by

$$\mathscr{S}(x,y,z,t;\tau) = c^2(t-\tau)^2 - [x-\xi(\tau)]^2 - [y-\eta(\tau)]^2 - [z-\zeta(\tau)]^2, \tag{2.107}$$

where ξ, η and ζ are smooth functions of the parameter τ. Then $1/\mathscr{S}$ is a wavefunction for any τ; it is a translated version of the wavefunction in (2.68). Then

$$u(x,y,z,t) = \int \frac{f(\tau)\,\mathrm{d}\tau}{\mathscr{S}(x,y,z,t;\tau)}$$

is a wavefunction for any reasonable choice of f and for constant integration limits.

For example, suppose we integrate around a simple closed contour in the complex τ-plane. Suppose further that this contour contains only one root $\tau = \tau_0(x,y,z,t)$ of $\mathscr{S} = 0$, a simple zero, so that

$$\mathscr{S}(x,y,z,t;\tau_0) = 0. \tag{2.108}$$

Then evaluating the residue at the simple pole shows that

$$u(x,y,z,t) = \frac{2\pi\mathrm{i}\, f(\tau_0)}{(\partial\mathscr{S}/\partial\tau)(\tau_0)} \tag{2.109}$$

is a wavefunction. Defining $L(\tau)$ by $cL(\tau) = -\frac{1}{2}\partial\mathscr{S}/\partial\tau$ and then dropping a constant factor of $-\pi\mathrm{i}/c$, we see that

$$u(x,y,z,t) = \frac{f(\tau_0)}{L(\tau_0)} \tag{2.110}$$

is a wavefunction, where

$$cL(\tau) = c^2(t-\tau) - [x-\xi(\tau)]\xi'(\tau) - [y-\eta(\tau)]\eta'(\tau) - [z-\zeta(\tau)]\zeta'(\tau). \tag{2.111}$$

Differentiating (2.108) partially with respect to x gives $cL(\tau_0)(\partial\tau_0/\partial x) = \xi(\tau_0) - x$. Similar calculations for y, z and t derivatives, followed by use of (2.108), show that τ_0 satisfies $\mathscr{H}\tau_0 = 0$, where $\mathscr{H}$ is defined by (2.82).

Suppose that a curve C is parametrised by

$$x = \xi(\tau), \quad y = \eta(\tau), \quad z = \zeta(\tau). \tag{2.112}$$

Let $\boldsymbol{\rho}(\tau) = (\xi(\tau), \eta(\tau), \zeta(\tau))$ and $\boldsymbol{v}(\tau) = (\xi'(\tau), \eta'(\tau), \zeta'(\tau))$. As τ varies, the point at $\boldsymbol{\rho}(\tau)$ moves along C with velocity $\boldsymbol{v}$. Let $\boldsymbol{M} = \boldsymbol{v}/c$ and suppose that the *Mach number* $M = |\boldsymbol{M}| < 1$: the motion is subsonic. Then, for any given $\boldsymbol{r} = (x,y,z)$ and t, there is a unique $\tau_0 < t$ satisfying (2.108); see [62, p. 116], [56, p. 309] and the calculations leading to (9.34). Moreover, τ_0 is an increasing function of t. In more detail, let $\boldsymbol{R}(\tau) = \boldsymbol{r} - \boldsymbol{\rho}(\tau)$, so that

$$\mathscr{S}(\boldsymbol{r},t;\tau) = c^2(t-\tau)^2 - R^2, \quad L(\tau) = c(t-\tau) - \boldsymbol{M}\cdot\boldsymbol{R}, \quad R = |\boldsymbol{R}|.$$

From (2.108), we have

$$\tau_0 = t - c^{-1}R(\tau_0) = t - c^{-1}\sqrt{[x-\xi(\tau_0)]^2 + [y-\eta(\tau_0)]^2 + [z-\zeta(\tau_0)]^2}, \tag{2.113}$$

an implicit equation for τ_0, given $\boldsymbol{r}$, t and $\boldsymbol{\rho}(\tau)$. Then the denominator in (2.110) can be expressed as

$$L(\tau_0) = (1-M_r)R \quad \text{evaluated at } \tau = \tau_0, \text{ where } \quad M_r = \boldsymbol{M}\cdot\boldsymbol{R}/R \tag{2.114}$$

is the component of $\boldsymbol{M}$ in the direction of $\boldsymbol{R}$. (Note that $\boldsymbol{R}/R$ is a unit vector, pointing from the source at $\boldsymbol{\rho}$ to the listener at $\boldsymbol{r}$.) As $M < 1$, we see from (2.114) that $|M_r| < 1$ and so $L(\tau_0) = 0$ only when $R(\tau_0) = 0$. This gives $\boldsymbol{r} = \boldsymbol{\rho}(\tau_0)$ and then (2.113) gives $t = \tau_0$. Hence the wavefunction (2.110) has a singularity that moves along the curve C. The quantity $\tau_0(x,y,z,t)$ is the *emission time*: a signal emitted at that time reaches the position (x,y,z) at time t.

If $v = 0$, choose a fixed τ_0, and then $L(\tau_0) = c(t - \tau_0) = R(\tau_0)$, whence

$$u(x,y,z,t) = \frac{f(t - c^{-1}R(\tau_0))}{R(\tau_0)},$$

which is the simple source (2.12).

If the point source moves at supersonic speeds, $v > c$ ($M > 1$), the situation is more complicated. For some discussion, see [62, p. 116], [245, p. 193] and [463, §5].

For a point source moving in a waveguide (such as the ocean), see [564].

There are other methods that can be used to solve the problem of a moving point source. Early approaches are discussed thoroughly in Schott's book [748]. Liénard's argument, which uses geometry and first principles, is given in many textbooks, such as [680, §19-1] and [293, §21-5]. For another method, see [799, §8.17]. Later, it became fashionable to start from a retarded volume potential (2.13) in which the strength

$$q(\boldsymbol{r},t) = \mathcal{Q}(t)\,\delta(\boldsymbol{r} - \boldsymbol{\rho}(t)),$$

where $\mathcal{Q}$ is specified and $\delta(\boldsymbol{r})$ is a Dirac delta. This approach goes back to Dirac himself [233, p. 150] and to Schwinger [751]. See, for example, [108, Chapter III], [451, §3.1], [644, §11.2], [440, §14.1], [245, §9.1], [56, §12.5] and [575, 160, 547, 228, 463].

3
Characteristics and Discontinuities

The wave equation (1.46) is the prototypical hyperbolic partial differential equation (PDE). Unlike with time-harmonic scattering, where solutions are governed by the Helmholtz equation (an elliptic PDE), solutions of the wave equation can be discontinuous. This means that the exterior domain B_e may be partitioned into subdomains by moving surfaces, with discontinuities across those surfaces.

The theory of linear hyperbolic PDEs is well developed. A prominent role is played by *characteristics*: What are they, how are they defined, and why are they of interest? For motivation, we can hardly do better than quote Courant & Hilbert:

> *The relevant fact, of great importance for wave propagation, is: Physically meaningful discontinuities of solutions occur only across characteristic surfaces (hence in this context such discontinuities are called wave fronts) and are propagated in these characteristics along bicharacteristic rays. This propagation is governed by a simple ordinary differential equation.*
>
> (Courant & Hilbert [203, p. 570])

This quotation is a little misleading because characteristics are not 'surfaces'; they are three-dimensional objects (hypersurfaces) in four-dimensional space-time, as we shall see in Section 3.1. We shall also discuss eikonal equations (Section 3.2), discontinuous solutions (Section 3.3) and weak solutions (Section 3.4). For acoustic problems, the underlying continuum mechanics puts a restriction on allowable discontinuities (Section 3.5). This restriction may not always be respected if one seeks weak solutions, so care is needed.

This chapter is an expanded and revised version of an earlier review paper [601]. A general reference for most of this chapter is the excellent book by Kosiński [494]. In particular, see [494, §17] for applications to acoustics. For those readers who prefer a distributional approach, see, for example, [283, 272, 273].

3.1 Characteristics

Locate a typical point in space by its position vector $\boldsymbol{r} = (x,y,z) = (x_1,x_2,x_3)$. Locate a typical point in space-time by $\mathbf{x} = (x_1,x_2,x_3,x_4) = (\boldsymbol{r},ct)$. We use two summation conventions. Repeated lower-case subscripts or superscripts are summed from 1 to 3, whereas repeated upper-case subscripts or superscripts are summed from 1 to 4. Thus, $|\boldsymbol{r}|^2 = x_i x_i$ and $|\mathbf{x}|^2 = x_I x_I$. The wave equation (1.46) can be written as

$$a_{ij} \frac{\partial^2 u}{\partial x_i \partial x_j} - \frac{\partial^2 u}{\partial x_4^2} = 0, \tag{3.1}$$

where $a_{ij} = \delta_{ij}$, the Kronecker delta. In general, if the quadratic form $\mathbb{Q}(\boldsymbol{\xi}) \equiv \mathbb{Q}(\xi_1, \xi_2, \xi_3) = a_{ij}\xi_i\xi_j$ is positive definite, the PDE (3.1) is said to be *hyperbolic* [322, §3.2], [524, p. 12]; for the wave equation, $\mathbb{Q}(\boldsymbol{\xi}) = \xi_i\xi_i = |\boldsymbol{\xi}|^2 > 0$ for all real $\boldsymbol{\xi} \neq \mathbf{0}$.

It is more convenient to write the second-order wave equation as a *first-order symmetric hyperbolic system*,

$$A_K \frac{\partial \mathbf{w}}{\partial x_K} = A_k \frac{\partial \mathbf{w}}{\partial x_k} + \frac{1}{c}\frac{\partial \mathbf{w}}{\partial t} = \mathbf{0} \tag{3.2}$$

where $\mathbf{w}$ is a column vector,

$$\mathbf{w} = \begin{pmatrix} w_1 \\ w_2 \\ w_3 \\ w_4 \end{pmatrix} = \begin{pmatrix} \partial u/\partial x \\ \partial u/\partial y \\ \partial u/\partial z \\ c^{-1}\partial u/\partial t \end{pmatrix} = \begin{pmatrix} v_1 \\ v_2 \\ v_3 \\ -p/(\rho c) \end{pmatrix}, \tag{3.3}$$

and A_K is a symmetric 4×4 matrix, $K = 1, 2, 3, 4$:

$$A_1 = \begin{pmatrix} 0 & 0 & 0 & -1 \\ 0 & 0 & 0 & 0 \\ 0 & 0 & 0 & 0 \\ -1 & 0 & 0 & 0 \end{pmatrix}, \quad A_2 = \begin{pmatrix} 0 & 0 & 0 & 0 \\ 0 & 0 & 0 & -1 \\ 0 & 0 & 0 & 0 \\ 0 & -1 & 0 & 0 \end{pmatrix}, \tag{3.4}$$

$$A_3 = \begin{pmatrix} 0 & 0 & 0 & 0 \\ 0 & 0 & 0 & 0 \\ 0 & 0 & 0 & -1 \\ 0 & 0 & -1 & 0 \end{pmatrix}, \quad A_4 = \begin{pmatrix} 1 & 0 & 0 & 0 \\ 0 & 1 & 0 & 0 \\ 0 & 0 & 1 & 0 \\ 0 & 0 & 0 & 1 \end{pmatrix}. \tag{3.5}$$

For a system equivalent to (3.2), see [303, §2.1.3]. If we denote the entries in A_K by a_{IJ}^K, we can write (3.2) as

$$a_{IJ}^K \frac{\partial w_J}{\partial x_K} = 0, \quad I = 1, 2, 3, 4. \tag{3.6}$$

Writing hyperbolic PDEs as first-order systems such as (3.2) is standard practice; see, for example, [322, §3.5], [94, Chapter 1], [275, §7.3] or [709, Appendix 2.I].

In the next few paragraphs, we follow closely [161, Chapter 5], where further details can be found; see also [881, §5.9] and [114, §2.3]. We take the defining property of characteristics to be that $\mathbf{w}$ need not be smooth across a characteristic, even though it satisfies the wave equation (3.2) elsewhere. To be more precise, write the equation of a characteristic $\mathscr{C}$ as $F(\mathbf{x}) = 0$. Denote a normal to $\mathscr{C}$ by $\mathbf{N} = (N_1, N_2, N_3, N_4)$; this 4-vector is parallel to the gradient of F, so that $N_I = \alpha\, \partial F/\partial x_I$ for some α. We seek those $\mathbf{N}(\mathbf{x})$ for which $\mathbf{w}(\mathbf{x})$ is not constrained to be smooth across $\mathscr{C}$.

Let $\mathbf{q} = (q_1, q_2, q_3, q_4)$ be an arbitrary 4-vector. Multiply (3.6) by q_I and sum over I, giving

$$q_I a_{IJ}^K \frac{\partial w_J}{\partial x_K} = 0. \tag{3.7}$$

This states that the directional derivative of w_J in the direction of the 4-vector

$$\mathbf{d}_J = \left(q_I a_{IJ}^1, q_I a_{IJ}^2, q_I a_{IJ}^3, q_I a_{IJ}^4\right) \tag{3.8}$$

is zero. We have four directions, one for each J:

$$\mathbf{d}_1 = (-q_4, 0, 0, q_1), \qquad \mathbf{d}_2 = (0, -q_4, 0, q_2),$$
$$\mathbf{d}_3 = (0, 0, -q_4, q_3), \qquad \mathbf{d}_4 = (-q_1, -q_2, -q_3, q_4).$$

Can we arrange that all four are perpendicular to a single direction $\mathbf{N}$? For this to happen, we require

$$q_I a_{IJ}^K N_K = 0, \quad J = 1, 2, 3, 4. \tag{3.9}$$

A consequence would be that the four 4-vectors $\mathbf{d}_J$ become linearly dependent.

At this stage, $\mathbf{q}$ is arbitrary. Our question reduces to seeking a non-zero $\mathbf{q}$ such that (3.9) holds, and this will be possible if the determinant of the 4×4 matrix with entries $(a_{IJ}^K N_K)$ is zero. Written out, using (3.4) and (3.5), the system (3.9) becomes

$$\begin{pmatrix} N_4 & 0 & 0 & -N_1 \\ 0 & N_4 & 0 & -N_2 \\ 0 & 0 & N_4 & -N_3 \\ -N_1 & -N_2 & -N_3 & N_4 \end{pmatrix} \begin{pmatrix} q_1 \\ q_2 \\ q_3 \\ q_4 \end{pmatrix} = \begin{pmatrix} 0 \\ 0 \\ 0 \\ 0 \end{pmatrix} \tag{3.10}$$

and setting the determinant to zero gives

$$N_4^2\{N_4^2 - (N_1^2 + N_2^2 + N_3^2)\} = 0. \tag{3.11}$$

One solution of (3.11) is $N_4 = 0$. Then (3.10) gives $q_4 = 0$ and $N_i q_i = 0$. As there is no time dependence, such characteristics are hypercylinders in space-time corresponding to arbitrary static surfaces in space.

Alternatively, from (3.11) we have

$$N_4^2 - (N_1^2 + N_2^2 + N_3^2) = 0, \tag{3.12}$$

which is [310, eqn (2.3.1)], for example. Written out in more detail, the system (3.10) becomes

$$N_4 q_1 = N_1 q_4, \quad N_4 q_2 = N_2 q_4, \quad N_4 q_3 = N_3 q_4, \quad N_4 q_4 = N_i q_i.$$

Then, ignoring constant factors, we can express the directions $\mathbf{d}_J$ in terms of the components of $\mathbf{N}$,

$$\mathbf{d}_1 = (N_4, 0, 0, -N_1), \quad \mathbf{d}_2 = (0, N_4, 0, -N_2),$$
$$\mathbf{d}_3 = (0, 0, N_4, -N_3), \quad \mathbf{d}_4 = (-N_1, -N_2, -N_3, N_4).$$

We note that, using (3.12), $N_i \mathbf{d}_i + N_4 \mathbf{d}_4 = \mathbf{0}$: we do have linear dependence.

3.2 The Eikonal Equations

As $N_I = \alpha\,\partial F/\partial x_I$, (3.12) gives the *eikonal equation* [709, p. 153]

$$\frac{\partial F}{\partial x_i}\frac{\partial F}{\partial x_i} = \left(\frac{\partial F}{\partial x}\right)^2 + \left(\frac{\partial F}{\partial y}\right)^2 + \left(\frac{\partial F}{\partial z}\right)^2 = \frac{1}{c^2}\left(\frac{\partial F}{\partial t}\right)^2. \tag{3.13}$$

For three more occurrences, see [248, Definition 10.1.1], [511, p. 148, eqn (1.7)] and [518, eqn (67.3)]. Luneburg [581, p. 23] calls (3.13) the *characteristic equation.* It is not a PDE for F; it is only supposed to be satisfied by those combinations of x, y, z and t for which $F(\mathbf{x}) = 0$.

However, it is perhaps more natural to write

$$F(\mathbf{x}) = x_4/c - \tau(x_1,x_2,x_3) = t - \tau(\boldsymbol{r}), \tag{3.14}$$

so that a characteristic $\mathscr{C}$ is defined by

$$t = \tau(\boldsymbol{r}), \quad \boldsymbol{r} \in \mathscr{H} \subset \mathbb{R}^3, \tag{3.15}$$

for some function τ and some spatial domain $\mathscr{H}$. Then, for fixed t, $\tau(\boldsymbol{r}) = t$ defines a surface in space. As t varies, this surface moves through space: it is called a *wavefront.*

Substituting (3.15) in (3.13) gives

$$\frac{\partial \tau}{\partial x_i}\frac{\partial \tau}{\partial x_i} = \frac{1}{c^2} \tag{3.16}$$

or, equivalently,

$$|\mathrm{grad}\,\tau| = c^{-1}. \tag{3.17}$$

These are also known as eikonal equations [310, eqn (2.3.3)], [248, eqn (10.5.2)], [105, eqn (5.1.6)], [275, p. 93]. For other occurrences, see [581, eqn (7.31)] and [445, eqn (3.34)], for example. Any solution of (3.17) may be called an eikonal function. According to Birkhoff, the 'eikonal function has had a curious history' [102, p. 9]. The word 'eikonal' itself is derived from the Greek word for 'image'. For some historical remarks, see [805, 102].

Unlike (3.13), solving (3.17) as a PDE generates characteristics. For more on this distinction, see [203, pp. 557–558].

We can now be more precise about $\mathbf{N}$. For $\mathbf{N}$ to be a *unit* 4-vector,

$$1 = N_I N_I = \alpha^2 \frac{\partial F}{\partial x_I}\frac{\partial F}{\partial x_I} = 2\frac{\alpha^2}{c^2},$$

using (3.14) and (3.16). Hence

$$\mathbf{N} = \pm\frac{1}{\sqrt{2}}\left(-c\frac{\partial \tau}{\partial x}, -c\frac{\partial \tau}{\partial y}, -c\frac{\partial \tau}{\partial z}, 1\right). \tag{3.18}$$

Then we take $+$ ($-$) if we want $\mathbf{N}$ to point towards the future (past). Note that we can write (3.18) as

$$\mathbf{N} = \pm(\boldsymbol{m}, 1)/\sqrt{2} \quad \text{where} \quad \boldsymbol{m} = -c\,\mathrm{grad}\,\tau \tag{3.19}$$

is a unit 3-vector due to (3.17).

A simple solution of the eikonal equation (3.16) is

$$\tau(\boldsymbol{r}) = t' + \widehat{\boldsymbol{e}} \cdot (\boldsymbol{r} - \boldsymbol{r}')/c, \tag{3.20}$$

where $(\boldsymbol{r}', ct')$ is an arbitrary fixed point in space-time and $\widehat{\boldsymbol{e}}$ is a constant unit 3-vector. This solution represents a plane moving through space at speed c in the direction of $\widehat{\boldsymbol{e}}$.

Further solutions of (3.16) are suggested by its evident spherical spatial symmetry. For example,

$$\tau(\boldsymbol{r}) = t' - |\boldsymbol{r} - \boldsymbol{r}'|/c \tag{3.21}$$

represents an expanding sphere, centred at $\boldsymbol{r}'$; the sphere has radius $c(t - t')$ at time $t > t'$. Similarly,

$$\tau(\boldsymbol{r}) = t' + |\boldsymbol{r} - \boldsymbol{r}'|/c \tag{3.22}$$

represents a contracting sphere of radius $c(t' - t)$ at time $t < t'$. Both (3.21) and (3.22) define cone-like structures (or *conoids*), $t = \tau(\boldsymbol{r})$, in space-time. They are often known as *light-cones*; the *forward* and *backward* light-cones are defined by (3.21) and (3.22), respectively.

3.3 Discontinuous Solutions

The calculations in Section 3.1 show that certain second-order derivatives of u, namely

$$N_I \frac{\partial w_J}{\partial x_I}, \quad J = 1, 2, 3, 4,$$

are unrestricted across characteristics: they can be discontinuous. (Recall the definition of w_J in (3.3).) John [445, §3.5] gives an alternative analysis of the same situation. However, it is more interesting to ask if u or first derivatives of u can be discontinuous across a characteristic.

Before doing that, we should clarify what we mean by 'discontinuous across a characteristic'. For a characteristic $\mathscr{C}$ defined by $F(\mathbf{x}) = 0$, we can consider the level sets of F defined by $F(\mathbf{x}) = F_0$, where F_0 is a constant. Then we can say that solutions for $\mathbf{x}$ when $F_0 > 0$ define one 'side' of $\mathscr{C}$, whereas solutions for $\mathbf{x}$ when $F_0 < 0$ define the other side. (Recall that $\mathscr{C}$ is a hypersurface in space-time.)

The situation is clearer when we define $\mathscr{C}$ by (3.15). The wavefront $t = \tau(\boldsymbol{r})$ defines a surface $\Gamma(t)$ moving through space as t varies. Denote the two sides of $\Gamma(t)$ by Γ^+ and Γ^-; see Fig. 3.1. The discontinuity (or jump) in some quantity g across $\Gamma(t)$ is denoted and defined by

$$[\![g]\!](\boldsymbol{r}, t) = g^+ - g^-, \quad \boldsymbol{r} \in \Gamma(t), \tag{3.23}$$

where $g^\pm$ is the limiting value of g when $\boldsymbol{r} \in \Gamma(t)$ is approached from the $\pm$ side.

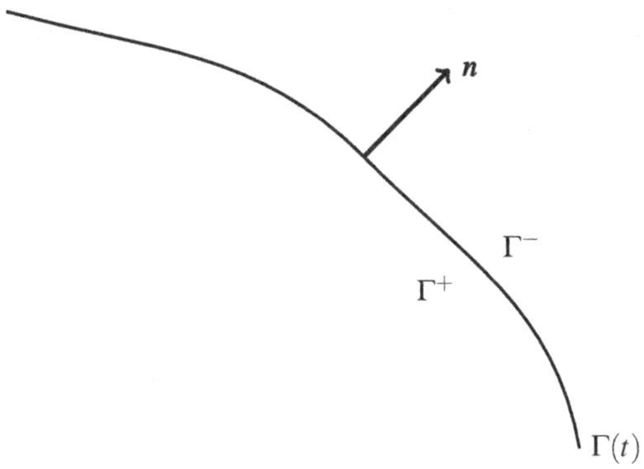

Figure 3.1 The wavefront Γ moves in the direction of the unit normal $\boldsymbol{n}$.

As t varies, $[\![g]\!]$ varies, giving

$$[\![g]\!](\boldsymbol{r}, \tau(\boldsymbol{r})) \equiv [\![g]\!](\boldsymbol{r}), \quad \boldsymbol{r} \in \mathscr{H} \subset \mathbb{R}^3, \tag{3.24}$$

with a slight abuse of notation. Thus (3.23) gives the jump across a surface $\Gamma(t)$ whereas (3.24) gives the jump across a characteristic.

A normal to Γ is grad τ, so that a unit normal $\boldsymbol{n}(\boldsymbol{r},t) = (n_1, n_2, n_3)$ is given by

$$n_i = \frac{\pm 1}{|\text{grad}\,\tau|}\frac{\partial \tau}{\partial x_i} = \pm c\frac{\partial \tau}{\partial x_i}, \quad i = 1,2,3,$$

using (3.17). Also, differentiating $t = \tau$ with respect to t gives

$$1 = \frac{\partial \tau}{\partial x_i}\frac{\mathrm{d}x_i}{\mathrm{d}t} = \pm\frac{1}{c}n_i\frac{\mathrm{d}x_i}{\mathrm{d}t}.$$

We conclude that the normal velocity of Γ is $\pm c$; the sign depends on the chosen direction of $\boldsymbol{n}$. This result was obtained by Love [572, §9] in 1904. He noted that his 'rather intricate analysis ... constitutes an abstract proof of the proposition that the velocity with which the wave-boundary advances is the velocity c .... From a physical point of view, this conclusion might be perhaps assumed' [572, p. 47]. Ten years later, Bateman [62, p. 21] wrote that (3.13) 'is the differential equation of the characteristics, it expresses that the moving boundary [the wavefront] moves normally to itself with the velocity' c.

At this point, let us fix our definitions. For a moving wavefront $\Gamma(t)$, we choose the unit normal $\boldsymbol{n}$ so that it points in the direction of motion and into the region denoted by Γ^- above (3.23); see Fig. 3.1. Thus (3.23) becomes

$$[\![g]\!] = (\text{value of } g \text{ just behind } \Gamma) - (\text{value of } g \text{ just ahead of } \Gamma). \tag{3.25}$$

This definition is convenient because in many applications the second term on the right-hand side of (3.25) is zero. Indeed, Love [572, §8] considered such a problem with Γ advancing into a quiescent region so that $u = 0$ ahead of Γ. He assumed further that u is continuous across Γ, $[\![u]\!]=0$, but first derivatives of u may be discontinuous. As $u=0$ at $\Gamma(t)$, we have

$$0=\frac{\mathrm{d}u}{\mathrm{d}t}=\frac{\partial u}{\partial t}+\frac{\partial u}{\partial x_i}\frac{\mathrm{d}x_i}{\mathrm{d}t}=\frac{\partial u}{\partial t}+cn_i\frac{\partial u}{\partial x_i}.$$

Thus

$$\frac{\partial u}{\partial n}+\frac{1}{c}\frac{\partial u}{\partial t}=0 \quad \text{on } \Gamma(t). \tag{3.26}$$

One year later, Love allowed motion ahead of Γ but with $[\![u]\!]=$ constant. He obtained [574, eqn (5)]

$$\left[\!\!\left[\frac{\partial u}{\partial n}\right]\!\!\right]+\frac{1}{c}\left[\!\!\left[\frac{\partial u}{\partial t}\right]\!\!\right]=0 \quad \text{on } \Gamma(t). \tag{3.27}$$

For a careful derivation of many such compatibility conditions (jump relations), see Chadwick & Powdrill [156] where it is also pointed out that (3.27) was derived by Hadamard [373, p. 101, eqn (48)] in his 1903 book on wave propagation.

In terms of pressure and velocity, (1.47) and (3.27) give

$$[\![p]\!]=\rho c\boldsymbol{n}\cdot[\![\boldsymbol{v}]\!] \quad \text{on } \Gamma(t). \tag{3.28}$$

This jump relation will appear later, in Section 3.5.

3.4 Weak Solutions and Discontinuities

3.4.1 Green's Formula

In order to study jumps across characteristics, we will use a four-dimensional version of Green's formula. To derive this, let Σ be a bounded region of space-time bounded by a hypersurface $\mathscr{S}$. The divergence theorem for such a region is

$$\int_\Sigma \frac{\partial V_I}{\partial x_I}\,\mathrm{d}\Sigma=\int_{\mathscr{S}} V_I N_I\,\mathrm{d}\mathscr{S}, \tag{3.29}$$

where $\mathbf{N}=(N_1,N_2,N_3,N_4)$ is a unit normal 4-vector to $\mathscr{S}$ pointing out of Σ and $\mathbf{V}=(V_1,V_2,V_3,V_4)$ is a continuously differentiable 4-vector field. Also the element of integration $\mathrm{d}\mathscr{S}$ is given by [310, eqn (2.2.3)]

$$|N_4|\,\mathrm{d}\mathscr{S}=\mathrm{d}V(\boldsymbol{r})=\mathrm{d}x_1\,\mathrm{d}x_2\,\mathrm{d}x_3, \quad N_4\neq 0. \tag{3.30}$$

If we suppose that $V_1=V_2=V_3=0$ in (3.29), we obtain (using $x_4=ct$)

$$\int_\Sigma \frac{\partial V_4}{\partial t}\,\mathrm{d}\Sigma=c\int_{\mathscr{S}} V_4 N_4\,\mathrm{d}\mathscr{S}.$$

Then put $V_4=u\,\partial v/\partial t$, repeat with u and v interchanged, and subtract the results to give

$$\int_\Sigma\left(u\frac{\partial^2 v}{\partial t^2}-v\frac{\partial^2 u}{\partial t^2}\right)\mathrm{d}\Sigma=c\int_{\mathscr{S}}\left(u\frac{\partial v}{\partial t}-v\frac{\partial u}{\partial t}\right)N_4\,\mathrm{d}\mathscr{S}. \tag{3.31}$$

Similarly, if we take $V_4 = 0$ in (3.29), we obtain

$$\int_\Sigma \frac{\partial V_i}{\partial x_i}\,\mathrm{d}\Sigma = \int_{\mathscr{S}} V_i N_i\,\mathrm{d}\mathscr{S}.$$

Then put $V_i = u\,\partial v/\partial x_i$, $i = 1,2,3$, repeat with u and v interchanged, and subtract the results to give

$$\int_\Sigma \left(u\nabla^2 v - v\nabla^2 u\right)\mathrm{d}\Sigma = \int_{\mathscr{S}} \left(u\frac{\partial v}{\partial x_i} - v\frac{\partial u}{\partial x_i}\right) N_i\,\mathrm{d}\mathscr{S}. \tag{3.32}$$

Hence, if we write the wave equation (1.46) as

$$\Box^2 u \equiv \nabla^2 u - c^{-2}\partial^2 u/\partial t^2 = 0,$$

we can combine (3.31) and (3.32) so as to obtain the four-dimensional version of Green's formula (see [310, eqn (3.2.3)])

$$\int_\Sigma \left(u\Box^2 v - v\Box^2 u\right)\mathrm{d}\Sigma = \int_{\mathscr{S}} \left(u\frac{\partial v}{\partial T} - v\frac{\partial u}{\partial T}\right)\mathrm{d}\mathscr{S}. \tag{3.33}$$

Here $\partial u/\partial T$ denotes a transverse derivative of u, defined by

$$\frac{\partial u}{\partial T} = N_i\frac{\partial u}{\partial x_i} - \frac{N_4}{c}\frac{\partial u}{\partial t} = T_I\frac{\partial u}{\partial x_I}, \tag{3.34}$$

say, where $\mathbf{T} = (T_1, T_2, T_3, T_4) = (N_1, N_2, N_3, -N_4)$. We have $N_I T_I = N_i N_i - N_4^2$, so that $\mathbf{T}$ is tangential to $\mathscr{S}$ if, and only if, $\mathscr{S}$ is a characteristic [310, §3.2]; here we have used (3.12) to define a characteristic.

3.4.2 Weak Solutions of the Wave Equation

Classical solutions of the wave equation are twice differentiable, but we expect that discontinuous solutions can occur and, in fact, such solutions are physically interesting. To handle them, we generalise the notion of 'solution'.

We say that u is a *weak solution* of the wave equation $\Box^2 u = 0$ in the space-time region Σ if

$$\int_\Sigma u\Box^2 v\,\mathrm{d}\Sigma = 0 \quad \text{for all } v \in V_\Sigma^0, \tag{3.35}$$

where V_Σ^0 is the set of all smooth *test functions* v with compact support contained in Σ (implying that $v \equiv 0$ in the vicinity of the boundary of Σ). Here, we are following Friedlander [310, pp. 42–45], Lax [524, p. 26] and Evans [275, §5.2.1] (where the set V_Σ^0 is denoted by $C_c^\infty(\Sigma)$). The basic idea was first outlined by Courant & Hilbert [202, pp. 469–470] in 1937.

If u is twice differentiable in Σ, we can use (3.33) to infer that

$$\int_\Sigma v\Box^2 u\,\mathrm{d}\Sigma = 0 \quad \text{for all } v \in V_\Sigma^0,$$

whence $\Box^2 u = 0$ in Σ. In other words, smooth weak solutions are classical solutions.

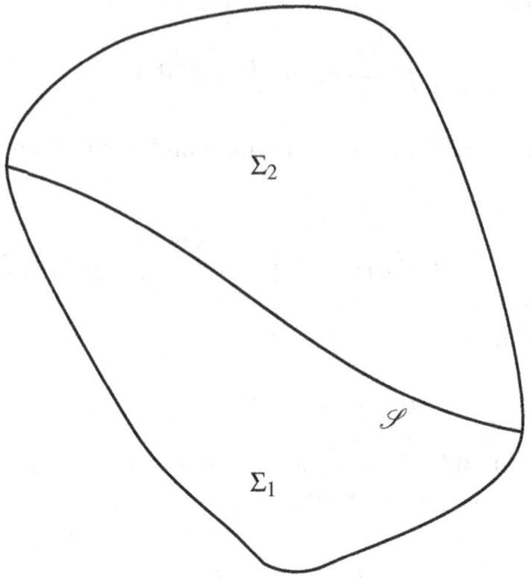

Figure 3.2 Two space-time regions, Σ_1 and Σ_2, separated by the hypersurface $\mathscr{S}$.

3.4.3 Discontinuous Weak Solutions

Suppose that u is discontinuous across some hypersurface, and that this hypersurface intersects a bounded space-time region Σ, splitting it into two sub-regions, Σ_1 and Σ_2. Let $\mathscr{S}$ denote the piece of the hypersurface inside Σ; it is the 'interface' between Σ_1 and Σ_2. See Fig. 3.2.

Clearly, if u is a weak solution of $\Box^2 u = 0$ in Σ, then it is also a classical solution in Σ_1 and Σ_2. To handle the interface, we write the definition (3.35) as

$$\int_{\Sigma_1} u\Box^2 v \,\mathrm{d}\Sigma + \int_{\Sigma_2} u\Box^2 v \,\mathrm{d}\Sigma = 0 \quad \text{for all } v \in V_\Sigma^0.$$

We use (3.33) twice, once in Σ_1 and once in Σ_2. As $\Box^2 u = 0$ in $\Sigma_1 \cup \Sigma_2$ and $v \equiv 0$ near the boundary of Σ (but not near $\mathscr{S}$), we are left with an integral over $\mathscr{S}$,

$$\int_{\mathscr{S}} \left([\![u]\!] \frac{\partial v}{\partial T} - v \left[\!\!\left[\frac{\partial u}{\partial T} \right]\!\!\right] \right) \mathrm{d}\mathscr{S} = 0 \quad \text{for all } v \in V_\Sigma^0, \tag{3.36}$$

where $[\![w]\!]$ is the jump in w across $\mathscr{S}$, that is, the difference in the values of w as a point in $\mathscr{S}$ is approached from Σ_1 and from Σ_2; this difference arises from the opposite directions of the outward pointing normals for Σ_1 and Σ_2.

Suppose first that $\mathscr{S}$ is not a characteristic. Then the transverse derivative, $\partial/\partial T$, is not a tangential derivative. This means that $\partial v/\partial T$ on $\mathscr{S}$ cannot be calculated from v on $\mathscr{S}$: they are independent and they can be chosen arbitrarily. We conclude from (3.36) that $[\![u]\!] = 0$: thus jump discontinuities can only occur across characteristics.

So, suppose next that $\mathscr{S}$ is a piece of a characteristic, $\mathscr{C}$. Then the transverse derivative is a tangential derivative. This means that $\partial v/\partial T$ on $\mathscr{C}$ can be calculated as a certain tangential derivative of v on $\mathscr{C}$. In more detail, define $\mathscr{C}$ by $t = \tau(\boldsymbol{r})$ for $\boldsymbol{r} \in \mathscr{H}$ where τ satisfies the eikonal equation (3.16); see Section 3.2. Then, given a function $w(\boldsymbol{r},t)$, its values on $\mathscr{C}$ are $w(\boldsymbol{r},\tau(\boldsymbol{r})) = \{w\}$, say, with $\boldsymbol{r} \in \mathscr{H}$; we use $\{\ \}$ to indicate evaluation on $\mathscr{C}$.

The chain rule gives

$$\frac{\partial\{w\}}{\partial x_i} = \left\{\frac{\partial w}{\partial x_i}\right\} + \left\{\frac{\partial w}{\partial t}\right\}\frac{\partial \tau}{\partial x_i}. \tag{3.37}$$

Hence the definition (3.34) gives

$$\begin{aligned}
\left\{\frac{\partial v}{\partial T}\right\} &= \left\{N_i\frac{\partial v}{\partial x_i} - \frac{N_4}{c}\frac{\partial v}{\partial t}\right\} = -cN_4\left\{\frac{\partial \tau}{\partial x_i}\frac{\partial v}{\partial x_i} + \frac{1}{c^2}\frac{\partial v}{\partial t}\right\} \\
&= -cN_4\left(\frac{\partial \tau}{\partial x_i}\left\{\frac{\partial v}{\partial x_i}\right\} + \frac{1}{c^2}\left\{\frac{\partial v}{\partial t}\right\}\right) \\
&= -cN_4\left(\frac{\partial \tau}{\partial x_i}\left(\frac{\partial\{v\}}{\partial x_i} - \left\{\frac{\partial v}{\partial t}\right\}\frac{\partial \tau}{\partial x_i}\right) + \frac{1}{c^2}\left\{\frac{\partial v}{\partial t}\right\}\right) \\
&= -cN_4\frac{\partial \tau}{\partial x_i}\frac{\partial\{v\}}{\partial x_i}
\end{aligned}$$

using (3.37) and the eikonal equation (3.16). A similar calculation gives

$$\left[\!\!\left[\frac{\partial u}{\partial T}\right]\!\!\right] = -cN_4\frac{\partial \tau}{\partial x_i}\frac{\partial[\![u]\!]}{\partial x_i}.$$

Hence, using (3.30), (3.36) becomes

$$\int_{\mathscr{H}}\left([\![u]\!]\frac{\partial \tau}{\partial x_i}\frac{\partial\{v\}}{\partial x_i} - \{v\}\frac{\partial \tau}{\partial x_i}\frac{\partial[\![u]\!]}{\partial x_i}\right)\frac{N_4}{|N_4|}\,\mathrm{d}V(\boldsymbol{r}) = 0. \tag{3.38}$$

Recall that the unit normal to a characteristic, $\mathbf{N}$, is given by (3.18); in particular, $N_4 = \pm 1/\sqrt{2}$. Therefore, if we assume that N_4 does not change sign on $\mathscr{C}$, we can delete the factor $N_4/|N_4|$ from the integrand in (3.38). Making this assumption, then, we obtain

$$\int_{\mathscr{H}}\left([\![u]\!]\frac{\partial \tau}{\partial x_i}\frac{\partial\{v\}}{\partial x_i} - \{v\}\frac{\partial \tau}{\partial x_i}\frac{\partial[\![u]\!]}{\partial x_i}\right)\mathrm{d}V(\boldsymbol{r}) = 0. \tag{3.39}$$

This holds for all smooth functions $\{v\}$ with support in $\mathscr{H}$. For such test functions, the divergence theorem gives

$$\int_{\mathscr{H}}[\![u]\!]\frac{\partial \tau}{\partial x_i}\frac{\partial\{v\}}{\partial x_i}\,\mathrm{d}V = -\int_{\mathscr{H}}\{v\}\frac{\partial}{\partial x_i}\left([\![u]\!]\frac{\partial \tau}{\partial x_i}\right)\mathrm{d}V.$$

Hence, as (3.39) holds for all test functions $\{v\}$, we obtain

$$0 = \frac{\partial}{\partial x_i}\left([\![u]\!]\frac{\partial \tau}{\partial x_i}\right) + \frac{\partial \tau}{\partial x_i}\frac{\partial[\![u]\!]}{\partial x_i} = 2\frac{\partial \tau}{\partial x_i}\frac{\partial[\![u]\!]}{\partial x_i} + [\![u]\!]\,\nabla^2\tau. \tag{3.40}$$

According to Friedlander [310, p. 45], this 'equation is called, following Luneburg, the *transport equation* associated with the wave fronts' (3.15). For other occurrences, see [308, eqn (5)], [127, eqn (7)], [310, eqn (3.3.7)] and [581, §11].

Equation (3.40) is a first-order PDE for $[\![u]\!]$ on $\mathscr{C}$. It is the differential equation mentioned in the quotation from Courant & Hilbert [203] on page 58.

To investigate the consequences of (3.40), we choose a function $\tau(\boldsymbol{r})$ that satisfies the eikonal equation. One simple choice is the plane-wave function (3.20); in particular, if we choose the unit vector $\widehat{\boldsymbol{e}}$ in the x_3-direction, we have

$$\tau(\boldsymbol{r}) = t' + \frac{x_3 - x_3'}{c}, \quad \frac{\partial \tau}{\partial x_i} = \frac{1}{c}\delta_{i3}, \quad \nabla^2 \tau = 0.$$

For this choice, (3.40) reduces to $\partial[\![u]\!]/\partial x_3 = 0$: $[\![u]\!]$ can be an arbitrary function of the in-plane variables $x = x_1$ and $y = x_2$ but it cannot depend on the out-of-plane variable $z = x_3$. To labour the point, we construct $[\![u]\!]$ from $u(x,y,z,t)$ with $t = \tau(x,y,z)$, giving $[\![u]\!]$ as a function of x, y and z; for a plane characteristic moving in the z-direction, $[\![u]\!]$ cannot change with z.

Next, suppose that we have a spherical wavefront with τ given by (3.21). Then, if we put $R = |\boldsymbol{r} - \boldsymbol{r}'|$, we obtain

$$\tau(\boldsymbol{r}) = t' + \frac{R}{c}, \quad \operatorname{grad}\tau = \frac{\boldsymbol{r} - \boldsymbol{r}'}{cR}, \quad \nabla^2 \tau = \frac{2}{cR},$$

and (3.40) reduces to

$$(x_i - x_i')\frac{\partial[\![u]\!]}{\partial x_i} + [\![u]\!] = 0. \tag{3.41}$$

If we define spherical polar coordinates by

$$x_1 - x_1' = R\sin\theta\cos\phi, \quad x_2 - x_2' = R\sin\theta\sin\phi, \quad x_3 - x_3' = R\cos\theta,$$

(3.41) simplifies to

$$R\frac{\partial[\![u]\!]}{\partial R} + [\![u]\!] = 0 \quad \text{whence} \quad [\![u]\!] = \frac{A(\theta,\phi)}{R}, \tag{3.42}$$

where A is an arbitrary function of θ and ϕ. Note that the wavefront is defined by $t = \tau$ whence $R = c(t - t')$ in (3.42).

We remark that the results of this section, including (3.42), are valid for any weak solutions of the wave equation such as the potential u, the pressure p or any Cartesian component of the velocity $\boldsymbol{v}$.

3.5 Jump Relations in Continuum Mechanics

Let us return to the quotation from Courant & Hilbert [203] on page 58. It refers to ‘physically meaningful discontinuities’; what does this mean? Love [572, p. 53] remarked that there may be conditions on the solution ‘imposed by the constitution of the medium or the nature of the disturbance, if it is to represent waves of a specified type transmitted through a specified medium’ with a footnote: ‘The importance of these conditions was emphasized by Dr. Larmor at the meeting at which the paper was communicated’ (in January 1903).

The conditions mentioned are derived within the general theory of continuum mechanics. For a thorough development, see [835, Part C]. For the special case of a compressible inviscid fluid, see, for example, [154, p. 118] or [368, §33.1]. In this context, there are two exact jump relations across a moving surface $\Gamma(t)$,

$$[\![\rho_{\text{ex}}(V - \boldsymbol{n}\cdot\boldsymbol{v}_{\text{ex}})]\!] = 0 \quad \text{and} \quad [\![\rho_{\text{ex}}(V - \boldsymbol{n}\cdot\boldsymbol{v}_{\text{ex}})\boldsymbol{v}_{\text{ex}} - p_{\text{ex}}\boldsymbol{n}]\!] = \boldsymbol{0}, \tag{3.43}$$

where ρ_{ex}, $\boldsymbol{v}_{\text{ex}}$ and p_{ex} denote the exact density, velocity and pressure, respectively, V is the normal velocity of Γ, and the unit normal to Γ, $\boldsymbol{n}$, points in the direction of motion. For a characteristic $\mathscr{C}$, Γ is defined by (3.15) and the normal velocity is $V = c$.

Now, in linear acoustics (see Section 1.1), we have $\rho_{\text{ex}} \simeq \rho + \varepsilon\rho_1$, $p_{\text{ex}} \simeq p_0 + \varepsilon p_1$ and $\boldsymbol{v}_{\text{ex}} \simeq \varepsilon\boldsymbol{v}_1$, where ρ and p_0 are constants and ε is a small parameter. Then we see that (3.43) is satisfied exactly at leading order in ε, whereas at first order in ε we obtain

$$c\,[\![\rho_1]\!] = \rho\,\boldsymbol{n}\cdot[\![\boldsymbol{v}_1]\!] \quad \text{and} \quad \rho c\,[\![\boldsymbol{v}_1]\!] = [\![p_1]\!]\boldsymbol{n}. \tag{3.44}$$

The (excess) pressure p is defined by $p_{\text{ex}} = p_0 + p$. Therefore, in linear acoustics, we have $p = \varepsilon p_1 = \varepsilon c^2\rho_1$, where the second equality comes from (1.22). We also define the velocity $\boldsymbol{v}$ by $\boldsymbol{v} = \varepsilon\boldsymbol{v}_1$. Then the first of (3.44) gives

$$[\![p]\!] = \rho c\,\boldsymbol{n}\cdot[\![\boldsymbol{v}]\!] \tag{3.45}$$

and the second gives

$$[\![p]\!]\boldsymbol{n} = \rho c\,[\![\boldsymbol{v}]\!]. \tag{3.46}$$

Evidently, (3.45) is the normal component of the vector equation (3.46). It is also equivalent to the condition found by Hadamard in 1903 [373] and by Love in 1905 [574]; see (3.28).

Let $\boldsymbol{t}$ be a tangent vector to Γ at $P \in \Gamma$. Then (3.46) gives $[\![\boldsymbol{v}]\!]\cdot\boldsymbol{t} = 0$: tangential components of the velocity are continuous across a wavefront Γ, whereas, by (3.45), the normal component of $\boldsymbol{v}$ has a jump proportional to the jump in the pressure across Γ. The wavefront Γ is sometimes called a *weak acoustic shock.*

These results go back to Christoffel [180]; see also [398], [470, p. 941] and [310, p. 45].

The jump relation (3.46) holds as $\Gamma(t)$ evolves. If we introduce the velocity potential u, so that, from (1.47), $\boldsymbol{v} = \operatorname{grad} u$ and $p = -\rho\,\partial u/\partial t$, (3.46) becomes

$$\left[\!\!\left[\frac{\partial u}{\partial t}\right]\!\!\right]\operatorname{grad}\tau = -[\![\operatorname{grad} u]\!], \quad \boldsymbol{r}\in\mathscr{H},$$

where we have used $\boldsymbol{n} = c\operatorname{grad}\tau$. But, from (3.37),

$$\left[\!\!\left[\frac{\partial u}{\partial t}\right]\!\!\right]\operatorname{grad}\tau = \operatorname{grad}[\![u]\!] - [\![\operatorname{grad} u]\!], \quad \boldsymbol{r}\in\mathscr{H}.$$

Comparing these two equations, we infer that $\operatorname{grad}[\![u]\!] = \boldsymbol{0}$ in $\mathscr{H}$, whence

$$[\![u]\!](\boldsymbol{r}) = \text{constant}, \quad \boldsymbol{r}\in\mathscr{H}. \tag{3.47}$$

As all physical quantities are obtained from appropriate derivatives of u, we may argue that we can take the constant to be zero. This would mean that the velocity potential is continuous across wavefronts, a result stated by Friedlander [310, p. 45], and then we could use the formula (3.26) at a wavefront. However, the constant in (3.47) may be determined by initial conditions, and then it may not be zero. We shall return to this point later.

When u is continuous across wavefronts, it can be advantageous to work with the potential u rather than the pressure p, for example. In general, p is not continuous across wavefronts; in fact, discontinuous pressure pulses are of great interest.

If we compare with the discussion of discontinuous weak solutions given in Section 3.4.3, we see that we must have $A = 0$ in (3.42) when u is required to be continuous across a wavefront. Similar formulas hold for p and the Cartesian components of $\boldsymbol{v}$, but then the coefficients corresponding to A will not vanish, in general.

If the constant in (3.47) is not zero, we see a mismatch between the physically justified jump condition (3.47) and the jump condition (3.42) obtained by seeking weak solutions.

4

Initial-Boundary Value Problems

Our study of time-domain scattering problems begins in this chapter. We consider a bounded obstacle $B \subset \mathbb{R}^3$ with smooth boundary S; this is the scatterer. Denote the unbounded region exterior to S by B_e. We are interested in solving the wave equation in B_e subject to initial and boundary conditions. This leads to a variety of initial-boundary value problems (IBVPs), also called 'mixed problems' by Hadamard [374, §23] and others; see [203, p. 471], [247], [165, Chapter 7], [408] and [524, Appendix D]. Some of these problems are formulated in Section 4.1. Related topics are acoustic energy (Section 4.2), incident sound pulses (Section 4.3) and compatibility conditions on the data (Section 4.4). Bateman's curious explicit solution for a sphere is given in Section 4.5. More elaborate IBVPs are described in Section 4.6; these include locally reacting boundaries, transmission problems, membranes and elastic shells. Section 4.7 is concerned with the forced (inhomogeneous) wave equation, with applications to aerodynamic sound. Finally, Section 4.8 gives a summary of available theoretical results covering existence, uniqueness and regularity for the major IBVPs.

4.1 Formulation of IBVPs

Formally, we seek a wavefunction $u(\mathbf{x})$ for $\mathbf{x} = (\boldsymbol{r}, ct) \in \Sigma = B_e \times \mathbb{R}^+$, where $\mathbb{R}^+$ denotes positive real numbers. Thus Σ is a semi-infinite hypercylinder in space-time. There is a boundary condition on the lateral boundary of Σ, $S \times \mathbb{R}^+$, and there are two initial conditions on the 'base' of Σ, $B_e \times \{0\}$. Of some interest will be the intersection of the lateral boundary and the base, $S \times \{0\} = \mathscr{E}$, say (the boundary S at $t = 0$), because this is where the boundary condition and the initial conditions may be in conflict (see Section 4.4). We call $\mathscr{E}$ the *space-time edge* or *corner* of Σ.

Returning to the formulation of IBVPs, suppose that, as usual, u is a velocity potential. Thus, from (1.47), the pressure $p = -\rho\, \partial u/\partial t$ and the velocity $\boldsymbol{v} = \operatorname{grad} u$. We seek a wavefunction $u(\mathbf{x}) \equiv u(P,t)$, where $P(x,y,z)$ is a typical point in B_e with position vector $\boldsymbol{r} = (x,y,z) = (x_1,x_2,x_3)$. The function $u(P,t)$ is subject to a boundary

Table 4.1 *Six initial-boundary value problems*

Problem	Initial conditions		Boundary condition on S
ID_0	$u(P,0)=u_0(P)$	$(\partial u/\partial t)(P,0)=u_1(P)$	$u(P,t)=0$
IP_0	$u(P,0)=u_0(P)$	$(\partial u/\partial t)(P,0)=u_1(P)$	$\partial u/\partial t=0$
IN_0	$u(P,0)=u_0(P)$	$(\partial u/\partial t)(P,0)=u_1(P)$	$\partial u/\partial n=0$
DI_0	$u(P,0)=0$	$(\partial u/\partial t)(P,0)=0$	$u(P,t)=d(P,t)$
PI_0	$u(P,0)=0$	$(\partial u/\partial t)(P,0)=0$	$-\rho\,\partial u/\partial t=p_S(P,t)$
NI_0	$u(P,0)=0$	$(\partial u/\partial t)(P,0)=0$	$\partial u/\partial n=v(P,t)$

condition when $P \in S$ and initial conditions at $t=0$. We take the latter as

$$u(P,0)=u_0(P) \quad \text{and} \quad \frac{\partial u}{\partial t}(P,0)=u_1(P) \quad \text{for all } P \in B_e, \tag{4.1}$$

where u_0 and u_1 are given functions.

We shall consider three choices for the boundary condition. Additional choices and problems are discussed in Section 4.6.

The first choice is the *Dirichlet condition*,

$$u(P,t)=d(P,t) \quad \text{for } P \in S \text{ and } t>0, \tag{4.2}$$

where d is a given function. The second choice is the *Neumann condition*,

$$\frac{\partial u}{\partial n}(P,t)=v(P,t) \quad \text{for } P \in S \text{ and } t>0, \tag{4.3}$$

where v is a given function and $\partial/\partial n$ denotes normal differentiation.

The Neumann condition (4.3) is physically realisable: the normal velocity is prescribed on S. However, the Dirichlet condition (4.2) does not have an obvious physical interpretation. For this reason, we also consider the 'pressure condition',

$$p(P,t)=-\rho\frac{\partial u}{\partial t}(P,t)=p_S(P,t) \quad \text{for } P \in S \text{ and } t>0, \tag{4.4}$$

where p_S is the prescribed pressure on S. At first sight, it appears that the solution satisfying (4.4) can be obtained from a time derivative of the solution to the Dirichlet problem, but we will see that the situation is not so straightforward. Note that p itself is a wavefunction, so IBVPs could be formulated directly in terms of p.

By linearity, the basic IBVPs outlined above can be reduced to problems with homogeneous boundary conditions or homogeneous initial conditions. This leads to six simpler problems, listed in Table 4.1.

The first three problems in Table 4.1 have zero boundary conditions, the last three have zero initial conditions. The literature is divided, with most authors preferring to work with zero initial conditions. In fact, there is no loss of generality in doing this, because we can construct a wavefunction satisfying the inhomogeneous initial conditions (4.1), and then subtract this wavefunction in order to obtain a new IBVP with zero initial conditions. The relevant construction is due to Poisson; see [516,

§287], [310, §1.6], [203, pp. 201–202], [881, §7.6], [197, §6.3], [518, §72] or [56, §12.2].

Much of the earlier mathematical literature assumes zero boundary conditions. For example, the whole of the famous book on *Scattering Theory* by Lax & Phillips [527] is concerned with Problem ID_0. Wilcox [886, 887] allows non-smooth S and he also considers Problem IN_0. The Wilcox and Lax–Phillips theories are abstract: their goal is not to actually construct solutions. Moreover, although different, it is known that their theories are equivalent [467]. (For extensions of these theories to electromagnetic problems, see [747] and [153, Chapter 6]; for elastic waves, see [769, 584]; and for fluid–solid interactions, see [51].)

Scattering theory originated with physics and it has a large and somewhat untidy literature, part of it concerned with abstract theory and part with specific physical problems. Wilcox [887] *gives ample references but he just mentions the work of Lax and Phillips, who treat the same problem in another way.* (Garding [324, p. 123])

There is also the work of Ladyzhenskaya, which goes back to the early 1950s. In her 1985 book, she discusses Problem ID_0 [511, §IV.3] and also the Robin problem with boundary condition $\partial u/\partial n + \sigma u = 0$ on S where $\sigma(P,t)$ is a prescribed function [511, §IV.5]. Taylor [817, §9.4] limits his discussion to Problem ID_0.

Prior to the work of Lax, Phillips and Wilcox, there is the book by Friedlander [310] on *Sound Pulses*. He also considers problems with zero boundary conditions but, in addition, he discusses scattering problems where a specified wave field (an incident 'sound pulse'; see Section 4.3) interacts with B; this leads to Problems DI_0 and NI_0 [310, pp. 8–9], with zero initial conditions and an inhomogeneous boundary condition. Bleistein [105, §5.4] mentions Problem DI_0. Ladyzhenskaya [511, p. 149] also mentions Problem DI_0 and a related Robin problem. She remarks that such IBVPs 'are more difficult, but are more frequently encountered in applications'. It is these problems, together with Problem PI_0, that will be our main concern.

4.2 Acoustic Energy

One advantage of assuming zero boundary conditions is that we have conservation of energy: energy is input via the initial conditions and there is no subsequent energy flux through the boundary S. Thus, writing in 1970, Morawetz referred to 'energy-conserving boundary conditions' [634, p. 663] and could state that 'for conservative systems we can prove rather easily the existence and uniqueness of solutions' [634, p. 661]. See also [165, Chapter 6, §6.18 and §7.5] and [275, §7.2.2.c]. For broader surveys, see [635, 247, 888].

4.2.1 Energy Conservation

The acoustic kinetic-energy density is $\frac{1}{2}\rho|\mathbf{v}|^2$ and the acoustic potential-energy density is $\frac{1}{2}p^2/(\rho c^2)$; see, for example, [559, §1.3] or [696, p. 39]. Then, using (1.47), the total acoustic energy in a domain $\mathscr{B} \subset \mathbb{R}^3$ is

$$E(t;\mathscr{B}) = \frac{1}{2}\rho \int_{\mathscr{B}} \mathbb{I}(\boldsymbol{r},t)\,\mathrm{d}V(\boldsymbol{r}) \tag{4.5}$$

where (recall the summation convention, Section 3.1)

$$\mathbb{I}(\boldsymbol{r},t) = \frac{\partial u}{\partial x_i}\frac{\partial u}{\partial x_i} + \frac{1}{c^2}\left(\frac{\partial u}{\partial t}\right)^2; \tag{4.6}$$

$\frac{1}{2}\rho\mathbb{I}$ is the total acoustic energy density. From (4.1),

$$E(0;\mathscr{B}) = \frac{1}{2}\rho \int_{\mathscr{B}} \left\{ \frac{\partial u_0}{\partial x_i}\frac{\partial u_0}{\partial x_i} + \frac{1}{c^2}u_1^2 \right\} \mathrm{d}V,$$

which we can assume is finite.

Subject to some mild conditions, E does not change: $E(t;\mathscr{B}) = E(0;\mathscr{B})$. This suggests seeking 'solutions with finite energy', called 'solutions wFE' by Wilcox [887]. For a fixed bounded domain $\mathscr{B}$, we have

$$\begin{aligned}
\frac{1}{\rho}\frac{\mathrm{d}E}{\mathrm{d}t} &= \int_{\mathscr{B}} \left(\frac{\partial u}{\partial x_i}\frac{\partial^2 u}{\partial x_i \partial t} + \frac{1}{c^2}\frac{\partial u}{\partial t}\frac{\partial^2 u}{\partial t^2} \right) \mathrm{d}V \\
&= \int_{\mathscr{B}} \left(\frac{\partial u}{\partial x_i}\frac{\partial^2 u}{\partial x_i \partial t} + \frac{\partial u}{\partial t}\frac{\partial^2 u}{\partial x_i \partial x_i} \right) \mathrm{d}V \\
&= \int_{\mathscr{B}} \frac{\partial}{\partial x_i}\left(\frac{\partial u}{\partial x_i}\frac{\partial u}{\partial t} \right) \mathrm{d}V = \int_{\partial\mathscr{B}} \frac{\partial u}{\partial n}\frac{\partial u}{\partial t}\,\mathrm{d}s,
\end{aligned} \tag{4.7}$$

where $\partial\mathscr{B}$ is the boundary of $\mathscr{B}$ and the normal vector on $\partial\mathscr{B}$ points out of $\mathscr{B}$. The integral over $\partial\mathscr{B}$ vanishes if u satisfies any of the homogeneous boundary conditions (on $\partial\mathscr{B}$) listed in the first three rows of Table 4.1, whence $E(t;\mathscr{B})$ is constant. Thus, these homogeneous conditions are indeed energy-conserving boundary conditions.

However, we are mainly interested in the exterior domain B_e, which is unbounded. To make progress, we assume that the functions u_0 and u_1 in (4.1) have compact support: we can assume that there is a large sphere of radius R containing both the support and the scatterer B (see Fig. 4.1). Then, at time t, $u \equiv 0$ outside a sphere of radius $R + ct$. We can then use the argument above with $\mathscr{B}$ taken as that part of B_e inside an even larger sphere, S_∞, large enough so that $u \equiv 0$ in a neighbourhood of S_∞. The surface integral in (4.7) will then be over $\partial\mathscr{B} = S \cup S_\infty$, and this will vanish.

4.2.2 Energy Inequality

To analyse energy further, we follow [310, Chapter 2, §4] and [506, §2.6.3]. Multiply the wave equation by $\partial u/\partial t$. Some calculation similar to the derivation of (4.7) gives

$$0 = -\frac{2}{c}\frac{\partial u}{\partial t}\Box^2 u = \frac{\partial V_I}{\partial x_I} \tag{4.8}$$

where

$$V_i = -\frac{2}{c}\frac{\partial u}{\partial t}\frac{\partial u}{\partial x_i}, \quad i = 1,2,3, \quad V_4 = \mathbb{I},$$

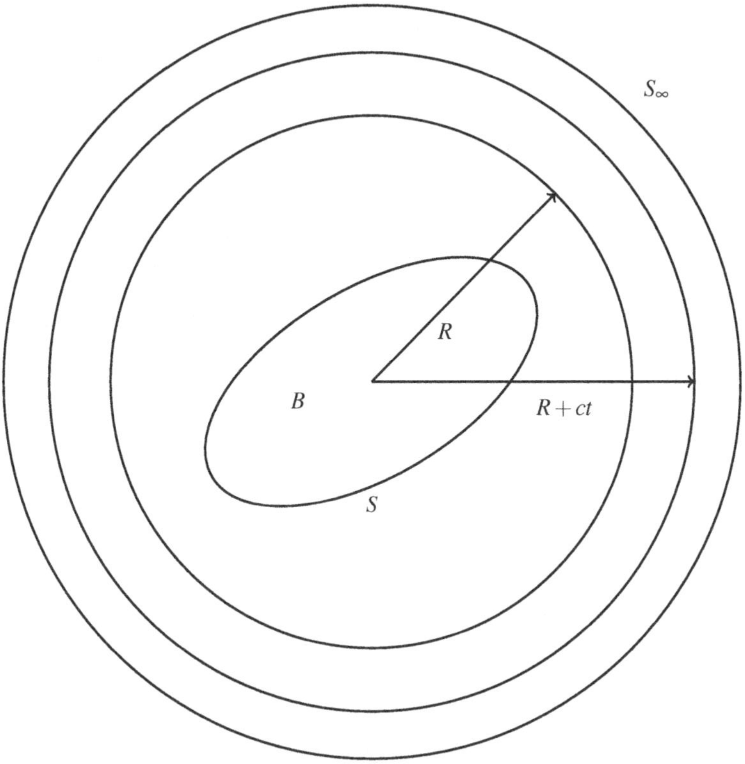

Figure 4.1 The domain $\mathscr{B}$ is bounded by the surface S and the sphere S_∞.

$\mathbb{I}$ is defined by (4.6) and we have used the notation defined in Section 3.1. Then, using the four-dimensional form of the divergence theorem, (3.29), we integrate (4.8) over a space-time region Σ and obtain

$$\int_{\mathscr{S}} V_I N_I \, \mathrm{d}\mathscr{S} = 0, \tag{4.9}$$

where Σ is bounded by the hypersurface $\mathscr{S}$ and $\mathbf{N}$ is a unit outward normal 4-vector to $\mathscr{S}$.

Next, we make a choice for Σ. Fix a time $t_0 \geq 0$ and a point $\boldsymbol{r}_0$. Let

$$\mathscr{B}_0(t) = \{\boldsymbol{r} \in \mathbb{R}^3 \; : \; |\boldsymbol{r} - \boldsymbol{r}_0| \leq ct\},$$

where $0 \leq t \leq t_0$. For fixed t, $\mathscr{B}_0(t)$ is a ball in $\mathbb{R}^3$ centred at $\boldsymbol{r}_0$ with radius ct; $\mathscr{B}_0(t)$ is a hyperplane in space-time.

Let $\mathscr{C}_0$ be a piece of the backward light-cone (3.22) defined by

$$\mathscr{C}_0 = \{\boldsymbol{r} \in \mathbb{R}^3, \, t \in \mathbb{R}_+ \; : \; t = t_0 - |\boldsymbol{r} - \boldsymbol{r}_0|/c, \; 0 \leq t \leq t_0\}.$$

Now choose Σ to be the space-time region bounded by the conoid $\mathscr{C}_0$ and two hyperplanes, $\mathscr{B}_0(t_0)$ and $\mathscr{B}_0(t_0-t)$, thus defining the boundary $\mathscr{S}$ in (4.9). We have $\mathbf{N}=(0,0,0,-1)$ on $\mathscr{B}_0(t_0)$ and $\mathbf{N}=(0,0,0,1)$ on $\mathscr{B}_0(t_0-t)$, with $\mathrm{d}\mathscr{S}=\mathrm{d}V$ on both (see (3.30)). Hence

$$\frac{1}{2}\rho\int_{\mathscr{B}_0(t_0)} V_I N_I\,\mathrm{d}\mathscr{S} = -E(t_0;\mathscr{B}_0(t_0)),$$
$$\frac{1}{2}\rho\int_{\mathscr{B}_0(t_0-t)} V_I N_I\,\mathrm{d}\mathscr{S} = E(t_0-t;\mathscr{B}_0(t_0-t)),$$

using (4.5). On $\mathscr{C}_0$, we can use (3.19), giving $\mathbf{N}=(\boldsymbol{m},1)/\sqrt{2}$ where $\boldsymbol{m}$ is a certain unit 3-vector (computable from (3.18) but not needed in detail here). Then, on $\mathscr{C}_0$,

$$\begin{aligned}V_I N_I &= -\frac{2}{c}\frac{\partial u}{\partial t}\frac{\partial u}{\partial x_i}N_i + \mathbb{I}N_4\\ &= \frac{1}{\sqrt{2}}\left\{\frac{\partial u}{\partial x_i}\frac{\partial u}{\partial x_i} - \frac{2}{c}\frac{\partial u}{\partial t}\frac{\partial u}{\partial x_i}m_i + \frac{1}{c^2}\left(\frac{\partial u}{\partial t}\right)^2\right\}\\ &= \frac{1}{\sqrt{2}}\left(\frac{\partial u}{\partial x_i} - \frac{m_i}{c}\frac{\partial u}{\partial t}\right)\left(\frac{\partial u}{\partial x_i} - \frac{m_i}{c}\frac{\partial u}{\partial t}\right)\end{aligned}$$

as $m_i m_i = |\boldsymbol{m}|^2 = 1$. Hence

$$\int_{\mathscr{C}_0} V_I N_I\,\mathrm{d}\mathscr{S} = \int_{\mathscr{C}_0}\left|\operatorname{grad} u - \frac{1}{c}\boldsymbol{m}\frac{\partial u}{\partial t}\right|^2\frac{\mathrm{d}\mathscr{S}}{\sqrt{2}} \geq 0.$$

Therefore (4.9) gives the energy inequality,

$$0 \leq E(t_0-t;\mathscr{B}_0(t_0-t)) \leq E(t_0;\mathscr{B}_0(t_0)), \quad 0\leq t\leq t_0. \tag{4.10}$$

As a simple consequence, suppose that $u(\boldsymbol{r},0)=0$ and $(\partial u/\partial t)(\boldsymbol{r},0)=0$ for all $\boldsymbol{r}\in\mathscr{B}_0(t_0)$, implying that $E(t_0;\mathscr{B}_0(t_0))=0$. Then (4.10) gives

$$E(t_0-t;\mathscr{B}_0(t_0-t))=0, \quad 0\leq t\leq t_0,$$

which implies that $u(\boldsymbol{r},t)=0$ for all space-time points $(\boldsymbol{r},ct)$ inside the space-time region Σ. This can be seen as a uniqueness theorem for the basic initial-value problem for $\square^2 u = 0$.

4.2.3 Energy Decay

Consider a bounded region $\mathscr{D}$ with $\mathscr{D}\subset B_\mathrm{e}\subset\mathbb{R}^3$. The acoustic energy in $\mathscr{D}$ is $E(t;\mathscr{D})$, (4.5). How does this quantity behave as a function of t, assuming we have energy-conserving boundary conditions? We expect that the energy generated by the initial conditions will be transported away to (spatial) infinity, as no energy is generated by the scatterer's boundary S, and so we expect that $E(t;\mathscr{D})\to 0$ as $t\to\infty$. This decay can be proved [526]. The rate of decay is known to depend on the shape of S. Suppose first that S is star-shaped (meaning that there is a point O inside S for which S can be defined by $r=r_0(\theta,\phi)$, where r_0 is a single-valued function and r, θ and ϕ are spherical polar coordinates centred at O: every point on S can be 'seen' from O).

For such surfaces, it can be proved that E decays exponentially: $E(t;\mathscr{D}) = O(e^{-\alpha t})$ as $t \to \infty$ for some $\alpha > 0$ [525, 633]. It is known that some restrictions on the shape of S are needed in order to secure exponential decay [634], but it is also known that the condition of star-shapedness is not necessary. For results in this direction, see [636] and [532, p. 278].

4.3 Incident Sound Pulses

An incident sound pulse is a wavefunction u_{inc}, and it can take various forms. A common class consists of *plane step pulses*, defined by

$$u_{\mathrm{inc}}(x,y,z,t) = w_{\mathrm{inc}}(t - [z - z_0]/c)\,\mathrm{H}(t - [z - z_0]/c), \tag{4.11}$$

where H is the Heaviside unit function ($\mathrm{H}(x) = 1$ for $x > 0$, $\mathrm{H}(x) = 0$ for $x < 0$), $w_{\mathrm{inc}}(t)$ is a specified function of one variable, and z_0 is a constant. As the name suggests, u_{inc} represents a plane step, a wavefront, propagating in the positive z-direction. Ahead of the front ($z > z_0 + ct$), $u_{\mathrm{inc}} \equiv 0$, and behind the front ($z < z_0 + ct$), $u_{\mathrm{inc}} = w_{\mathrm{inc}}(t - [z - z_0]/c)$. It is usual to choose z_0 so that the wavefront first reaches S at $t = 0$. For example, if S is a sphere of radius a, centred at the origin, we would choose $z_0 = -a$.

The wavefront $\Gamma(t)$ is at $z = z_0 + ct$ and the unit normal to Γ, $\boldsymbol{n}$, points in the $+z$ direction. Using the definition (3.25), we have $[\![u_{\mathrm{inc}}]\!] = w_{\mathrm{inc}}(0)$, a constant, and

$$[\![p]\!] = \rho c\, w'_{\mathrm{inc}}(0) = \rho c\, \boldsymbol{n} \cdot [\![\boldsymbol{v}]\!],$$

in agreement with the jump relation (3.46).

Plane step pulses propagating in other directions are easily constructed. Similar to (2.3), let $\widehat{\boldsymbol{e}}$ be a constant unit vector. Then

$$u(x,y,z,t) = f(t - \widehat{\boldsymbol{e}} \cdot \boldsymbol{r}/c) \tag{4.12}$$

is a wavefunction, for any smooth (twice differentiable) function $f(\xi)$. It represents a plane wave propagating in the direction of $\widehat{\boldsymbol{e}}$. Introducing an appropriate Heaviside function, as in (4.11), gives a plane step pulse with discontinuities across the wavefront.

Returning to (4.11), various choices for w_{inc} have been made in the literature. These include

$$w_{\mathrm{inc}}(t) = \mathscr{U}_0, \qquad w_{\mathrm{inc}}(t) = \mathscr{U}_0 \cos \omega_0 t,$$
$$w_{\mathrm{inc}}(t) = \mathscr{U}_0\, \omega_0 t, \quad w_{\mathrm{inc}}(t) = \mathscr{U}_0 \sin \omega_0 t,$$

where $\mathscr{U}_0$ and ω_0 are constants. The first two of these have $w_{\mathrm{inc}}(0) = \mathscr{U}_0 \neq 0$, whereas the second two have $w_{\mathrm{inc}}(0) = 0$; often, the latter two are preferable because they satisfy $[\![u_{\mathrm{inc}}]\!] = 0$. Another choice with this property is Friedlander's 'favourite incident pulse' [253, p. 282], $w_{\mathrm{inc}}(t) = \mathscr{U}_0\, \omega_0 t\, e^{-\omega_0 t}$.

If we want $w_{\text{inc}}(0) = 0$ and $w'_{\text{inc}}(0) = 0$, we can choose

$$w_{\text{inc}}(t) = 2\sin\omega_0 t - \sin 2\omega_0 t. \tag{4.13}$$

Equation (4.11) can be generalised slightly to

$$u_{\text{inc}}(x,y,z,t) = w_{\text{inc}}(\xi)\{\mathrm{H}(\xi) - \mathrm{H}(\xi - \xi_0)\}, \tag{4.14}$$

where $\xi = t - [z - z_0]/c$ and ξ_0 is a positive constant. This gives a pulse of finite width, $c\xi_0$, with $u_{\text{inc}} \equiv 0$ ahead of the first front at $z = z_0 + ct$ and behind the second front at $z = z_0 + ct - c\xi_0$. For example, if we combine (4.13) with $\xi_0 = 2\pi/\omega_0$, we obtain a 'wavelet' with $w_{\text{inc}}(t) = 0$ and $w'_{\text{inc}}(t) = 0$ at both $t = 0$ and $t = \xi_0$ [443, eqn (8.87)].

Rather than plane pulses, we can also consider *spherical step pulses*, generated by a simple source located at a point $\boldsymbol{r}'$, (2.12). Thus

$$u_{\text{inc}}(\boldsymbol{r},t) = R^{-1} w_{\text{inc}}(t - [R - R_0]/c)\,\mathrm{H}(t - [R - R_0]/c), \tag{4.15}$$

where $R = |\boldsymbol{r} - \boldsymbol{r}'|$, R_0 is a constant and $w_{\text{inc}}(t)$ is a specified function of one variable. There is a spherical wavefront at $R = R_0 + ct$, with $u_{\text{inc}} \equiv 0$ ahead of the front. Again, it is usual to choose R_0 so that the wavefront first reaches S at $t = 0$. For example, if S is a sphere of radius a, centred at the origin, and we place the source at $(x,y,z) = (0,0,z')$ with $z' > 0$, we would choose $R_0 = |z' - a|$. Note that we could place the source *inside* S if we want to generate exact solutions of IBVPs; these can be useful for testing numerical schemes [356].

4.4 Compatibility Conditions

The solution to an elliptic [PDE] *is weakly singular in the corners of the domain unless the forcing and boundary data are special. These 'corner' singularities are well known. ... It is less well known that hyperbolic* [PDEs] *are equally prone to singularities in the corners of the space-time domain—that is to say, at the spatial boundaries at the initial time, t = 0. Unless the initial conditions, boundary data and forcing satisfy* [certain] *'compatibility' conditions, the kth spatial derivative of u will be unbounded at the spatial boundary for some finite order k. ... In the absence of damping,* [these] *weak singularities propagate away from the boundary and persist forever.* (Boyd & Flyer [124, p. 281])

The compatibility conditions mentioned have been well studied; Boyd & Flyer [124] refer to the early work of Ladyzhenskaya (see [511, p. 165]) and to papers by Rauch & Massey [710] and Smale [781], among others. A later paper by Temam [819] is also relevant. Some authors are interested mainly in computational aspects [124, 819], others in analytical results on smoothness of solutions [710, 781]. For the one-dimensional wave equation, see [445, p. 7] and [734, p. 12].

For simple examples, consider Problems ID_0 and DI_0, and examine the behaviour near the space-time corner $\mathscr{E}$, where $t = 0$ and $P \in S$. For Problem ID_0, we have the boundary condition $u(P,t) = 0$ for $P \in S$ and the initial condition $u(P,0) = u_0(P)$ for

$P \in B_e$. If we want both of these to hold at $\mathscr{E}$, then $u_0(P) = 0$ for $P \in S$; this is a compatibility condition. Similarly, for Problem DI_0, the simplest compatibility condition is $d(P,0) = 0$, $P \in S$. If these conditions are not satisfied, u will be discontinuous across characteristics passing through $\mathscr{E}$, in general.

Higher-order compatibility conditions have been worked out. For example, Boyd & Flyer [124, Theorem 1] give formulas for Problems ID_0 and IN_0. The underlying question is often: What do we have to do at $\mathscr{E}$ to make the solution smoother? In our applications, we have some kind of incident wave (such as a pressure step pulse), with known properties, and we want to calculate the scattered waves: we do not have the luxury of being able to make the solution smoother by adjusting the conditions at $\mathscr{E}$. In addition, we also have to ensure that physical constraints across wavefronts are satisfied.

4.5 Bateman's IBVP for a Sphere

Consider Problem DI_0 for a sphere of radius a. Thus, in terms of spherical polar coordinates, r, θ and ϕ, the problem is to determine $u(r,\theta,\phi,t)$ subject to zero initial conditions and the boundary condition

$$u(a,\theta,\phi,t) = d(\theta,\phi),$$

where the given function d does not depend on t. Bateman [65, 66] showed that the solution of this problem is

$$u(r,\theta,\phi,t) = (a/r)\overline{d}(r,\theta,\phi,t),$$

where $\overline{d}$ is the mean value of d around that circle on the sphere $r = a$ whose points are all at a distance ct from the point at (r,θ,ϕ); if there is no such circle, $u = 0$.

To see that u is a wavefunction, suppose that $d(\theta,\phi) = Y_n^m(\theta,\phi)$, where Y_n^m is a spherical harmonic of degree n. In an earlier paper, Bateman [63] (see also [67, §7.42]) had shown that

$$\overline{d}(r,\theta,\phi,t) = P_n(\cos\alpha)Y_n^m(\theta,\phi),$$

where P_n is a Legendre polynomial and the angle $\alpha(r,t)$ is defined by

$$c^2t^2 = r^2 + a^2 - 2ar\cos\alpha.$$

Then, writing $\Box^2 u$ as (2.6) and substituting $u = (a/r)\overline{d}$, we obtain

$$\begin{aligned}\Box^2 u = \frac{aY_n^m}{r}\bigg\{&\left(\left[\frac{\partial}{\partial r}\cos\alpha\right]^2 - \frac{1}{c^2}\left[\frac{\partial}{\partial t}\cos\alpha\right]^2\right)P_n''(\cos\alpha)\\ &+\left(\frac{\partial^2}{\partial r^2}\cos\alpha - \frac{1}{c^2}\frac{\partial^2}{\partial t^2}\cos\alpha\right)P_n'(\cos\alpha) - \frac{n(n+1)}{r^2}P_n(\cos\alpha)\bigg\}\\ = -ar^{-3}Y_n^m&\left\{\sin^2\alpha\, P_n''(\cos\alpha) - 2\cos\alpha\, P_n'(\cos\alpha) + n(n+1)\,P_n(\cos\alpha)\right\},\end{aligned}$$

and this vanishes because of the differential equation satisfied by P_n, (2.34). For some extensions of Bateman's results, see a paper by Erdélyi [261].

4.6 Further Problems and Boundary Conditions

Up to now, we have considered three different boundary conditions on the surface S in which u, $\partial u/\partial n$ or $\partial u/\partial t$ is prescribed. Many other choices for the boundary condition on S could be made. For example, Hagstrom [379, eqn (2.3)] considers

$$\alpha \frac{\partial u}{\partial t} + \beta \frac{\partial u}{\partial n} + \gamma u = f \quad \text{on } S, \tag{4.16}$$

where α, β, γ and f are given. For (4.16) with $\gamma = 0$, see [531, 375]. In [94, §7.1.1], the authors consider an IBVP posed in a half-space, $z > 0$, with

$$\frac{\partial u}{\partial t} + \beta \frac{\partial u}{\partial z} + \boldsymbol{a} \cdot \operatorname{grad} u = 0 \quad \text{on } z = 0,$$

where $\boldsymbol{a}$ is a constant vector.

More complicated situations arise if waves can pass through S (leading to various transmission problems) or if S itself can move. We describe a few of these situations below.

A variant of (4.16), involving the boundary values of $\nabla^2 u$, is

$$\nabla^2 u + \beta \frac{\partial u}{\partial n} + \gamma u = 0 \quad \text{on } S. \tag{4.17}$$

This is called the *Wentzell boundary condition* after the author of a 1959 paper [857] (although several different spellings of his name are common). For more information, see [320] and [347].

In another variant of (4.16), the function f is adjusted (controlled) so as to minimise some measure of the scattered field. For such 'furtivity problems' [591], the scatterer may be described as 'active' or 'smart' [289].

4.6.1 Locally Reacting Surfaces

Love [574, p. 94] considered the following problem:

Sphere Vibrating Radially in Air.

The sphere will be treated as an elastic membrane of mass M and surface density σ, which is maintained nearly at a definite radius a by springs. It will be supposed that, in the absence of the air, the frequency of vibration of the sphere would be $\omega_0/2\pi$. If ρ is the density of the air, a the radius of the sphere when in equilibrium under the pressure of the air, $a+\xi$ its radius at time t, p the excess of pressure above that in equilibrium, the equation of motion of the sphere is

$$M(\ddot{\xi} + \omega_0^2 \xi) = -4\pi a^2 p,$$

where dots denote differentiation with respect to t. ... [Use $M = 4\pi a^2 \sigma$ and $p = -\rho\, \partial u/\partial t$.] The above equation may be written

$$\sigma(\ddot{\xi} + \omega_0^2 \xi) = \rho\, \partial u/\partial t \quad \text{at } r = a. \tag{4.18}$$

[The velocity potential u] must satisfy the [wave] equation outside the sphere and the condition

$$\dot{\xi} = \frac{\partial u}{\partial r} \quad \text{at } r = a. \tag{4.19}$$

Love solved the problem exploiting spherical symmetry using the basic solution (2.10) and methods developed in Chapter 6. Bromwich [130, §7] solved the same problem using what we would recognise now as Laplace transforms.

The surface envisaged by Love behaves in a simple (and perhaps unrealistic) manner: 'pushing the surface at one point does not move it elsewhere' [696, p. 110] or 'the motion, normal to the surface, of one portion of the surface is ... independent of the motion of any other part of the' surface [644, p. 260]. Such surfaces are called *locally reacting surfaces* [125, §3.4], [126, §2.3]. Their study in the mathematical literature goes back to a short paper by Beale & Rosencrans [77]; they used

$$m(P)\,\frac{\partial^2 \xi}{\partial t^2} + d(P)\,\frac{\partial \xi}{\partial t} + k(P)\,\xi = \rho\,\frac{\partial u}{\partial t} \quad \text{and} \quad \frac{\partial \xi}{\partial t} = \frac{\partial u}{\partial n} \quad \text{for } P \in S \tag{4.20}$$

and $t > 0$, where m, d and k are prescribed, $\xi(P,t)$ represents the outward displacement of S and, as usual, the normal vector on S points outwards into B_e. Evidently, (4.20) generalises (4.18) and (4.19). For more information, see [76] and [320, §2].

The local nature of (4.18) and (4.20) is clear because we do not see any tangential derivatives of ξ: any elastic properties of S are not included. If such properties are important, we can model S as a thin elastic shell; see Section 4.6.4. For a simpler problem, we may be able to model S (or a piece of S) as a thin membrane; see Section 4.6.3.

Equations (4.20) have been used in simple models of the basilar membrane in the inner ear [545, eqns (3) and (4)], [468, §20.2.2]. This membrane is a moving interface between two fluid domains, with the right-hand side of $(4.20)_1$ replaced by the jump in the acoustic pressure across the membrane; see also Section 4.6.3. Actually, the 'name "basilar membrane" is misleading, as it is not a true membrane [because] when it is cut, the edges do not retract. Thus it is not under tension; resistance to movement comes from the bending elasticity' [468, p. 948]. Consequently, more elaborate models have been developed; see, for example, [790, 436] and [659].

4.6.2 Transmission Problems

Suppose that the interior of S, B, is occupied by another acoustic medium, with density ρ_0 and sound speed c_0. The motion in B is determined from a velocity potential u_0 that satisfies the wave equation $\nabla^2 u_0 = c_0^{-2}\, \partial^2 u_0/\partial t^2$. The potentials u in B_e and u_0 in B are connected by a pair of transmission conditions on the interface S; these are usually continuity of pressure and continuity of normal velocity:

$$\rho \frac{\partial u}{\partial t} = \rho_0 \frac{\partial u_0}{\partial t} \quad \text{and} \quad \frac{\partial u}{\partial n} = \frac{\partial u_0}{\partial n} \quad \text{on } S.$$

The resulting *transmission problem* is appropriate for scattering by a blob of fluid surrounded by a different fluid.

There are analogous problems of *fluid–solid interaction*, in which B is filled with an elastic solid. Then the elastodynamic displacement vector $\boldsymbol{u}$ satisfies (1.91) in B. The usual relations on S are continuity of normal velocity, zero tangential tractions (because the exterior fluid is inviscid) and a balance between the exterior presssure and the normal traction. The resulting fluid–solid interaction problem has received an increasing amount of attention recently (see [860, 423, 424, 48, 555, 342]). Replacing the solid domain by a thin elastic shell leads to problems that have been studied extensively because of their application to metal structures submerged in water; see Section 4.6.4.

4.6.3 Membrane Problems

Suppose that M is a flat membrane occupying a portion of the plane $z = 0$; M could be the skin of a drum [714] (so that M is a portion of S) or it could be set into a rigid baffle [503, 504]. Let $w(x,y,t)$ be the out-of-plane displacement of the membrane in the z-direction. It satisfies a forced two-dimensional wave equation,

$$T\left(\frac{\partial^2 w}{\partial x^2} + \frac{\partial^2 w}{\partial y^2}\right) - \rho_{\mathrm{m}} \frac{\partial^2 w}{\partial t^2} = [\![p]\!], \tag{4.21}$$

where T is the membrane tension (force per unit length), ρ_{m} is the mass per unit area of the membrane and $[\![p]\!]$ is the jump in the acoustic pressure across the membrane,

$$[\![p]\!](x,y,t) = p(x,y,0+,t) - p(x,y,0-,t). \tag{4.22}$$

The wave speed in the membrane is $\sqrt{T/\rho_{\mathrm{m}}}$. In addition, the vertical components of the velocity must be continuous across the membrane,

$$\frac{\partial u}{\partial z} = \frac{\partial w}{\partial t} = \frac{\partial u_0}{\partial z} \quad \text{at } z = 0, \tag{4.23}$$

where u and u_0 are the velocity potentials above ($z > 0$) and below the membrane, respectively. This simple model has served as a prototype for acoustic interactions with wave-bearing boundaries.

4.6.4 Elastic Shells

Suppose that the surface S is a thin elastic shell, with compressible fluid both inside and outside. The equations of motion of such a shell are very complicated, in general; see, for example, [498, 665, 666, 856]. Instead of writing down these equations in full generality, we consider one very special case, axisymmetric motions of a spherical shell. In terms of spherical polar coordinates (r,θ,ϕ) (see (2.5)), the midsurface of the sphere is at $r=a$ and there is no dependence on the azimuthal angle ϕ. The displacement of the midsurface has the form

$$\boldsymbol{u}(\theta,t)=u_r(\theta,t)\,\widehat{\boldsymbol{r}}+u_\theta(\theta,t)\,\widehat{\boldsymbol{\theta}}.$$

The governing equations for u_r and u_θ are as follows:

$$\begin{aligned}(1+\beta^2)&\left[\frac{\partial^2 u_\theta}{\partial\theta^2}+\frac{\partial u_\theta}{\partial\theta}\cot\theta-\left(\nu+\cot^2\theta\right)u_\theta\right]-\beta^2\frac{\partial^3 u_r}{\partial\theta^3}\\ &-\beta^2\frac{\partial^2 u_r}{\partial\theta^2}\cot\theta+\left[1+\nu+\beta^2(\nu+\cot^2\theta)\right]\frac{\partial u_r}{\partial\theta}=\frac{a^2}{c_p^2}\frac{\partial^2 u_\theta}{\partial t^2},\end{aligned}\tag{4.24}$$

$$\begin{aligned}\beta^2&\frac{\partial^3 u_\theta}{\partial\theta^3}+2\beta^2\frac{\partial^2 u_\theta}{\partial\theta^2}\cot\theta-\left[(1+\nu)(1+\beta^2)+\beta^2\cot^2\theta\right]\frac{\partial u_\theta}{\partial\theta}\\ &+\left[\beta^2(2-\nu+\cot^2\theta)-(1+\nu)\right]u_\theta\cot\theta-\beta^2\frac{\partial^4 u_r}{\partial\theta^4}-2\beta^2\frac{\partial^3 u_r}{\partial\theta^3}\cot\theta\\ &+\beta^2(1+\nu+\cot^2\theta)\frac{\partial^2 u_r}{\partial\theta^2}-\beta^2(2-\nu+\cot^2\theta)\frac{\partial u_r}{\partial\theta}\cot\theta\\ &-2(1+\nu)u_r=\frac{a^2}{c_p^2}\left(\frac{\partial^2 u_r}{\partial t^2}+\frac{[\![p]\!]}{\rho_s h}\right).\end{aligned}\tag{4.25}$$

In these equations, the solid shell has thickness h, Poisson's ratio ν, mass density ρ_s, and compressional wavespeed c_p (see Section 1.6.2). The parameter $\beta^2=\frac{1}{12}(h/a)^2$. The motion of the shell is driven by $[\![p]\!]$, the jump in the acoustic pressure $p(r,\theta,t)$ across the midsurface of the shell, defined by

$$[\![p]\!](\theta,t)=p(a+,\theta,t)-p(a-,\theta,t).$$

Equations (4.24) and (4.25) can be found in [457, §7.17]; see also [392, eqns (7) and (8)] and [258, eqns (9) and (10)]. (In line 8 of [258, eqn (10)], replace $(1-\nu)$ by $(1+\nu)$.) In addition, the normal components of the velocity must be continuous across the shell, whence

$$\frac{\partial u}{\partial r}=\frac{\partial u_r}{\partial t}=\frac{\partial u_0}{\partial r}\quad\text{at } r=a,\tag{4.26}$$

where u is the (total) velocity potential outside the shell and u_0 is the velocity potential inside the shell.

Equations (4.24) and (4.25) apply to a thin elastic spherical shell of constant thickness h, taking extensional and bending stresses into account. Ignoring the latter leads to the equations for a thin elastic spherical *membrane*. These equations are obtained by putting $\beta^2=0$, giving

$$\frac{\partial^2 u_\theta}{\partial\theta^2}+\frac{\partial u_\theta}{\partial\theta}\cot\theta-\left(\nu+\cot^2\theta\right)u_\theta+(1+\nu)\frac{\partial u_r}{\partial\theta}=\frac{a^2}{c_p^2}\frac{\partial^2 u_\theta}{\partial t^2},\tag{4.27}$$

$$\frac{\partial u_\theta}{\partial\theta}+u_\theta\cot\theta+2u_r=-\frac{a^2}{(1+\nu)c_p^2}\left(\frac{\partial^2 u_r}{\partial t^2}+\frac{[\![p]\!]}{\rho_s h}\right).\tag{4.28}$$

Modifications to these equations have been developed. For example, Kuo et al. [509] incorporate prestrain so as to model motions of inflated balloons. Grinfeld [358] has derived equations that allow for thickness variations, an essential feature in the modelling of soap-bubble oscillations.

For *spheroidal* elastic shells, see [394] and references therein.

4.6.5 Hydrodynamic Problems

The governing equations for small-amplitude water waves in the time domain were given in Section 1.6.3. Thus the velocity potential $u(x,y,z,t)$ satisfies Laplace's equation in the water together with the free-surface boundary condition at $z=0$, (1.93).

The simplest IBVPs arise when there are no objects in the water so that we have $\nabla^2 u = 0$ everywhere in the half-space $z<0$. The appropriate initial conditions are imposed on the mean free surface,

$$u(x,y,0,0) = u_0(x,y) \quad \text{and} \quad (\partial u/\partial t)(x,y,0,0) = u_1(x,y), \tag{4.29}$$

where u_0 and u_1 are specified functions. (Linearity implies that we can solve one problem with $u_0 \equiv 0$ and one with $u_1 \equiv 0$.) Finding $u(x,y,z,t)$ is known as the *Cauchy–Poisson problem* [516, §255]. The problem can be solved by introducing Fourier transforms in the horizontal directions (or, equivalently, Hankel transforms). For some details, see [785, §32.6], [795, §6.4] and [534, §50].

Suppose now that there is a rigid object B immersed in the water; the object has wetted surface S. There are two cases to consider. In the first (and simpler) case, the motion of B is prescribed, which means that $\partial u/\partial n$ is given on S. For some analysis of this problem, see [299, 315] and Section 10.7. For impulsive motions of B, starting from rest, one may use power series in t [626].

Perhaps more interesting is the case where B represents a floating rigid body (such as a ship); its motion has to be determined by solving the relevant equations of motion together with the equations for u. The resulting IBVP has been formulated and studied in several papers, such as [444, 872, 873, 557, 609].

4.7 The Forced Wave Equation

Consider the forced wave equation,

$$\square^2 u = -q(\boldsymbol{r},t) \tag{4.30}$$

where $\square^2 u = \nabla^2 u - c^{-2}\partial^2 u/\partial t^2$ and q is specified. We assume that q is compactly supported in space and time,

$$q(\boldsymbol{r},t) = 0 \quad \text{for } r = |\boldsymbol{r}| > a, \text{ for } t \le 0 \text{ and for } t > t_0, \tag{4.31}$$

where a and t_0 are positive constants. (These assumptions can be relaxed.) For forward problems, we seek $u(\boldsymbol{r},t)$ subject to zero initial conditions (Section 4.7.1). For inverse problems, the goal is to determine q from a knowledge of u (Section 4.7.4).

We remark that there is a growing literature on stochastic versions of (4.30) in which q is a random field of some kind; see, for example, [213] and [179, Chapter 5].

4.7.1 Forward Problem

In the simplest case, we are given q and we seek $u(\boldsymbol{r},t)$ everywhere in space for $t > 0$. This field could be regarded as the incident field u_{inc} when there is an obstacle present.

We look for a causal solution of (4.30), meaning that $u(\boldsymbol{r},t) \equiv 0$ for all $\boldsymbol{r}$ when $t \leq 0$. This implies that we seek a *particular* solution of (4.30); all non-trivial solutions of the homogeneous equation $\Box^2 u = 0$ are inadmissible here.

A Laplace transform of (4.30) with respect to t gives

$$\nabla^2 U - (s/c)^2 U = -Q(\boldsymbol{r},s), \tag{4.32}$$

where $U(\boldsymbol{r},s) = \mathscr{L}\{u(\boldsymbol{r},t)\}$ and $Q(\boldsymbol{r},s) = \mathscr{L}\{q(\boldsymbol{r},t)\}$. (See Chapter 5 for more details on the use of $\mathscr{L}$; in particular, $\mathscr{L}\{u\}$ is defined by (5.1).)

We can solve (4.32) using a *volume potential*; these are defined by

$$W(P) = W(\boldsymbol{r}) = \int_B \varphi(\boldsymbol{r}') \frac{\mathrm{e}^{-\lambda R}}{4\pi R}\, \mathrm{d}\boldsymbol{r}',$$

where B is a bounded domain, φ is a given function, λ is a constant and $R = |\boldsymbol{r} - \boldsymbol{r}'|$. Then [190, §8.2], [599, Appendix H]

$$(\nabla^2 - \lambda^2) W = \begin{cases} 0, & P \in B_{\mathrm{e}}, \\ -\varphi(P), & P \in B, \end{cases}$$

where B_{e} is the region exterior to B. Hence

$$U(\boldsymbol{r},s) = \int Q(\boldsymbol{r}',s) \frac{\mathrm{e}^{-(s/c)R}}{4\pi R}\, \mathrm{d}\boldsymbol{r}', \tag{4.33}$$

where the integration is over all of $\mathbb{R}^3$. Making use of (1.74) and (4.31), we have $Q(\boldsymbol{r},s)\mathrm{e}^{-sR/c} = \mathscr{L}\{q(\boldsymbol{r},t - R/c)\}$, and so we can invert (4.33) to obtain

$$u(\boldsymbol{r},t) = \int q(\boldsymbol{r}',t - R/c)\, \frac{\mathrm{d}\boldsymbol{r}'}{4\pi R}, \tag{4.34}$$

which expresses u as a retarded volume potential; see (2.13). This formula can be found in many places, such as [871, §63], [311, eqn (3.15)], [445, eqn (2.29)], [230, eqn (4.1)], [205, eqn (2.9)], [245, eqn (7.8)], [422, eqn (1.6.6)], [303, eqn (2.24)], [271, eqn (16.35)] and [229, §1.3.1]. Multipole expansions of (4.34) can be developed; see [405] for details and [234, §2.3] for an application. For applications of (4.34) to moving sources in the context of photoacoustics, see [36].

As an alternative, given that we have to solve (4.32) for U everywhere, we can use a three-dimensional Fourier transform $\mathscr{F}_3$, defined by

$$\widehat{U}(\boldsymbol{\xi},s) = \mathscr{F}_3\{U(\boldsymbol{r},s)\} = \int U(\boldsymbol{r},s)\,\mathrm{e}^{-\mathrm{i}\boldsymbol{\xi}\cdot\boldsymbol{r}}\, \mathrm{d}\boldsymbol{r}.$$

Fourier transforming (4.32) gives

$$(\xi^2 + (s/c)^2)\,\widehat{U}(\boldsymbol{\xi},s) = \widehat{Q}(\boldsymbol{\xi},s)$$

where $\xi = |\boldsymbol{\xi}|$. Hence

$$\widehat{U}(\boldsymbol{\xi},s) = \frac{c^2}{s^2+\xi^2c^2}\,\widehat{Q}(\boldsymbol{\xi},s) = c^2\,\mathscr{L}\left\{\frac{\sin(\xi ct)}{\xi c}\right\}\mathscr{L}\{\widehat{q}(\boldsymbol{\xi},t)\},$$

where $\widehat{q}(\boldsymbol{\xi},t) = \mathscr{F}_3\{q(\boldsymbol{r},t)\}$. Inverting, using the convolution theorem for Laplace transforms,

$$\widehat{u}(\boldsymbol{\xi},t) = c^2\int_0^t \frac{\sin(\xi c[t-\tau])}{\xi c}\,\widehat{q}(\boldsymbol{\xi},\tau)\,\mathrm{d}\tau, \tag{4.35}$$

where $\widehat{u}(\boldsymbol{\xi},t) = \mathscr{F}_3\{u(\boldsymbol{r},t)\}$; for a similar formula, see [111, eqn (67)].

Inverting $\mathscr{F}_3$ in (4.35) gives

$$u(\boldsymbol{r},t) = \frac{c^2}{(2\pi)^3}\int \mathrm{e}^{\mathrm{i}\boldsymbol{\xi}\cdot\boldsymbol{r}}\int_0^t \frac{\sin(\xi c[t-\tau])}{\xi c}\int q(\boldsymbol{r}',\tau)\,\mathrm{e}^{-\mathrm{i}\boldsymbol{\xi}\cdot\boldsymbol{r}'}\,\mathrm{d}\boldsymbol{r}'\,\mathrm{d}\tau\,\mathrm{d}\boldsymbol{\xi}. \tag{4.36}$$

Formally, we may hope to change the order of integration, giving

$$u(\boldsymbol{r},t) = \int_0^t\int \mathscr{G}(\boldsymbol{r}-\boldsymbol{r}',t-\tau)\,q(\boldsymbol{r}',\tau)\,\mathrm{d}\boldsymbol{r}'\,\mathrm{d}\tau \tag{4.37}$$

where

$$\mathscr{G}(\boldsymbol{r},t) = \frac{c^2}{(2\pi)^3}\int \frac{\sin(\xi ct)}{\xi c}\,\mathrm{e}^{\mathrm{i}\boldsymbol{\xi}\cdot\boldsymbol{r}}\,\mathrm{d}\boldsymbol{\xi}. \tag{4.38}$$

However the triple integral defining $\mathscr{G}$ is divergent, although it may be expressed in terms of Dirac deltas. Indeed, we may interpret $\mathscr{G}$ as the *causal fundamental solution* of the wave equation, which means that $\mathscr{G}(\boldsymbol{r}-\boldsymbol{r}',t-\tau)$ solves the forced wave equation (4.30) with $q(\boldsymbol{r},t) = \delta(\boldsymbol{r}-\boldsymbol{r}')\,\delta(t-\tau)$ together with the causal requirement that $\mathscr{G}(\boldsymbol{r}-\boldsymbol{r}',t-\tau) \equiv 0$ for $t<\tau$. This fundamental solution is given by

$$\mathscr{G}(\boldsymbol{r},t) = (4\pi r)^{-1}\,\delta(t-r/c). \tag{4.39}$$

See, for example, [643, eqn (11.3.4)], [248, eqn (10.2.20)], [56, eqn (11.2.1)], [421, eqn (1.7.20)], [271, eqn (16.33)], [229, eqn (1.20a)], [541, §5.3] and (9.31).

Inspection of (4.39) shows that $\mathscr{G}(\boldsymbol{r},t) = 0$ for $r \neq ct$. We can interpret $\mathscr{G}$ as a simple source (2.11), with a spherical singular wavefront propagating away from $r=0$ with speed c.

The formula (4.37) with (4.38) can be found in [229, eqns (1.19) and (1.33)] and [829, eqns (2)–(4)]. For many other integral representations for $u(\boldsymbol{r},t)$, see [230].

4.7.2 The Limiting Amplitude Principle

Suppose that time-harmonic forcing is switched on at $t=0$, giving

$$q(\boldsymbol{r},t) = \tilde{q}(\boldsymbol{r})\,\mathrm{H}(t)\sin\omega t.$$

Then, we might expect that, after transients have died out, a time-harmonic solution persists, one that would have been obtained by solving a corresponding time-harmonic (frequency domain) problem (which includes the Sommerfeld radiation condition) directly. This idea is known as the *limiting amplitude principle*. The earliest papers and proofs are Russian; see Eidus [255] for a review. Morawetz [631, 632] has two well-known papers on this topic; see also [137, 522] and [271, §19.5]. For an example of an IBVP with a time-harmonic boundary condition, see (6.20). The limiting amplitude principle may also be used computationally, to solve frequency-domain problems by solving a related time-domain problem instead [502].

4.7.3 Aerodynamic Sound

Lighthill's 'acoustic analogy' gives a method for estimating aerodynamic sound production [560, 563]. 'It is based on decoupling the acoustic problem from the fluid dynamic one: the velocity and pressure fields, obtained through a separate numerical simulation, are used as source terms in an inhomogeneous wave equation whose solution reconstructs the noise' [184, p. 1]: there is no 'significant back-reaction of the sound produced on the flow field itself' [560, p. 565]. In its simplest form, the governing equations are reduced exactly to a forced wave equation (4.30) with a complicated forcing q. This can then be solved using (4.34). In his admirable historical survey, Morfey summarises as follows:

> *Most of the recent progress in understanding the sound produced by blades in airflows has been based on the acoustic analogy proposed by Lighthill* [560] ... [in which] *the pressure* p *(or density* ρ*) is formally regarded as a small-amplitude sound field, driven in a uniform fluid at rest. Sources of sound are formally identified with non-zero values of the quantity* q. ... *Primitive versions of the acoustic analogy were given earlier by Rayleigh* [711, §296] *and Lamb* [516, §291]. ... *Whereas* [they] *restricted their analyses to small disturbances, the essential step later taken by Lighthill was to incorporate the non-linear terms which describe sound generation by turbulence. This was accomplished by evaluating* q*, the equivalent source distribution, from the general equations of fluid motion.* (Morfey [637, p. 593])

Lighthill's approach has been extended in many directions. For example, Curle [212] showed how to include the effects of solid bodies, using the Kirchhoff formula (Section 9.1). For more information, see [234, §2.1], [205, §3], [245, §7.4], [294], [422, Chapter 2], [420, Chapter 4], [22, Chapter 9] and [345, Chapter 4]. For criticisms of the acoustic analogy, see [208, 290] and [865, §3].

4.7.4 Inverse Source Problems

Suppose that u is an expanding wave, meaning that $u(\boldsymbol{r},t) = 0$ for $r > ct + a$ and some constant a. The corresponding far-field pattern, f_0, is defined by (2.18):

$$f_0(\widehat{\boldsymbol{r}},t) = \lim_{r\to\infty} \{ru(\boldsymbol{r},t + r/c)\}.$$

Representing u using (4.34), and making use of $R = r - \hat{\boldsymbol{r}} \cdot \boldsymbol{r}' + O(r^{-1})$ as $r \to \infty$, we obtain

$$f_0(\hat{\boldsymbol{r}},t) = \frac{1}{4\pi} \int q(\boldsymbol{r}', t + \hat{\boldsymbol{r}} \cdot \boldsymbol{r}'/c)\, \mathrm{d}\boldsymbol{r}',$$

which is [311, eqn (3.16)], [387, eqn (5.48)] and [229, eqn (1.51)]. Friedlander proved that if $f_0(\hat{\boldsymbol{r}},t) = 0$ for all directions $\hat{\boldsymbol{r}}$ and for all $0 \le t \le t_0$, then $u(\boldsymbol{r},t) = 0$ in $r \ge a$ for $(r-a)/c \le t \le (r-a)/c + t_0$; see [311, Theorem 1] and [312, top of p. 387]. In other words, there is at most one expanding wave u with a given far-field pattern f_0.

However, the same cannot be said about the source function q. Friedlander notes [311, top of p. 60] that there are 'non-radiating source fields' q such that the corresponding u (found by solving (4.30)) vanishes for $r \ge a$. Later [314, p. 552], he gave a simple argument to show that non-radiating source fields exist: take a function $v(\boldsymbol{r},t)$ with the same support as in (4.31), then $q = \Box^2 v$ is non-radiating. In fact, this argument had been given earlier by Westervelt [877, p. 200]: 'It should be obvious that the part of the source function which is equal to the d'Alembertian of anything will contribute nothing to the scattered radiation field'. The same argument is repeated by Bleistein & Cohen [106], who then show how uniqueness of q can be restored by placing certain restrictions on the allowable q. See also [645] and [229, Chapter 5] for further information on inverse source problems.

4.8 Existence and Uniqueness Results

We have already indicated that problems with zero boundary conditions are well covered in the literature. (See text at the beginning of Section 4.2.) For problems in acoustics, where p and $\boldsymbol{v}$ are specified at $t = 0$, see Leis [541, Chapter 7].

Next, we summarise the literature on existence and uniqueness results for Problems DI_0 and NI_0, problems with zero initial conditions. Although there is earlier work [402], we start with the papers by Kreiss [499] and Sakamoto [733] from 1970. A problem is posed in a wedge-shaped region in space-time, with $x > 0$, $t > 0$, all y and all z, so that the spatial domain is a half-space $x > 0$ with a flat boundary, Γ, at $x = 0$ on which there is a boundary condition. The dependence on t is removed using a Laplace transform, and the dependence on y and z is removed using Fourier transforms parallel to Γ. The resulting theory was worked out for first-order systems by Kreiss [499] and for higher-order scalar problems by Sakamoto [733]. For expositions and extensions, see [165, Chapter 7], [734], [94, Part II], [369, Chapter 9] and [256].

The situation with more realistic spatial geometries (such as with our exterior domain B_e) is more complicated:

A closer view of what is going on can be obtained using space-time Fourier transformation. For this, one has to assume that Γ is flat. Then all the operators are convolutions and as such are represented by multiplication operators in Fourier space. If Γ is not flat but smooth, then *the results for the flat case describe the principal part of the operators. To construct a complete analysis, one has to consider lower order terms coming from coordinate transformations and localizations. Whereas this is a well-known technique in the elliptic and parabolic cases, namely part of the calculus of pseudodifferential operators, it has so far prevented the construction of a completely satisfactory theory for the hyperbolic case.*

(Costabel [198, §2.4])

For constructions of this kind, see [165, Chapter 7], [520, 500] and [369, Chapter 10]. In particular, the review by Lasiecka & Triggiani [520] gives a useful summary, which we shall use below.

An alternative approach was taken by Bamberger & Ha Duong [42, 43]. They use a Laplace transform in t leading to an elliptic boundary value problem for $U = \mathscr{L}\{u\}$, the Laplace transform of u. A weak solution U is sought in the Sobolev space $H^1(B_e)$, and it is represented with layer potentials (Section 9.4). Inversion leads to a study of time-domain boundary integral equations. For a detailed study, see the book by Sayas [742]. A similar approach has been used for penetrable scatterers [367].

4.8.1 Function Spaces and Notation

Let $\mathbb{T}$ denote the time interval, $0 < t \leq T$, with $T > 0$. When $t = 0$ is included, we write $\overline{\mathbb{T}}$; this distinction is relevant when compatibility conditions are required. We are interested in solving the wave equation in the hypercylinder $\Sigma \equiv B_e \times \mathbb{T}$. Let $\overline{B_e} = B_e \cup S$ and $\overline{\Sigma} = \overline{B_e} \times \overline{\mathbb{T}}$. As we are interested here in IBVPs with zero initial conditions, we can often take $T = \infty$ (in which case we write $\mathbb{R}_+$ or $\overline{\mathbb{R}_+}$ for the time interval) and we do not need to worry about B_e being unbounded: we have $u(\boldsymbol{r},t) = 0$ for sufficiently large $|\boldsymbol{r}|$.

Next, we need sets of functions mapping time to function spaces involving spatial variables. Thus 'we are going to consider u not as a function of $\boldsymbol{r}$ and t together, but rather as a mapping of t into a space of functions of $\boldsymbol{r}$' [275, §7.1.1.b]. For example, $C(\overline{\mathbb{T}};X)$ denotes the set of continuous functions of t, $w(\boldsymbol{r},t) : \overline{\mathbb{T}} \to X$, where X might be $L^2(B_e)$ or a Sobolev space $H^s(B_e)$ [833, §§24–26], [275, §5], [271, §13], [231]. Spaces $L^2(\mathbb{T};X)$ and $H^s(\mathbb{T};X)$ are defined similarly [833, §39], [275, §5.9.2]. Lubich [578] introduced the space $H_0^s(\mathbb{T};X)$ in which the underlying space, $H_0^s(\mathbb{T})$, contains the restrictions of those functions $f \in H^s(\mathbb{R})$ to $\mathbb{T}$ with $f(t) \equiv 0$ for $t < 0$; functions in this space are Laplace transformable. When $s = m$, a positive integer, the functions $f \in H_0^m(\mathbb{T})$ satisfy $f^{(r)}(0) = 0$, $r = 0, 1, \ldots, m-1$, where $f^{(m)}$ is the mth derivative of f [578, eqn (2.5)].

One case of special interest occurs when $X = C(\overline{B_e})$, because $C(\overline{\mathbb{T}};C(\overline{B_e})) = C(\overline{\Sigma})$, the set of continuous functions on $\overline{\Sigma}$. This case is of interest because, if u is a velocity potential, then u must be continuous everywhere.

Bamberger & Ha Duong [42, 43] use different spaces. To describe these, we start with some standard definitions from distribution theory. The space $\mathscr{D}(\mathbb{R})$ consists of all infinitely differentiable complex-valued functions on $\mathbb{R}$ with compact support. Functions in $\mathscr{D}(\mathbb{R})$ are called *test functions*. A continuous linear functional

$f : \mathscr{D}(\mathbb{R}) \to \mathbb{C}$ is called a *distribution* or a *generalised function*. It is called *regular* if it is *generated* by a locally integrable function f using the formula [357, Theorem 1.14]

$$\langle f, \varphi \rangle = \int_{-\infty}^{\infty} f(t)\, \varphi(t)\, \mathrm{d}t \quad \text{for all } \varphi \in \mathscr{D}(\mathbb{R}). \tag{4.40}$$

Note that it is common to use the same letter for both the name of the function and the name of the distribution. The space of all distributions is denoted by $\mathscr{D}'(\mathbb{R})$.

The *Schwartz space* $\mathscr{S}(\mathbb{R})$ consists of all infinitely differentiable complex-valued functions $\varphi(t)$ ($t \in \mathbb{R}$) that, together with all their derivatives, decay faster than the inverse of any polynomial as $|t| \to \infty$; Lighthill [562, §2.1] refers to such φ as *good functions*. Every $\varphi \in \mathscr{S}(\mathbb{R})$ has a Fourier transform $\mathscr{F}\{\varphi\} = \int_{-\infty}^{\infty} \varphi(t)\, \mathrm{e}^{\mathrm{i}\omega t}\, \mathrm{d}t$, with $\mathscr{F}\{\varphi\} \in \mathscr{S}$: the Fourier transform of a good function is a good function [452, Theorem 2.4], [357, Proposition 3.10], [271, §10.1].

A continuous linear functional $f : \mathscr{S}(\mathbb{R}) \to \mathbb{C}$ is called a *tempered distribution*. It is regular if it is generated by a function f using the formula (4.40) for all $\varphi \in \mathscr{S}(\mathbb{R})$ [357, Theorem 3.18]. Evidently, this definition requires some conditions on the integrability of $f(t)$ and on its growth as $|t| \to \infty$. The space of all tempered distributions is denoted by $\mathscr{S}'(\mathbb{R})$; we have $\mathscr{D} \subset \mathscr{S}$ and $\mathscr{S}' \subset \mathscr{D}'$ [271, §9].

The (generalised) Fourier transform of a tempered distribution $f \in \mathscr{S}'(\mathbb{R})$ is another tempered distribution $\tilde{f} \in \mathscr{S}'(\mathbb{R})$ defined by the formula $\langle \tilde{f}, \varphi \rangle = \langle f, \mathscr{F}\{\varphi\} \rangle$ for all $\varphi \in \mathscr{S}(\mathbb{R})$ [750, eqn (V, 2; 11)], [357, Definition 3.20], [271, Definition 10.1].

We extend these notions to Laplace transforms. The standard definition for ordinary functions, $\mathscr{L}\{f\} = \int_0^\infty f(t)\, \mathrm{e}^{-st}\, \mathrm{d}t$, involves $f(t)$ for $t > 0$. If $f(t)$ grows slower than an exponential, $\mathscr{L}\{f\}$ exists for $\operatorname{Re} s > 0$. In addition, we usually stipulate that f is a *causal function* (1.82), $f(t) = 0$ for $t < 0$; thus, the support of f, $\operatorname{supp} f = \overline{\mathbb{R}_+}$. To generalise, we need to define the support of a distribution $f \in \mathscr{D}'(\mathbb{R})$. We say that $f = 0$ on an open set $U \subset \mathbb{R}$ if $\langle f, \varphi \rangle = 0$ for any $\varphi \in \mathscr{D}(\mathbb{R})$ with $\operatorname{supp} \varphi$ in U. If U_m is the largest open set on which $f = 0$, we define $\operatorname{supp} f = \mathbb{R} \setminus U_\mathrm{m}$, which is a closed set [271, §6]. Hence we can define $f \in \mathscr{D}'(\mathbb{R})$ to be causal if $\operatorname{supp} f = \overline{\mathbb{R}_+}$, and we then write $f \in \mathscr{D}'_+$. Similarly, we can define $f \in \mathscr{S}'(\mathbb{R})$ to be a causal tempered distribution, in which case we write $f \in \mathscr{S}'_+$.

In order to define the Laplace transform of a causal distribution $f \in \mathscr{D}'_+$, consider

$$\mathscr{L}\{f\} = \int_0^\infty f(t)\, \mathrm{e}^{-st}\, \mathrm{d}t = \int_{-\infty}^{\infty} f(t)\, \mathrm{e}^{-st}\, \mathrm{d}t = \mathscr{F}\{\mathrm{e}^{-\xi t} f(t)\},$$

where we have used the causal nature of f and we have put $s = \xi - \mathrm{i}\omega$. Formally, the left-hand side of this equation defines $\mathscr{L}\{f\}$ for an ordinary function f. The right-hand side defines $\mathscr{L}\{f\}$ when $f \in \mathscr{D}'_+$, but only if we can find a real number ξ for which $\mathrm{e}^{-\xi t} f \in \mathscr{S}'$ [833, eqn (43.3)]. Of course, if f is already tempered, $f \in \mathscr{S}'_+$, we can put $\xi = 0$; the exponential convergence factor is redundant. For more details, see [901, §8.3], [452, Chapter 12], [750, §VI.2] and [833, §43].

The reader may wonder why the Laplace transform is now more important to us than the Fourier transform. The reason is that, if we content ourselves with dealing with distributions *which vanish for $t < 0$, then we can apply the Laplace transform to far more distributions than we can the Fourier transform—the Laplace transform makes sense for distributions which grow exponentially at $+\infty$, whereas only the tempered ones are Fourier-transformable.*

(Treves [833, p. 419])

Next, as we did before, we move from $\mathbb{C}$-valued mappings to X-valued mappings, where X is a space involving spatial variables. Thus, an X-valued distribution is a continuous linear mapping from $\mathscr{D}(\mathbb{R})$ to X. The set of all such distributions is denoted by $\mathscr{D}'(\mathbb{R};X)$. Similarly, a tempered X-valued distribution is a continuous linear mapping from $\mathscr{S}(\mathbb{R})$ to X. The set of all such tempered distributions is denoted by $\mathscr{S}'(\mathbb{R};X)$. See [833, §39.3].

A *causal tempered distribution with values in* X is a tempered X-valued distribution f such that $\langle f, \varphi \rangle = 0$ for all those $\varphi \in \mathscr{S}(\mathbb{R})$ with $\varphi(t) = 0$ for $t < 0$. Sayas [742, p. 22] denotes the space of all such distributions by CT(X); it is denoted by $\mathscr{S}'_+(X)$ in [42, 377]. Elements of this space include any function in $L^2(\mathbb{R}_+;X)$ and those $w(\boldsymbol{r},t) \in C(\overline{\mathbb{R}_+};X)$ satisfying $\|w(\cdot,t)\|_X \le \mathscr{C}(1+t^m)$ for $t \ge 0$ where $\mathscr{C}$ is a constant and m is a positive integer [742, §2.1]. The causal and tempered nature of distributions in CT(X) means that their Laplace transforms can be defined (without introducing exponential factors) [742, §2.3]. In [42], the notation

$$L'(X) = \{f \in \mathscr{D}'_+ : \mathrm{e}^{-\xi t} f \in \mathscr{S}'_+(X)\} \tag{4.41}$$

is introduced. This is the set of all Laplace-transformable causal X-valued distributions; the same set is denoted by $LT(\xi,X)$ in [377, 722], by $\mathscr{L}'_+(X)$ in [167] and by $\mathscr{L}'_\xi(\mathbb{R}_+,X)$ in [375, 144]. Certain subspaces of $L'(X)$, denoted here by $\mathscr{H}^s_\xi(\mathbb{R}_+,X)$, were also introduced in [42, §3]; when $s = m$, a non-negative integer,

$$\mathscr{H}^m_\xi(\mathbb{R}_+,X) = \{f \in L'(X) : \mathrm{e}^{-\xi t} f \in H^m(\mathbb{R};X)\}. \tag{4.42}$$

4.8.2 Problem DI$_0$

We seek a solution u of the wave equation with zero initial conditions and a Dirichlet boundary condition ($u = d$ on S). From [520, §2], we have the following results:

D1. Suppose that $d \in L^2(\mathbb{T};L^2(S)) \equiv L^2(S \times \mathbb{T})$ (without any compatibility conditions on d at $t = 0$). Then the unique solution of Problem DI$_0$ is in the space $C(\overline{\mathbb{T}};L^2(B_\mathrm{e}))$.

D2. Suppose that d is smoother, $d \in C(\overline{\mathbb{T}};H^{1/2}(S)) \cap H^1(\mathbb{T};L^2(S))$, together with the compatibility condition, $d \equiv 0$ on S when $t = 0$. Then the solution is smoother, $u \in C(\overline{\mathbb{T}};H^1(B_\mathrm{e}))$.

D3. Suppose that d is even smoother, $d \in C(\overline{\mathbb{T}};H^{3/2}(S)) \cap H^2(\mathbb{T};L^2(S))$, together with two compatibility conditions, $d \equiv 0$ and $\partial d/\partial t \equiv 0$ on S when $t = 0$. Then the solution is smoother, $u \in C(\overline{\mathbb{T}};H^2(B_\mathrm{e}))$.

Now, functions in H^1 need not be continuous, whereas functions in H^2 must be continuous; this is an example of the Sobolev imbedding theorem. In detail, if $f \in$

$H^s(B_e)$ with $s > \frac{3}{2}$, then $f \in C(\overline{B_e})$ [426, Theorem 4.1.6]. Thus, imposing two compatibility conditions, $d = 0$ and $\partial d/\partial t = 0$ on S (case D3 above), is sufficient to obtain a continuous solution of Problem DI_0.

Bamberger & Ha Duong [42, Problem (I), p. 407] start by making assumptions similar to those in case D2 above: $d \in C(\overline{\mathbb{R}_+};H^{1/2}(S))$ together with one compatibility condition, $d \equiv 0$ on S when $t = 0$. They go on to prove that if $d \in L'(H^{1/2}(S))$ (see (4.41) for the definition of $L'(X)$) then the unique solution of Problem DI_0 is in $L'(H^1(B_e))$ [42, Théorème 1]; no compatibility conditions are mentioned. For summaries (and interpretations) of their results, see [578, §2.3] and [278, §1]. With our notation, Lubich [578, p. 370] states: 'For smooth compatible boundary data d there exists a unique smooth solution u with $u(\cdot,t) \in H^1(B_e)$ for all t', although the phrases 'smooth solution' and 'compatible boundary data' are not made precise. Solutions for smoother d were also studied in [42, Théorème 2] using the subspaces $\mathscr{H}^s_\xi$, (4.42); see [167, Theorem 4(c)] and [278, §1].

We note that the practical significance of the exponential weight appearing in the definition of $L'(X)$, (4.41), is unclear. It does cause difficulties in the error analysis of space-time variational methods [449, p. 139]; see Section 10.3.2.

4.8.3 Problem NI_0

Compared with the Dirichlet problem, 'the situation is drastically different in the Neumann case' [520, p. 105]. For this problem, we seek a solution u of the wave equation with zero initial conditions and a Neumann boundary condition ($\partial u/\partial n = v$ on S). From [520, §3], we have the following results:

N1. Suppose that $v \in L^2(S \times \mathbb{T})$ (without any compatibility conditions on v at $t = 0$). Then the unique solution of Problem NI_0 is in $C(\overline{\mathbb{T}};H^\alpha(B_e))$. In general, the parameter $\alpha = \frac{3}{5} - \varepsilon$ (where $\varepsilon > 0$ is arbitrary) but $\alpha = \frac{2}{3}$ when S is a sphere.

N2. Suppose that v is smoother, $v \in C(\overline{\mathbb{T}};H^{\alpha-1/2}(S)) \cap H^1(\mathbb{T};L^2(S))$, together with the compatibility condition, $v \equiv 0$ on S when $t = 0$. Then the solution is smoother, $u \in C(\overline{\mathbb{T}};H^{\alpha+1}(B_e))$. We have $\alpha + 1 = \frac{3}{2} + \left(\frac{1}{10} - \varepsilon\right)$. Therefore, if we have $0 < \varepsilon < \frac{1}{10}$, the Sobolev imbedding theorem shows that our solution of Problem NI_0 is in $C(\overline{\Sigma})$.

Bamberger & Ha Duong [43, Problem (P_+), p. 598] state Problem NI_0 assuming that $v \equiv 0$ on S for $t \le 0$ (which includes the compatibility condition in case N2 above). They try to incorporate this condition into a modified definition of the space $L'(X)$ (see (4.41)): '$L'(X)$ désigne les distributions sur $\mathbb{R}$, à valeurs dans X, nulles pour $t \le 0$ et majorées à $t \to +\infty$ par une exponentielle en t' [43, p. 601]. This definition of $L'(X)$ is then used in their definition of $\mathscr{H}^s_\xi(\mathbb{R}_+,X)$, (4.42). Their result [43, Théorème 1] is: given $v \in \mathscr{H}^{s+1}_\xi(\mathbb{R}_+,H^{-1/2}(S))$, Problem NI_0 has a unique solution $u \in \mathscr{H}^s_\xi(\mathbb{R}_+,H^1(B_e))$, where $s \in \mathbb{R}$. Chappell's result [162, p. 1587] is: given $v \in H^{s+1}_0(\mathbb{T};H^{-1/2}(S))$, 'there exists a unique solution to the IBVP' $u \in H^s_0(\mathbb{T};H^1(B_e))$, where H_0^s is a Lubich space [578] (see Section 4.8.1); when $s = m$, a non-negative integer, the conditions on v imply that v and its first m time derivatives must vanish at $t = 0$. These results agree when $s = 0$, but it is unclear how they compare when $s > 0$, especially with respect to compatibility conditions. Moreover, increasing s does not affect the spatial smoothness beyond lying in $H^1(B_e)$, which does not agree with the regularity result in case N2 above. Interestingly, Falletta et al. [279, §2] want C^2 solutions based on 'sufficiently smooth compatible data'; in particular [279, eqn (2.1)] 'at least the following conditions' are assumed: $v = \partial v/\partial t = \partial^2 v/\partial t^2 = 0$ at $t = 0$ for all $P \in S$.

4.8.4 Summary

We have tried to give a summary of available theoretical results concerning two basic IBVPs with zero initial conditions. There are some results for other problems, such as those involving the forced wave equation (Section 4.7) or Maxwell's equations (Section 1.6.1). For the boundary condition $\partial u/\partial n + \beta\, \partial u/\partial t = f$ on S, see [375].

One gets the impression that further refinement of the theoretical developments is desirable. Physical applications suggest that we want to know: Given properties of the boundary data (such as d and v), when can we assert that the solution is continuous and piecewise smooth for all $t > 0$ and everywhere in B_e? We shall investigate this question in Section 6.1.4, using explicit calculations for simple problems with spherical symmetry.

5
Use of Laplace Transforms

It is natural to use Laplace transforms to solve initial-value problems. Their use converts the wave equation into the modified Helmholtz equation. The earliest use of transform methods for wave generation by a pulsating sphere is probably that in Jeffreys' 1927 book [442, §6.3] on *Operational Methods*.

However, there is a difficulty: What is the effect of possible discontinuities across wavefronts? Can we ignore such discontinuities? We investigate these questions in this chapter; see also [601, §8]. We also introduce Fourier transforms (Section 5.6) and relate them to Laplace transforms and to frequency-domain problems. Finally, we give a brief discussion on Laguerre transforms (Section 5.7).

5.1 Formal Treatment

Basic properties of Laplace transforms were collected in Section 1.4, with an emphasis on functions of one variable. Here, our focus is on wavefunctions $u(\boldsymbol{r},t)$, solutions of the wave equation (1.46). Define the Laplace transform of $u(\boldsymbol{r},t)$ with respect to t by

$$U(\boldsymbol{r},s) = \mathscr{L}\{u\} = \int_0^\infty u(\boldsymbol{r},t)\,\mathrm{e}^{-st}\,\mathrm{d}t. \tag{5.1}$$

Applying $\mathscr{L}$ to (1.46) gives

$$\nabla^2 U - (s/c)^2 U = 0, \tag{5.2}$$

where (for simplicity) we have assumed zero initial conditions and we have not worried about any possible discontinuities in u or its partial derivatives. Equation (5.2) is the *modified Helmholtz equation*, an elliptic PDE. If it can be solved together with the Laplace-transformed boundary condition (see Section 5.4), we can then invert to obtain u from U,

$$u(\boldsymbol{r},t) = \mathscr{L}^{-1}\{U\} = \frac{1}{2\pi\mathrm{i}}\int_{\mathrm{Br}} U(\boldsymbol{r},s)\,\mathrm{e}^{st}\,\mathrm{d}s, \tag{5.3}$$

where Br is the Bromwich contour in the complex s-plane. This method has been used extensively for spherical boundaries (see Sections 6.2 and 7.2).

5.2 Laplace Transform of Discontinuous Functions

Let us investigate the effects of discontinuities. Thus suppose there is a surface of discontinuity at $t = \tau(\boldsymbol{r})$. Then, starting from (5.1), and splitting the range of integration,

$$\mathscr{L}\{u\} = \int_0^\tau u(\boldsymbol{r},t)\,\mathrm{e}^{-st}\,\mathrm{d}t + \int_\tau^\infty u(\boldsymbol{r},t)\,\mathrm{e}^{-st}\,\mathrm{d}t,$$

we differentiate to obtain

$$\begin{aligned}\frac{\partial}{\partial x_i}\mathscr{L}\{u\} &= \int_0^\tau \frac{\partial u}{\partial x_i}\mathrm{e}^{-st}\,\mathrm{d}t + u(\boldsymbol{r},\tau-)\,\mathrm{e}^{-s\tau}\frac{\partial \tau}{\partial x_i} + \int_\tau^\infty \frac{\partial u}{\partial x_i}\mathrm{e}^{-st}\,\mathrm{d}t - u(\boldsymbol{r},\tau+)\,\mathrm{e}^{-s\tau}\frac{\partial \tau}{\partial x_i}\\ &= \mathscr{L}\left\{\frac{\partial u}{\partial x_i}\right\} + [\![u]\!]\,\mathrm{e}^{-s\tau}\frac{\partial \tau}{\partial x_i},\end{aligned} \tag{5.4}$$

where $u(\boldsymbol{r},\tau\pm) = \lim_{\varepsilon\to 0} u(\boldsymbol{r},\tau\pm\varepsilon^2)$ and $[\![u]\!] = u(\boldsymbol{r},\tau-) - u(\boldsymbol{r},\tau+)$, in agreement with (3.25). Hence, replacing u by $\partial u/\partial x_i$,

$$\begin{aligned}\mathscr{L}\{\nabla^2 u\} &= \frac{\partial}{\partial x_i}\mathscr{L}\left\{\frac{\partial u}{\partial x_i}\right\} - \left[\!\!\left[\frac{\partial u}{\partial x_i}\right]\!\!\right]\frac{\partial \tau}{\partial x_i}\mathrm{e}^{-s\tau}\\ &= \frac{\partial}{\partial x_i}\left(\frac{\partial U}{\partial x_i} - [\![u]\!]\,\mathrm{e}^{-s\tau}\frac{\partial \tau}{\partial x_i}\right) - \left[\!\!\left[\frac{\partial u}{\partial x_i}\right]\!\!\right]\frac{\partial \tau}{\partial x_i}\mathrm{e}^{-s\tau}\\ &= \nabla^2 U - \left(\frac{\partial [\![u]\!]}{\partial x_i}\frac{\partial \tau}{\partial x_i} + [\![u]\!]\,\nabla^2\tau - s[\![u]\!]\frac{\partial \tau}{\partial x_i}\frac{\partial \tau}{\partial x_i} + \left[\!\!\left[\frac{\partial u}{\partial x_i}\right]\!\!\right]\frac{\partial \tau}{\partial x_i}\right)\mathrm{e}^{-s\tau}.\end{aligned} \tag{5.5}$$

For time derivatives, we have

$$\begin{aligned}\mathscr{L}\left\{\frac{\partial u}{\partial t}\right\} &= \int_0^\tau \frac{\partial u}{\partial t}\mathrm{e}^{-st}\,\mathrm{d}t + \int_\tau^\infty \frac{\partial u}{\partial t}\mathrm{e}^{-st}\,\mathrm{d}t\\ &= \left[u\mathrm{e}^{-st}\right]_0^\tau + s\int_0^\tau u\mathrm{e}^{-st}\,\mathrm{d}t + \left[u\mathrm{e}^{-st}\right]_\tau^\infty + s\int_\tau^\infty u\mathrm{e}^{-st}\,\mathrm{d}t\\ &= s\mathscr{L}\{u\} - u_0 + [\![u]\!]\,\mathrm{e}^{-s\tau}\end{aligned} \tag{5.6}$$

and

$$\begin{aligned}\mathscr{L}\left\{\frac{\partial^2 u}{\partial t^2}\right\} &= s\mathscr{L}\left\{\frac{\partial u}{\partial t}\right\} - u_1 + \left[\!\!\left[\frac{\partial u}{\partial t}\right]\!\!\right]\mathrm{e}^{-s\tau}\\ &= s^2\mathscr{L}\{u\} - su_0 - u_1 + \left(s[\![u]\!] + \left[\!\!\left[\frac{\partial u}{\partial t}\right]\!\!\right]\right)\mathrm{e}^{-s\tau},\end{aligned} \tag{5.7}$$

where we have used the initial conditions (4.1).

Equations (5.4)–(5.7) can be found in a paper by Chadwick & Powdrill [155]. They show the effects of discontinuities across $t = \tau(\boldsymbol{r})$; if there are no discontinuities, all the terms involving $[\![\cdot]\!]$ are absent, and we obtain standard formulas.

5.3 Laplace Transform of Wavefunctions

Suppose that u is a wavefunction. Then applying $\mathscr{L}$ to the wave equation (1.46), using (5.5) and (5.7), gives

$$\nabla^2 U - (s/c)^2 U = f_{\text{ic}}(\boldsymbol{r},s) + f(\boldsymbol{r},s)\,\mathrm{e}^{-s\tau(\boldsymbol{r})}, \tag{5.8}$$

where the initial conditions are contained in $f_{\text{ic}} = -[su_0(\boldsymbol{r}) + u_1(\boldsymbol{r})]/c^2$ and

$$f = \frac{\partial [\![u]\!]}{\partial x_i}\frac{\partial \tau}{\partial x_i} + [\![u]\!]\,\nabla^2\tau - s[\![u]\!]\left(\frac{\partial \tau}{\partial x_i}\frac{\partial \tau}{\partial x_i} - \frac{1}{c^2}\right) + \left[\!\!\left[\frac{\partial u}{\partial x_i}\right]\!\!\right]\frac{\partial \tau}{\partial x_i} + \frac{1}{c^2}\left[\!\!\left[\frac{\partial u}{\partial t}\right]\!\!\right].$$

The chain rule gives (3.37), which implies

$$\frac{\partial [\![u]\!]}{\partial x_i} = \left[\!\!\left[\frac{\partial u}{\partial x_i}\right]\!\!\right] + \left[\!\!\left[\frac{\partial u}{\partial t}\right]\!\!\right]\frac{\partial \tau}{\partial x_i}.$$

Using this to eliminate $[\![\partial u/\partial x_i]\!]$, we obtain

$$f = 2\frac{\partial [\![u]\!]}{\partial x_i}\frac{\partial \tau}{\partial x_i} + [\![u]\!]\,\nabla^2\tau - \left(s[\![u]\!] + \left[\!\!\left[\frac{\partial u}{\partial t}\right]\!\!\right]\right)\left(\frac{\partial \tau}{\partial x_i}\frac{\partial \tau}{\partial x_i} - \frac{1}{c^2}\right).$$

Now, we know that discontinuities can only occur across characteristics. It follows that τ satisfies the eikonal equation (3.16) thus simplifying f to

$$f = 2\frac{\partial [\![u]\!]}{\partial x_i}\frac{\partial \tau}{\partial x_i} + [\![u]\!]\,\nabla^2\tau, \tag{5.9}$$

which does not depend on the transform variable s.

5.4 Laplace Transform of Boundary Conditions

Write $u(\boldsymbol{r},t) \equiv u(P,t)$ and $U(\boldsymbol{r},s) = U(P,s)$. As in Section 4.1, we impose a boundary condition at $P \in S$. The simplest is the Dirichlet condition (4.2): $u(P,t) = d(P,t)$ for $P \in S$ and $t > 0$. Hence

$$U(P,s) = D(P,s), \quad P \in S, \tag{5.10}$$

where $D = \mathscr{L}\{d\}$ and we have assumed that d is a continuous function of t.

The pressure condition (4.4) requires that $-\rho\,\partial u/\partial t = p_S(P,t)$ for $P \in S$ and $t > 0$. Applying $\mathscr{L}$, using (5.6), gives

$$sU(P,s) - u(P,0) + [\![u]\!]\,\mathrm{e}^{-s\tau(P)} = -\rho^{-1}P_S(P,s), \quad P \in S, \tag{5.11}$$

where $P_S(P,s) = \mathscr{L}\{p_S\}$ and $\tau(P) = \tau(\boldsymbol{r})$. We are free to choose $u(P,0) = u_0(P)$ for $P \in S$. Doing this ensures that $[\![u]\!] \equiv 0$ and then (5.11) simplifies.

For the Neumann condition (4.3), $\partial u/\partial n(P,t) = v(P,t)$, we use (5.4) and obtain

$$\frac{\partial U}{\partial n} = V(P,s) + [\![u]\!]\,\mathrm{e}^{-s\tau(P)}\frac{\partial \tau}{\partial n}, \quad P \in S, \tag{5.12}$$

where $V = \mathscr{L}\{v\}$ and we have assumed that v is a continuous function of t. Again, we are free to choose $u(P,0) = u_0(P)$ for $P \in S$, thus ensuring that $[\![u]\!] \equiv 0$ with a corresponding simplification to (5.12).

5.5 Synthesis

At this stage, we have found that $U = \mathscr{L}\{u\}$ satisfies (5.8), an inhomogeneous form of the modified Helmholtz equation. The inhomogeneous term is $f_{\rm ic} + f$ where the first part, $f_{\rm ic}$, comes from the initial conditions; it is absent when we have zero initial conditions. The second part, f, is defined by (5.9); it involves the (unknown) jump across wavefronts, $[\![u]\!]$.

Ideally, we would like to have $[\![u]\!] = 0$, because this would simplify both the PDE for U and the boundary condition on U.

If we are satisfied with a weak solution, then we know that $[\![u]\!]$ satisfies the transport equation (3.40). This immediately yields $f \equiv 0$: for weak solutions, we can ignore the presence of discontinuities and (5.2) obtains. On the other hand, we know that weak solutions need not respect the physical constraints across wavefronts (see Section 3.5), so care is needed.

If we are not satisfied with a weak solution, we can start by noting that the physical constraints lead to (3.47), $[\![u]\!] = \text{constant}$. This reduces (5.9) to $f = [\![u]\!]\,\nabla^2\tau$, which is non-zero in general. However, if we can arrange that $[\![u]\!] = 0$, then $f \equiv 0$ and, again, we can ignore the presence of discontinuities. Indeed, in most cases, this can be done: for the Dirichlet problem, we require the compatibility condition $d(P,0) = u_0(P)$, $P \in S$; for the pressure and Neumann conditions, we are free to set $[\![u]\!] = 0$. It is worth emphasising that even though u is continuous, p, $\mathbf{v}$, and derivatives of u can be discontinuous.

As far as we know, all published applications of the Laplace transform to scattering problems have been based on the formal method (where all effects of possibly non-zero jumps $[\![u]\!]$ are ignored), as outlined in Section 5.1. Later, we shall describe some of these applications in detail; for problems with spherical symmetry, see Section 6.2.

As an aside, we note that there is some literature on the use of Laplace transforms for other problems involving discontinuities, weak solutions and generalised functions (distribution theory). We mention a few papers [327, 112, 35, 78]. For textbook treatments in the context of distribution theory, see [901, §8.3], [452, Chapter 12], [750, §VI.2] and [833, §43].

5.6 Fourier Transforms

Fourier transforms provide a clear connection between time-domain and frequency-domain problems. We describe that connection here. As with Laplace transforms, methods based on Fourier transforms suffer from the same potential difficulty with discontinuous functions; we ignore that difficulty here.

5.6.1 Basic Definitions and Strategy

Define the (temporal) Fourier transform of $u(\boldsymbol{r},t)$ by

$$\widehat{u}(\boldsymbol{r},\omega) = \mathscr{F}\{u\} = \int_{-\infty}^{\infty} u(\boldsymbol{r},t)\,\mathrm{e}^{\mathrm{i}\omega t}\,\mathrm{d}t, \tag{5.13}$$

where ω is real (although this restriction may be relaxed). Evidently, the integral will converge only if $u(\boldsymbol{r},t)$ behaves suitably as $t \to \pm\infty$.

The corresponding inverse Fourier transform is

$$u(\boldsymbol{r},t) = \mathscr{F}^{-1}\{\widehat{u}\} = \frac{1}{2\pi}\int_{-\infty}^{\infty} \widehat{u}(\boldsymbol{r},\omega)\,\mathrm{e}^{-\mathrm{i}\omega t}\,\mathrm{d}\omega. \tag{5.14}$$

In our applications, u is real, and then (5.13) implies

$$\overline{\widehat{u}(\boldsymbol{r},\omega)} = \widehat{u}(\boldsymbol{r},-\omega), \tag{5.15}$$

where the overbar denotes complex conjugation. This fact simplifies (5.14) to

$$u(\boldsymbol{r},t) = \mathscr{F}^{-1}\{\widehat{u}\} = \frac{1}{\pi}\int_{0}^{\infty} \mathrm{Re}\left\{\widehat{u}(\boldsymbol{r},\omega)\,\mathrm{e}^{-\mathrm{i}\omega t}\right\}\mathrm{d}\omega, \tag{5.16}$$

an integral over positive frequencies ω. This formula is the basic connection between frequency-domain solutions $\widehat{u}(\boldsymbol{r},\omega)$ and time-domain solutions $u(\boldsymbol{r},t)$.

To exploit this connection, the following strategy emerges naturally. Write $\widehat{u} = \widehat{u}_{\mathrm{inc}} + \widehat{u}_{\mathrm{sc}}$, where $\widehat{u}_{\mathrm{inc}}(\boldsymbol{r},\omega)$ is the Fourier transform of the given incident field and $\widehat{u}_{\mathrm{sc}}(\boldsymbol{r},\omega)$ is the unknown scattered field. Applying $\mathscr{F}$ to the wave equation shows that the governing PDE in the exterior domain B_{e} is the Helmholtz equation,

$$(\nabla^2 + (\omega/c)^2)\widehat{u} = 0. \tag{5.17}$$

In addition, there will be a boundary condition on the scatterer (for example, $\widehat{u} = 0$ on S) and the *Sommerfeld radiation condition* at infinity [189, §3.2], [599, §1.3.1]

$$r\left(\frac{\partial \widehat{u}_{\mathrm{sc}}}{\partial r} - \mathrm{i}\frac{\omega}{c}\widehat{u}_{\mathrm{sc}}\right) \to 0 \quad \text{as } r = |\boldsymbol{r}| \to \infty \tag{5.18}$$

uniformly in all directions; here, it is assumed that ω is real and positive. The radiation condition (5.18) implies that $\widehat{u}_{\mathrm{sc}}$ behaves as $r^{-1}\mathrm{e}^{\mathrm{i}\omega r/c}$ as $r \to \infty$; with the time dependence $\mathrm{e}^{-\mathrm{i}\omega t}$ (see (5.16)), we see that the waves are propagating away from the scatterer. Next, solve the frequency-domain boundary-value problem for $\widehat{u}_{\mathrm{sc}}$ using any convenient method (see [599] for various possibilities). Finally, use the inverse Fourier transform to recover u from $\widehat{u}$; for some discussion of numerical aspects, see [443, §8.2] and [19].

The approach outlined above is convenient if one has reliable software available for solving frequency-domain scattering problems. Use of a quadrature rule to approximate the inversion integral (5.16) implies solving scattering problems at a finite number of frequencies ω. Moreover, these problems are uncoupled, so that the result is a parallel-in-frequency method [610]. This basic method has been used by Klaseboer et al. [487]; they solve a boundary integral equation to obtain $\widehat{u}_{\mathrm{sc}}$. Anderson et al. [19] adopt a similar approach, except they use a smoothly time-windowed form of (5.13), leading to an efficient numerical scheme. The basic method has also been used for scattering by spheres, where the frequency-domain problems can be solved by separation of variables; see Section 7.2.4 for references.

Computationally, the main issues with implementing the basic method concern accuracy and efficiency. These will be discussed later, in Section 10.9.

5.6.2 A Time-Domain Radiation Condition

There is a time-domain version of the Sommerfeld radiation condition (5.18). To derive it, use (5.13) and consider

$$\frac{\partial \widehat{u}}{\partial r} - \mathrm{i}\frac{\omega}{c}\widehat{u} = \int_{-\infty}^{\infty}\left(\frac{\partial u}{\partial r} - \mathrm{i}\frac{\omega}{c}u\right)\mathrm{e}^{\mathrm{i}\omega t}\,\mathrm{d}t = \int_{-\infty}^{\infty}\left(\frac{\partial u}{\partial r} + \frac{1}{c}\frac{\partial u}{\partial t}\right)\mathrm{e}^{\mathrm{i}\omega t}\,\mathrm{d}t$$

after an integration by parts. Denote the left-hand side of this equation by $F(\omega)$ and denote the integrand on the right-hand side by $f(t)\,\mathrm{e}^{\mathrm{i}\omega t}$. Then Parseval's relation [785, §3.6] gives

$$\begin{aligned}\int_{-\infty}^{\infty}\{f(t)\}^2\,\mathrm{d}t &= \frac{1}{2\pi}\int_{-\infty}^{\infty} f(t)\int_{-\infty}^{\infty} F(\omega)\,\mathrm{e}^{-\mathrm{i}\omega t}\,\mathrm{d}\omega\,\mathrm{d}t \\ &= \frac{1}{2\pi}\int_{-\infty}^{\infty} F(\omega)\,F(-\omega)\,\mathrm{d}\omega = \frac{1}{\pi}\int_{0}^{\infty} F(\omega)\,F(-\omega)\,\mathrm{d}\omega \\ &= \frac{1}{\pi}\int_{0}^{\infty} |F(\omega)|^2\,\mathrm{d}\omega,\end{aligned}$$

after use of (5.15). Multiplying by r^2, we obtain

$$\int_{-\infty}^{\infty} r^2\left(\frac{\partial u}{\partial r} + \frac{1}{c}\frac{\partial u}{\partial t}\right)^2\mathrm{d}t = \frac{1}{\pi}\int_{0}^{\infty} r^2\left|\frac{\partial \widehat{u}}{\partial r} - \mathrm{i}\frac{\omega}{c}\widehat{u}\right|^2\mathrm{d}\omega.$$

Thus, imposing the Sommerfeld radiation condition (5.18) for every positive frequency ω is equivalent to imposing

$$\int_{-\infty}^{\infty} r^2\left(\frac{\partial u}{\partial r} + \frac{1}{c}\frac{\partial u}{\partial t}\right)^2\mathrm{d}t \to 0 \quad \text{as } r\to\infty, \text{ uniformly in all directions.} \tag{5.19}$$

This *time-domain Sommerfeld radiation condition* is used in [580, §2.2] (but note the sign error in [580, eqn (19)]).

5.6.3 Further Remarks

When solving an IBVP, we can suppose that $u(\boldsymbol{r},t) = 0$ for all $t < 0$, thus simplifying (5.13) to

$$\widehat{u}(\boldsymbol{r},\omega) = \int_{0}^{\infty} u(\boldsymbol{r},t)\,\mathrm{e}^{\mathrm{i}\omega t}\,\mathrm{d}t, \tag{5.20}$$

which looks like a Laplace transform. Indeed, if we put $\omega = \mathrm{i}s$, we obtain

$$U(\boldsymbol{r},s) \equiv \widehat{u}(\boldsymbol{r},\mathrm{i}s) = \int_0^\infty u(\boldsymbol{r},t)\,\mathrm{e}^{-st}\,\mathrm{d}t = \mathscr{L}\{u\}.$$

The same substitution converts the inversion formula (5.14) into (5.3).

For scattering problems involving an incident sound pulse (with no disturbance ahead of the pulse), we can readily formulate an IBVP, and then it is appropriate to use Laplace transforms. However there are also scattering problems in which the incident field does not have a well-defined front. For example, a popular choice is a Gaussian pulse, such as

$$u_{\mathrm{inc}}(\boldsymbol{r},t) = \mathscr{U}_0 \exp\{-\omega_0^2(t-[z-z_0]/c)^2\}, \tag{5.21}$$

where $\mathscr{U}_0$, ω_0 and z_0 are constants. The corresponding Fourier transform is easily calculated:

$$\widehat{u}_{\mathrm{inc}}(\boldsymbol{r},\omega) = (\mathscr{U}_0/\omega_0)\sqrt{\pi}\,\mathrm{e}^{-\omega^2/(4\omega_0^2)}\,\mathrm{e}^{\mathrm{i}(z-z_0)\omega/c}.$$

This represents a plane wave (because it is a multiple of $\mathrm{e}^{\mathrm{i}(\omega/c)z}$) in the frequency domain. The scattering of such a wave by an obstacle is the basic problem in time-harmonic scattering theory.

Generalising (5.21), suppose that u_{inc} is a plane wave (4.12),

$$u_{\mathrm{inc}}(\boldsymbol{r},t) = f(t-\widehat{\boldsymbol{e}}\cdot\boldsymbol{r}/c). \tag{5.22}$$

Then, assuming that $\widehat{f}(\omega) = \mathscr{F}f$ exists,

$$\widehat{u}_{\mathrm{inc}}(\boldsymbol{r},\omega) = \widehat{f}(\omega)\exp\{\mathrm{i}(\omega/c)\widehat{\boldsymbol{e}}\cdot\boldsymbol{r}\}, \tag{5.23}$$

which is another plane wave in the frequency domain with amplitude $\widehat{f}(\omega)$.

In some circumstances, $\widehat{u}(\boldsymbol{r},\omega)$ can be written as a Herglotz wavefunction [190, §3.3],

$$\widehat{u}(\boldsymbol{r},\omega) = \int_\Omega g(\widehat{\boldsymbol{s}},\omega)\,\mathrm{e}^{\mathrm{i}(\omega/c)\boldsymbol{r}\cdot\widehat{\boldsymbol{s}}}\,\mathrm{d}\Omega(\widehat{\boldsymbol{s}}), \tag{5.24}$$

where Ω is the unit sphere and g is a function defined on Ω. Given ω, $\widehat{u}(\boldsymbol{r},\omega)$ satisfies (5.17) for all $\boldsymbol{r} \in \mathbb{R}^3$ if $g \in L^2(\Omega)$. The far-field behaviour of Herglotz wavefunctions (5.24) has been studied; see [603] and references therein.

Inverting the Fourier transform using (5.14) gives

$$u(\boldsymbol{r},t) = \frac{1}{2\pi}\int_{-\infty}^{\infty}\int_\Omega g(\widehat{\boldsymbol{s}},\omega)\,\mathrm{e}^{\mathrm{i}\{(\omega/c)\boldsymbol{r}\cdot\widehat{\boldsymbol{s}}-\omega t\}}\,\mathrm{d}\Omega(\widehat{\boldsymbol{s}})\,\mathrm{d}\omega. \tag{5.25}$$

Integral representations of this kind have been used for incident fields [610, 591].

5.7 Laguerre Transforms

Another relative of the Laplace transform is the *Laguerre transform*, defined by

$$U_n(\boldsymbol{r},s) = \mathscr{L}_n\{u\} = \int_0^\infty u(\boldsymbol{r},t)\,\mathrm{e}^{-st}L_n(st)\,\mathrm{d}t, \tag{5.26}$$

where $L_n \equiv L_n^0$ is a Laguerre polynomial [661, §18.3]. By definition,

$$L_n(x) = \frac{e^x}{n!}\frac{d^n}{dx^n}\left(x^n e^{-x}\right),$$

so that $L_0(x) = 1$ and $\mathscr{L}_0 = \mathscr{L}$, the Laplace transform. However, unlike with $\mathscr{L}$, s plays a minor role: it is a (positive) scaling parameter. Indeed, inverting Laguerre transforms exploits the fact that Laguerre polynomials form a complete orthonormal set with respect to a weighted inner product. Thus, if $f(t)$ is such that $\int_0^\infty |f(t)|^2 e^{-t}\,dt$ is finite, then we can expand f as

$$f(t) = s\sum_{n=0}^{\infty} f_n L_n(st)$$

with Laguerre–Fourier coefficients given by

$$f_n = \mathscr{L}_n\{f\} = \int_0^\infty f(t)\,e^{-st}\,L_n(st)\,dt.$$

For basic properties and simple applications of the Laguerre transform, see [608, 469] and [224, Chapter 16]. Applications to IBVPs for the wave equation were made by Chapko & Kress [159]; see also [177, §3.3]. For example, Chapko & Kress [159, Theorem 2.1] show that if $u(\boldsymbol{r},t)$ is a smooth (twice differentiable) solution of Problem DI_0, with $u = d$ on S, then $U_n(\boldsymbol{r},s) = \mathscr{L}_n\{u\}$ (see (5.26)) solves

$$\nabla^2 U_n - (s/c)^2 U_n = (s/c)^2 \sum_{m=0}^{n-1} (n-m+1) U_m \quad \text{in } B_e, \tag{5.27}$$

with $U_n = d_n$ on S, where d_n is the nth Laguerre–Fourier coefficient of the Dirichlet data d. (The right-hand side of (5.27) is replaced by 0 when $n = 0$.) Thus, the IBVP has been reduced to a sequence of coupled BVPs for the forced modified Helmholtz equation. 'A drawback of this method is that it may be necessary to use fairly large numbers of terms ($n \geq 20$ [159]) to get good convergence and this requires storing many terms to obtain the approximations' [177, p. 10]. It is also unclear how to choose the scaling parameter s.

6

Problems with Spherical Symmetry

In this chapter, we consider IBVPs for a sphere of radius a. We assume that the forcing is such that the waves generated have spherical symmetry, which means that u depends on r and t only. This is a strong assumption but it permits exact solutions, and these solutions are revealing. We pay attention to allowable discontinuities.

We start in Section 6.1 by constructing solutions using simple sources at the centre of the sphere. Then, in Section 6.2, we solve the same problems using Laplace transforms.

6.1 Use of Simple Sources

Because of the assumed spherical symmetry, the wave equation reduces to (2.9), and we can write down the general solution. Thus, from (2.10),

$$u(r,t) = r^{-1}\{f(r-ct)+g(r+ct)\}, \tag{6.1}$$

where f and g are arbitrary smooth functions. These are to be determined using initial and boundary conditions. Specifically, we seek $u(r,t)$ in $\mathscr{Q}$, a quarter of the rt-plane where $r > a$ and $t > 0$. There is a boundary condition at $r = a$ for $t > 0$, and two initial conditions at $t = 0$ for $r > a$. There is a characteristic of the PDE (2.9) emanating from the corner at $(r,t) = (a,0)$ along the straight line $r = a+ct$, $t > 0$. Denote this line by $\mathscr{W}$ as it corresponds to the wavefront; solutions may be discontinuous across $\mathscr{W}$. See Fig. 6.1 for a sketch of the rt-plane. In three-dimensional space, there is a spherical wavefront, $\Gamma(t)$, at $r = a+ct$. The space-time edge or corner $\mathscr{E}$ is at $r = a$, $t = 0$; it is $\Gamma(0)$.

We use the wavefront $\mathscr{W}$ to partition the quadrant $\mathscr{Q} = \mathscr{Q}^+ \cup \mathscr{Q}^- \cup \mathscr{W}$, where

$$\mathscr{Q}^+ = \{(r,t) : a < r < a+ct,\ t > 0\} \quad \text{and} \quad \mathscr{Q}^- = \{(r,t) : r > a+ct,\ t > 0\}; \tag{6.2}$$

thus $\mathscr{Q}^+$ is the region behind the wavefront and $\mathscr{Q}^-$ is the region ahead of the wavefront. The boundary condition is imposed on one edge of $\mathscr{Q}^+$ and the initial conditions are imposed on one edge of $\mathscr{Q}^-$. See Fig. 6.1.

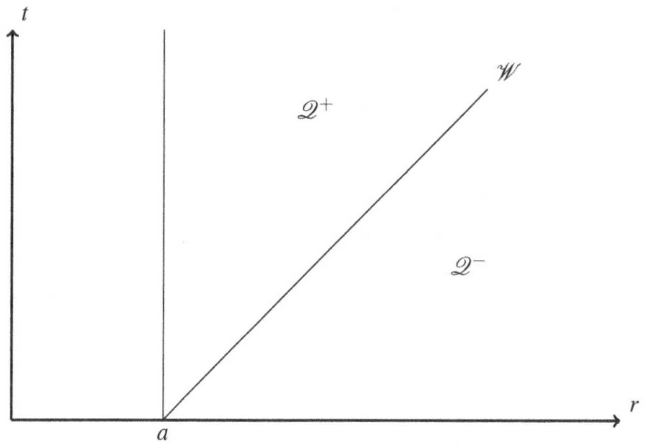

Figure 6.1 The quadrant $\mathscr{Q} = \mathscr{Q}^+ \cup \mathscr{Q}^- \cup \mathscr{W}$ in the rt-plane.

The initial conditions are (4.1), which become

$$u(r,0) = u_0(r), \quad \frac{\partial u}{\partial t}(r,0) = u_1(r), \quad r > a,$$

where u_0 and u_1 are given. When combined with (6.1), they give (cf. (1.52))

$$f(r) + g(r) = ru_0(r), \quad -f'(r) + g'(r) = (r/c)u_1(r), \quad r > a$$

whence, for $r > a$,

$$f(r) = \frac{1}{2} ru_0(r) - \frac{1}{2c}\int_a^r \xi u_1(\xi)\,\mathrm{d}\xi + C, \tag{6.3}$$

$$g(r) = \frac{1}{2} ru_0(r) + \frac{1}{2c}\int_a^r \xi u_1(\xi)\,\mathrm{d}\xi - C, \tag{6.4}$$

where C is an arbitrary constant. Hence (6.1) gives

$$u(r,t) = \frac{1}{2r}\{(r+ct)u_0(r+ct) + (r-ct)u_0(r-ct)\} + \frac{1}{2cr}\int_{r-ct}^{r+ct} \xi u_1(\xi)\,\mathrm{d}\xi \tag{6.5}$$

for $r > a + ct$: the solution in this domain, $\mathscr{Q}^-$, does not depend on the boundary condition because information from the boundary at $r = a$ first reaches the wavefront $\Gamma(t)$ after travelling at speed c for time t.

The solution behind the wavefront in $\mathscr{Q}^+$ is determined using both the boundary condition and the initial conditions, and it splits as

$$u(r,t) = u_{\mathrm{bc}}(r,t) + u_{\mathrm{ic}}(r,t), \quad a < r < a + ct. \tag{6.6}$$

We calculate u for Dirichlet, pressure and Neumann boundary conditions separately.

6.1.1 Dirichlet Boundary Condition

If we take the Dirichlet condition (4.2),

$$u(a,t) = d(t), \quad t > 0, \tag{6.7}$$

where $d(t) \equiv d(P,t)$ is given when P is on the sphere, and use it in (6.1), we obtain

$$ad(t) = f(a-ct) + g(a+ct), \quad t > 0, \tag{6.8}$$

whence $f(\xi) = ad([a-\xi]/c) - g(2a-\xi)$ for $\xi < a$. Hence (6.1) and (6.4) give (6.6) in which

$$u_{\text{bc}}(r,t) = (a/r)d(t - [r-a]/c), \tag{6.9}$$

$$u_{\text{ic}}(r,t) = \frac{1}{2r}\{(r+ct)u_0(r+ct) - (2a-r+ct)u_0(2a-r+ct)\} + \frac{1}{2cr}\int_{2a-r+ct}^{r+ct} \xi u_1(\xi)\,\mathrm{d}\xi. \tag{6.10}$$

Note that $u_{\text{bc}}(a,t) = d(t)$ and $u_{\text{ic}}(a,t) = 0$, thus verifying (6.7).

Across the wavefront at $r = a + ct$, we find that the jump in u (defined by (3.25)) is

$$[\![u]\!] = (a/r)\{d(0) - u_0(a)\},$$

where we have used (6.5), (6.6), (6.9) and (6.10). We note that this result, $[\![u]\!] = A/r$, is consistent with the known jump behaviour of weak solutions, (3.42). However, the physical constraint is much stronger: it requires that $[\![u]\!]$ be constant as the wavefront evolves; see (3.47). Thus we must impose the consistency condition

$$u_0(a) = d(0).$$

In fact, this condition gives more, namely $[\![u]\!] = 0$.

A similar argument shows that $d(t)$ must be continuous for $t > 0$, otherwise unphysical discontinuities will be induced in u.

6.1.2 Pressure Boundary Condition

Next, consider the pressure condition (4.4),

$$p(a,t) = -\rho\frac{\partial u}{\partial t}(a,t) = p_a(t), \quad t > 0, \tag{6.11}$$

where $p_a(t) \equiv p_S(P,t)$ is given when P is on the sphere. Use of (6.1) gives

$$ap_a(t) = \rho c\left\{f'(a-ct) - g'(a+ct)\right\}, \quad t > 0.$$

An integration then gives

$$f(\xi) = -\frac{a}{\rho}\int_0^{(a-\xi)/c} p_a(\eta)\,\mathrm{d}\eta - g(2a-\xi) + aA, \quad \xi < a, \tag{6.12}$$

where A is an arbitrary constant. Hence (6.1) and (6.4) give (6.6) in which $u_{\text{ic}}(r,t)$ is given by (6.10) again and

$$u_{\text{bc}}(r,t) = -\frac{a}{r\rho}\int_0^{t-(r-a)/c} p_a(\eta)\,\mathrm{d}\eta + \frac{a}{r}A, \quad a < r < a+ct. \tag{6.13}$$

Note that $-\rho(\partial u_{\text{bc}}/\partial t) = p_a(t)$ and $\partial u_{\text{ic}}/\partial t = 0$ when $r = a$, which confirms (6.11).

Across the wavefront at $r = a + ct$, we obtain

$$[\![u]\!] = (a/r)\{A - u_0(a)\}.$$

For this to be independent of r, in accordance with the physical constraint (3.47), we must take the constant $A = u_0(a)$, whence $[\![u]\!] = 0$.

We also find $[\![p]\!] = (a/r)\{\rho u_1(a) + p_a(0)\}$. Thus we obtain $[\![p]\!] = 0$ if $p_a(0) = -\rho u_1(a)$. If this condition is not satisfied, the pressure will jump across the wavefront even though the potential does not. Similarly, if $p_a(t)$ is not continuous for $t > 0$, points of discontinuity in $p_a(t)$ will induce admissible discontinuities in $p(r,t)$ but no discontinuities in $u(r,t)$.

Of course, jumps in p across wavefronts are what we expect in more general cases, such as when a plane pressure pulse is scattered by a sphere.

6.1.3 Neumann Boundary Condition

Instead of (6.7) or (6.11), we can take the Neumann boundary condition (4.3),

$$\frac{\partial u}{\partial r}(a,t) = v(t), \quad t > 0, \tag{6.14}$$

where $v(t) \equiv v(P,t)$ is given. Substitution of (6.1) in (6.14) gives a first-order differential equation for f,

$$f'(\xi) - a^{-1}f(\xi) = h(\xi), \quad \xi < a,$$

where

$$h(\xi) = av([a-\xi]/c) - g'(2a-\xi) + a^{-1}g(2a-\xi)$$

and g is given by (6.4). Solving gives

$$f(\xi)\mathrm{e}^{-\xi/a} = A_1 - \int_\xi^a h(\eta)\mathrm{e}^{-\eta/a}\,\mathrm{d}\eta, \quad \xi < a, \tag{6.15}$$

where A_1 is an arbitrary constant. The piece of h containing v generates u_{bc} in (6.6):

$$u_{\text{bc}}(r,t) = -\frac{a}{r}\mathrm{e}^{(r-ct)/a}\int_{r-ct}^a \mathrm{e}^{-\eta/a}v([a-\eta]/c)\,\mathrm{d}\eta, \quad a < r < a+ct; \tag{6.16}$$

it is straightforward to check that u_{bc} satisfies (6.14). Substituting the other piece of h in (6.15) gives

$$\begin{aligned} f(\xi)\mathrm{e}^{-\xi/a} &= A_1 + \int_\xi^a g'(2a-\eta)\,\mathrm{e}^{-\eta/a}\,\mathrm{d}\eta - \frac{1}{a}\int_\xi^a g(2a-\eta)\,\mathrm{e}^{-\eta/a}\,\mathrm{d}\eta \\ &= A_2 + g(2a-\xi)\,\mathrm{e}^{-\xi/a} - \frac{2}{a}\int_\xi^a g(2a-\eta)\,\mathrm{e}^{-\eta/a}\,\mathrm{d}\eta, \end{aligned} \tag{6.17}$$

after an integration by parts; A_2 $(= A_1 - \mathrm{e}^{-1}g(a))$ is an arbitrary constant. We use (6.17) with $\xi = r - ct$, and then (6.1) and (6.6) give

$$\begin{aligned} u_{\text{ic}}(r,t) = &\frac{A_2}{r}\,\mathrm{e}^{(r-ct)/a} + \frac{1}{r}\{g(r+ct) + g(2a-r+ct)\} \\ &- \frac{2}{ar}\mathrm{e}^{(r-ct)/a}\int_{r-ct}^a g(2a-\eta)\,\mathrm{e}^{-\eta/a}\,\mathrm{d}\eta. \end{aligned} \tag{6.18}$$

For the last term, we have, using (6.4),

$$\begin{aligned} \int_\xi^a g(2a-\eta)\,\mathrm{e}^{-\eta/a}\,\mathrm{d}\eta &= \frac{1}{2}\int_\xi^a (2a-\eta)u_0(2a-\eta)\,\mathrm{e}^{-\eta/a}\,\mathrm{d}\eta - C\int_\xi^a \mathrm{e}^{-\eta/a}\,\mathrm{d}\eta \\ &\quad + \frac{1}{2c}\int_\xi^a \mathrm{e}^{-\eta/a}\int_a^{2a-\eta} s u_1(s)\,\mathrm{d}s\,\mathrm{d}\eta \\ &= \frac{1}{2\mathrm{e}^2}\int_a^{2a-\xi} s\,\mathrm{e}^{s/a}u_0(s)\,\mathrm{d}s - aC\mathrm{e}^{-\xi/a} + \frac{aC}{\mathrm{e}} \\ &\quad + \frac{a}{2c}\mathrm{e}^{-\xi/a}\int_a^{2a-\xi} s u_1(s)\,\mathrm{d}s - \frac{a}{2c\mathrm{e}^2}\int_a^{2a-\xi} s\,\mathrm{e}^{s/a}u_1(s)\,\mathrm{d}s. \end{aligned}$$

Hence, using (6.4) again, (6.18) gives

$$\begin{aligned} u_{\text{ic}}(r,t) = &\frac{A}{r}\,\mathrm{e}^{(r-ct)/a} + \frac{1}{2r}\{(r+ct)u_0(r+ct) + (2a-r+ct)u_0(2a-r+ct)\} \\ &+ \frac{1}{2cr}\int_{2a-r+ct}^{r+ct} \xi u_1(\xi)\,\mathrm{d}\xi \\ &- \frac{\mathrm{e}^{(r-ct)/a}}{r\mathrm{e}^2}\int_a^{2a-r+ct} \xi\,\mathrm{e}^{\xi/a}\left(\frac{u_0(\xi)}{a} - \frac{u_1(\xi)}{c}\right)\mathrm{d}\xi \end{aligned} \tag{6.19}$$

for $a < r < a + ct$, where A $(= A_2 - 2C/\mathrm{e})$ is an arbitrary constant. Direct calculation verifies that $\partial u_{\text{ic}}/\partial r = 0$ on $r = a$.

Note that the first term in (6.19), $(A/r)\mathrm{e}^{(r-ct)/a}$, is a wavefunction that satisfies the homogeneous boundary condition, $\partial u/\partial r = 0$ on $r = a$. This feature is discussed briefly by John [445, p. 15].

Across the wavefront at $r = a + ct$, we find that $[\![u]\!] = A\mathrm{e}/r$, after using (6.5), (6.6), (6.16) and (6.19). Then the physical constraint (3.47) implies that we must take $A = 0$, whence $[\![u]\!] = 0$. Note that seeking a weak solution would lead to non-uniqueness because the term involving A is admissible.

Table 6.1 *The six initial-boundary value problems in Table 4.1 for a sphere of radius a. The wavefront* Γ *is at* $r = a + ct$.

Problem	Behind Γ, in $\mathscr{Q}^+$	Ahead of Γ, in $\mathscr{Q}^-$	Physical constraint
ID_0	$u = u_{\mathrm{ic}}$, defined by (6.10)	u is given by (6.5)	$u_0(a) = 0$
IP_0	$u = u_{\mathrm{ic}} + aA/r$, u_{ic} from (6.10)	u is given by (6.5)	$u_0(a) = A$
IN_0	$u = u_{\mathrm{ic}}$, defined by (6.19)	u is given by (6.5)	$A = 0$ in (6.19)
DI_0	$u = u_{\mathrm{bc}}$, defined by (6.9)	$u \equiv 0$	$d(0) = 0$
PI_0	$u = u_{\mathrm{bc}}$, defined by (6.13)	$u \equiv 0$	$A = 0$ in (6.13)
NI_0	$u = u_{\mathrm{bc}} + (A/r)\mathrm{e}^{(r-ct)/a}$, with u_{bc} defined by (6.16)	$u \equiv 0$	$A = 0$

6.1.4 Discussion and Special Cases

In Table 6.1, we collect the results for the six simpler problems listed in Table 4.1. In each case, there is a wavefront $\Gamma(t)$ at $r = a + ct$, and the physical constraint (3.47) implies that $[\![u]\!] = 0$ across Γ.

The pressure and velocity fields are easily calculated. For example, let us calculate the pressure jump across the wavefront when there are zero initial conditions. For Problem DI_0, (6.9) gives $[\![p]\!] = -\rho(a/r)d'(0)$, so that $[\![p]\!] = 0$ if $d'(0) = 0$ in addition to $d(0) = 0$. For Problem PI_0, $[\![p]\!] = (a/r)p_a(0)$. For Problem NI_0, (6.16) gives $[\![p]\!] = -\rho(a/r)v(0)$.

Next, we calculate the jump in the normal velocity across the wavefront for Problem DI_0. We find that

$$[\![\partial u/\partial r]\!] = -(a/[cr])\,d'(0) \quad \text{at } r = a + ct.$$

Therefore imposing $d(0) = d'(0) = 0$ leads to a smoother solution u; additional smoothness can be induced by requiring higher derivatives of $d(t)$ to vanish at $t = 0$.

One interesting feature of the solutions derived is that we did not impose any condition far from the sphere: there is no analogue of the Sommerfeld radiation condition. However, for problems with zero initial conditions, we can start by writing

$$u(r,t) = r^{-1} f(r - ct),$$

with an outgoing wavefunction. The initial conditions are satisfied by taking $f(\xi) = 0$ for all $\xi > a$, whence $u \equiv 0$ in $\mathscr{Q}^-$. The solution in $\mathscr{Q}^+$ requires $f(\xi)$ for $a - ct < \xi < a$. The boundary condition will give $f(a - ct)$ for $t > 0$, that is, $f(\xi)$ for $\xi < a$. For Problem DI_0, $f(a - ct) = ad(t)$, leading to $u = u_{\mathrm{bc}}$, as in (6.9). For Problem PI_0, we find that f is given by (6.12) with $g \equiv 0$ therein, leading to $u = u_{\mathrm{bc}}$ as in (6.13). For Problem NI_0, we find that $u = (A/r)\mathrm{e}^{(r-ct)/a} + u_{\mathrm{bc}}$ with u_{bc} given by (6.16). Thus we recover all the results listed in the last three rows of Table 6.1.

Let us consider two examples of time-harmonic forcing with zero initial conditions. For Problem DI_0 with $d(t) = \sin \omega t$, (6.9) gives

$$u(r,t) = (a/r)\sin\Omega \quad \text{for } a < r < a + ct,$$

where $\Omega = \omega t - k(r-a)$ and $k = \omega/c$. Thus, for fixed r, time-harmonic oscillations are seen as soon as the wavefront has passed. On the other hand, for Problem NI_0 with $v(t) = \sin \omega t$, (6.16) gives

$$u(r,t) = \frac{a^2}{r\{1+(ka)^2\}}\left(ka\cos\Omega - \sin\Omega - ka\,\mathrm{e}^{(r-a-ct)/a}\right) \quad \text{for } a < r < a + ct, \tag{6.20}$$

so that time-harmonic oscillations are achieved exponentially fast. This is an example of the *limiting amplitude principle*; see Section 4.7.2. The corresponding pressure on the sphere is

$$p(a,t) = -\rho \left.\frac{\partial u}{\partial t}\right|_{r=a} = \frac{\rho\omega a}{1+(ka)^2}\left(ka\sin\omega t + \cos\omega t - \mathrm{e}^{-ct/a}\right). \tag{6.21}$$

This solution has been used [607, eqn (B8)] to test a numerical scheme based on the convolution quadrature method; see Section 10.4.

The literature on spherically symmetric pulsations of a sphere, using the general solution (6.1), is extensive, going back to Lord Rayleigh [711, §279] and Pierre Duhem [249, pp. 235–237]. See also Love [572, 574], Lamb [516, §286], Morse [642, §27], Barnes & Anderson [53], John [445, p. 15], Morawetz [635, p. 29], Lighthill [559, §1.11], Pierce [696, §4-1] and Barbone & Crighton [50, §2.1]. Whitham [881, §7.3] solves what he calls the 'balloon problem', with $p(r,0) = 0$ for $r > a$ and $p(r,0) = P$, a constant, for $0 \le r < a$. Initially, there is no motion and then the balloon is burst at $t = 0$. The initial conditions are $u(r,0) = 0$ and $\partial u/\partial t = -(P/\rho)\,\mathrm{H}(a-r)$. For large-amplitude oscillations of a spherical bubble, see [472, 473].

For similar treatments of spherical cavities in a solid, see the 1960 review by Hopkins [419, §5] and [350, Chapter 2].

6.2 Use of Laplace Transforms

The problems solved in Section 6.1 can also be solved using Laplace transforms. We do that here, paying attention to the effects of the expanding wavefront; see also [601, §9].

For simplicity, let us take zero initial conditions so that $f_{\mathrm{ic}} = 0$ in (5.8). We expect a wavefront at $t = \tau(r) = (r-a)/c$, with $[\![u]\!]$ constant. From (5.8), (5.9) and the spherically symmetric form $\nabla^2 u = r^{-1}(\partial^2/\partial r^2)(ru)$, we obtain

$$\frac{\partial^2(rU)}{\partial r^2} - \frac{s^2}{c^2}(rU) = [\![u]\!]\frac{\partial^2(r\tau)}{\partial r^2} = \frac{2}{c}[\![u]\!], \tag{6.22}$$

where $U = \mathscr{L}\{u\}$. Solving this equation,

$$rU(r,s) = \mathscr{A}(s)\,\mathrm{e}^{-sr/c} + \mathscr{B}(s)\,\mathrm{e}^{sr/c} - (2c/s^2)[\![u]\!],$$

where $\mathscr{A}$ and $\mathscr{B}$ are arbitrary. We take $\mathscr{B} = 0$ because we do not want solutions that grow exponentially with r, whence

$$rU(r,s) = \mathscr{A}(s)\,\mathrm{e}^{-sr/c} - (2c/s^2)[\![u]\!]. \tag{6.23}$$

6.2.1 Dirichlet Boundary Condition

For Problem DI_0, we apply the Laplace transform of the Dirichlet boundary condition (6.7), namely (5.10) at $r = a$, whence

$$rU(r,s) = aD(s)\mathrm{e}^{-s(r-a)/c} + (2c/s^2)\left(\mathrm{e}^{-s(r-a)/c} - 1\right)[\![u]\!].$$

Inverting, using $\mathscr{L}\{f(t-b)\mathrm{H}(t-b)\} = \mathrm{e}^{-sb}\mathscr{L}\{f(t)\}$, gives

$$\begin{aligned} u(r,t) &= (a/r)d(t-[r-a]/c)\,\mathrm{H}(t-[r-a]/c) \\ &\quad + (2c/r)\{(t-[r-a]/c)\mathrm{H}(t-[r-a]/c) - t\}\,[\![u]\!]. \end{aligned} \tag{6.24}$$

There is a wavefront at $r = a + ct$. Behind the wavefront (in $\mathscr{Q}^+$, see (6.2)),

$$u(r,t) = (a/r)d(t-[r-a]/c) + (2/r)(a-r)[\![u]\!].$$

Ahead of the wavefront (in $\mathscr{Q}^-$), $u(r,t) = -(2ct/r)[\![u]\!]$. Combining these two equations so as to calculate $[\![u]\!]$ gives

$$[\![u]\!] = (a/r)d(0).$$

But $[\![u]\!]$ is required to be constant, whence $d(0) = [\![u]\!] = 0$. Thus (6.24) reduces to

$$u(r,t) = (a/r)d(t-[r-a]/c)\,\mathrm{H}(t-[r-a]/c), \tag{6.25}$$

in agreement with the solution of Problem DI_0 obtained in Section 6.1.4 (see Table 6.1). We observe that, for this problem, the term on the right-hand side of (6.22) is, in fact, zero.

What would have happened if we simply applied the Laplace transform $\mathscr{L}$ to the governing equation for $u(r,t)$, (2.9), oblivious to the possible presence of discontinuities? Proceeding formally, we would obtain

$$\frac{\partial^2(rU)}{\partial r^2} - \frac{s^2}{c^2}(rU) = 0$$

with solution $rU(r,s) = \mathscr{A}(s)\mathrm{e}^{-sr/c}$. Applying the boundary condition (6.7) gives $U(a,s) = D(s)$. This determines $\mathscr{A}$ whence $rU(r,s) = aD(s)\mathrm{e}^{-s(r-a)/c}$. Inverting this formula gives precisely (6.25). Thus, for this problem, formal application of $\mathscr{L}$ gives the correct result. However, subsequent inspection of (6.25) might suggest that u is allowed to be discontinuous across wavefronts; it is not.

6.2.2 Pressure Boundary Condition

Next, consider Problem PI_0. The Laplace transform of the pressure boundary condition (6.11) gives (5.11), which becomes $sU(a,s) = -\rho^{-1}P_a(s)$ with $P_a = \mathscr{L}\{p_a\}$; recall that, for this problem, we can assume that $[\![u]\!] = 0$. Then (6.23) gives $s\mathscr{A}\,\mathrm{e}^{-sa/c} = -(a/\rho)P_a$ and

$$rU(r,s) = -\frac{a}{\rho s}P_a(s)\,\mathrm{e}^{-s(r-a)/c} = -\frac{a}{\rho}\mathrm{e}^{-s(r-a)/c}\mathscr{L}\left\{\int_0^t p_a(\eta)\,\mathrm{d}\eta\right\}.$$

Thus, inverting,

$$u(r,t) = -\frac{a}{r\rho}\mathrm{H}(t-[r-a]/c)\int_0^{t-(r-a)/c} p_a(\eta)\,\mathrm{d}\eta, \tag{6.26}$$

in agreement with the solution of Problem PI_0 obtained in Section 6.1.4 (see Table 6.1).

6.2.3 Neumann Boundary Condition

For Problem NI_0, we apply the Laplace transform of the Neumann boundary condition (6.14), namely (5.12), which becomes $\partial U/\partial r = V + [\![u]\!]/c$ at $r = a$. Using this in (6.23) gives

$$\mathscr{A}(s) = \frac{\mathrm{e}^{sa/c}}{sa+c}\left(\frac{2c^2}{s^2}[\![u]\!] - a^2[\![u]\!] - a^2cV(s)\right),$$

whence

$$\begin{aligned} rU(r,s) &= -\frac{a^2cV(s)}{sa+c}\,\mathrm{e}^{-s(r-a)/c} + [\![u]\!]\left(\frac{2c^2 - s^2a^2}{(sa+c)s^2}\,\mathrm{e}^{-s(r-a)/c} - \frac{2c}{s^2}\right) \\ &= -acV(s)\mathscr{L}\{\mathrm{e}^{-ct/a}\}\mathrm{e}^{-s(r-a)/c} - 2[\![u]\!]\mathscr{L}\{ct\} \\ &\quad + [\![u]\!]\mathscr{L}\{a\mathrm{e}^{-ct/a} + 2ct - 2a\}\,\mathrm{e}^{-s(r-a)/c}. \end{aligned} \tag{6.27}$$

The convolution theorem (1.76) gives

$$V(s)\mathscr{L}\{\mathrm{e}^{-ct/a}\} = \mathscr{L}\left\{\int_0^t v(\eta)\,\mathrm{e}^{-c(t-\eta)/a}\,\mathrm{d}\eta\right\}.$$

Hence, inverting,

$$\begin{aligned} u(r,t) = &-\frac{ac}{r}\mathrm{H}(t-[r-a]/c)\,\mathrm{e}^{(r-ct)/a}\int_0^{t-(r-a)/c} v(\eta)\,\mathrm{e}^{(c\eta-a)/a}\,\mathrm{d}\eta \\ &+ r^{-1}\left\{(a\mathrm{e}^{(r-a-ct)/a} + 2ct - 2r)\mathrm{H}(t-[r-a]/c) - 2ct\right\}[\![u]\!]. \end{aligned}$$

Behind the wavefront at $r = a + ct$, this solution simplifies to

$$u(r,t) = -\frac{ac}{r}\,\mathrm{e}^{(r-ct)/a}\int_0^{t-(r-a)/c} v(\eta)\,\mathrm{e}^{(c\eta-a)/a}\,\mathrm{d}\eta + \frac{1}{r}\left\{a\mathrm{e}^{(r-a-ct)/a} - 2r\right\}[\![u]\!], \tag{6.28}$$

whereas ahead of the wavefront $u(r,t) = -(2ct/r)[\![u]\!]$. Combining these two equations so as to calculate $[\![u]\!]$ gives $[\![u]\!] = 0$. Hence we find agreement with the solution of Problem NI_0 obtained in Section 6.1.4 (see Table 6.1 and (6.16)).

Summarising, as $[\![u]\!] = 0$, (6.27) and (6.28) reduce to

$$U(r,s) = -\frac{a^2 V(s)}{r(1+sa/c)}\,\mathrm{e}^{-s(r-a)/c} \qquad \text{and} \tag{6.29}$$

$$u(r,t) = -\frac{ac}{r}\,\mathrm{e}^{-cT/a}\mathrm{H}(T)\int_0^T v(\eta)\,\mathrm{e}^{\eta c/a}\,\mathrm{d}\eta, \tag{6.30}$$

respectively, with $T = t-(r-a)/c$.

7
Scattering by a Sphere

Nevertheless, a word of caution is in order. The sphere with its high degree of symmetry is not a good pointer to what will happen to asymmetric bodies. ... Getting the answers right for a sphere is a necessary but not sufficient condition for the validation of a numerical procedure.
(Jones [454, p. 552])

This chapter is concerned with the scattering of an incident wave, a sound pulse, by a sphere. In general, the solution u will not be spherically symmetric. (That simpler situation was studied in detail in Chapter 6.)

The standard method is to use Laplace transforms, as outlined in Section 5.1. Thus $U = \mathscr{L}\{u\}$ satisfies the modified Helmholtz equation, and then this equation is solved by the method of separation of variables in spherical polar coordinates. Typical solutions have the form $k_n(sr/c)\,\mathrm{e}^{st}\,Y_n^m$, where k_n is a modified spherical Bessel function, Y_n^m is a spherical harmonic and s is the Laplace transform variable. After the boundary condition has been applied, U is inverted to obtain u. It is assumed throughout that $[\![u]\!] = 0$ and that u satisfies zero initial conditions. We develop the method in Section 7.2; see also [600, §3].

In Section 7.4, we use an alternative method [600, §4] based on the integral representation (2.36). Typical solutions have the form $r^{-1}P_n(ct/r)Y_n^m$, where P_n is a Legendre polynomial.

Other topics covered in this chapter include residual potential methods and non-reflecting boundary kernels (Section 7.3), moving spheres (Section 7.5) and scattering by a spheroid (Section 7.6).

7.1 Preliminaries

We consider a sphere of radius a. For scattering problems, we suppose that there is an incident plane pulse defined by (4.11),

$$u_{\mathrm{inc}}(z,t) = w_{\mathrm{inc}}(t - [z+a]/c)\,\mathrm{H}(t - [z+a]/c), \tag{7.1}$$

where H is the Heaviside unit function and $w_{\text{inc}}(t)$ is specified; see Section 4.3 for some examples. The total field is $u + u_{\text{inc}}$. As the incident pulse does not reach the sphere until $t = 0$, u satisfies zero initial conditions.

There are other possibilities for the incident field, such as a simple source (2.12) located at a point outside the sphere. Alternatively, we could consider radiation problems where waves are generated by prescribed forcing of the sphere.

7.2 Use of Laplace Transforms

As we have assumed that u is continuous with zero initial conditions, it follows that $U(\boldsymbol{r},s) = \mathscr{L}\{u(\boldsymbol{r},t)\}$ satisfies the modified Helmholtz equation, $\nabla^2 U = (s/c)^2 U$. Separating variables in spherical polar coordinates, (r,θ,ϕ), we find solutions

$$i_n(sr/c)\,Y_n^m(\theta,\phi) \quad \text{and} \quad k_n(sr/c)\,Y_n^m(\theta,\phi),$$

where i_n and k_n are modified spherical Bessel functions and $Y_n^m(\theta,\phi)$ is a spherical harmonic; see Section 2.3.1. We discard the solutions containing $i_n(sr/c)$ because of their exponential growth with r. For simplicity, we also assume that the incident wave is axisymmetric about the z-axis. Then we can write

$$U(\boldsymbol{r},s) = U(r,\theta,s) = \sum_{n=0}^{\infty} B_n(s)\,k_n(sr/c)\,P_n(\cos\theta), \tag{7.2}$$

where P_n is a Legendre polynomial and the functions $B_n(s)$ $(n = 0,1,2,\ldots)$ are to be determined from the boundary condition on the sphere. Then, inverting $\mathscr{L}$,

$$u(r,\theta,t) = \sum_{n=0}^{\infty} u_n(r,t)\,P_n(\cos\theta), \tag{7.3}$$

where

$$u_n(r,t) = \frac{1}{2\pi \mathrm{i}} \int_{\mathrm{Br}} B_n(s)\,k_n(sr/c)\,\mathrm{e}^{st}\,\mathrm{d}s \tag{7.4}$$

and Br is the Bromwich contour in the complex s-plane.

7.2.1 Dirichlet Boundary Condition

For the axisymmetric Dirichlet problem, the boundary condition is

$$u(a,\theta,t) = d(\theta,t), \quad 0 \le \theta \le \pi, \quad t > 0. \tag{7.5}$$

To impose it, let us expand d similarly to (7.3),

$$d(\theta,t) = \sum_{n=0}^{\infty} d_n(t)\,P_n(\cos\theta); \tag{7.6}$$

orthogonality of Legendre polynomials gives

$$d_n(t) = \frac{2n+1}{2} \int_0^{\pi} d(\theta,t) P_n(\cos\theta)\sin\theta\,\mathrm{d}\theta. \tag{7.7}$$

Applying $\mathscr{L}$ to $u_n(a,t) = d_n(t)$ gives $B_n(s)\,k_n(sa/c) = D_n(s)$, with $D_n(s) = \mathscr{L}\{d_n\}$. Hence (7.3) and (7.4) give $u(r,\theta,t)$ with

$$u_n(r,t) = \frac{1}{2\pi \mathrm{i}} \int_{\mathrm{Br}} D_n(s)\, \frac{k_n(sr/c)}{k_n(sa/c)}\, \mathrm{e}^{st}\, \mathrm{d}s. \tag{7.8}$$

Recall that (see (2.30))

$$\frac{2}{\pi} x^{n+1} \mathrm{e}^x\, k_n(x) = \theta_n(x) = \sum_{j=0}^{n} \frac{(2n-j)!\, x^j}{2^{n-j}\, j!\, (n-j)!}, \tag{7.9}$$

a polynomial in x of degree n known as a *reverse Bessel polynomial*. For example, $\theta_0(x) = 1$, $\theta_1(x) = x+1$ and $\theta_2(x) = x^2+3x+3$. Much is known about these polynomials; see [437, §4.10], [661, §18.34] and the book by Grosswald [364]. For example, all zeros of $\theta_n(x)$ are simple [364, p. 75] and they are in the left half of the complex x-plane. We denote them by $\beta_{n,m}$ with $m = 1,2,\ldots,n$: $\theta_n(\beta_{n,m}) = 0$ with $\operatorname{Re}\beta_{n,m} < 0$. Asymptotically, for large n, the zeros lie on a certain convex arc, symmetric about the real axis, meeting the imaginary axis at $\pm \mathrm{i}n$ and crossing the real axis at $-n\zeta_0$ with $\zeta_0 \simeq 0.66$. This result is due to Olver [673, p. 354]; for the convex arc, rotate any of the following figures clockwise by $\pi/2$: [673, Fig. 15], [3, Fig. 9.6], [661, Fig. 10.21.6]. For a plot of the zeros of θ_{10} and θ_{11}, see [356, Fig. 5]. For early tabulations of $\beta_{n,m}$, see [878, Table I], [354, Table I] and [355, Table I].

In terms of θ_n, (7.8) becomes

$$u_n(r,t) = \frac{1}{2\pi \mathrm{i}} \left(\frac{a}{r}\right)^{n+1} \int_{\mathrm{Br}} D_n(s)\, \frac{\theta_n(sr/c)}{\theta_n(sa/c)}\, \mathrm{e}^{s(t-[r-a]/c)}\, \mathrm{d}s. \tag{7.10}$$

In particular, as $\theta_0 = 1$ and $\mathscr{L}\{d_0(t-b)\mathrm{H}(t-b)\} = \mathrm{e}^{-sb} D_0(s)$, the spherically symmetric component is

$$u_0(r,t) = (a/r)\, d_0(T)\mathrm{H}(T) \quad \text{with} \quad T = t-(r-a)/c, \tag{7.11}$$

where H is the Heaviside unit function. Equation (7.11) agrees with (6.25).

Motivated by (7.10) and (7.11), define $\Psi_n(r,s)$ by

$$\Psi_n(r,s) = \frac{a^n\, \theta_n(sr/c)}{r^n\, \theta_n(sa/c)} - 1 \tag{7.12}$$

so that (7.10) becomes

$$u_n(r,t) = (a/r)\{d_n(T)\mathrm{H}(T) + w_n(r,t)\} \tag{7.13}$$

with

$$w_n(r,t) = \frac{1}{2\pi \mathrm{i}} \int_{\mathrm{Br}} D_n(s)\, \Psi_n(r,s)\, \mathrm{e}^{sT}\, \mathrm{d}s. \tag{7.14}$$

We shall describe two approaches for determining u_n from (7.13) and (7.14).

First Approach: Use the Convolution Theorem

We start by noticing that $x^{-n}\theta_n(x) \to 1$ as $x \to \infty$ (see (7.9)). Hence, from (7.12), $\Psi_n(r,s) \to 0$ as $s \to \infty$, implying that there is a function $\psi_n(r,t)$ with $\mathscr{L}\{\psi_n\} = \Psi_n$. Then, by the convolution theorem (1.76),

$$D_n(s)\Psi_n(r,s) = \mathscr{L}\left\{ \int_0^t d_n(t')\, \psi_n(r,t-t')\,\mathrm{d}t' \right\}$$

and hence (7.13) becomes

$$u_n(r,t) = \frac{a}{r}\left\{ d_n(T) + \int_0^T d_n(t')\, \psi_n(r,T-t')\,\mathrm{d}t' \right\}\mathrm{H}(T). \tag{7.15}$$

This is [356, eqn (15)]. To use (7.15), we need $d_n(t)$ (which is defined by (7.7) as an integral of the boundary data $d(\theta,t)$) and $\psi_n(r,t)$ (which is defined by inverting (7.12)).

To obtain ψ_n from Ψ_n, we can use a partial-fraction expansion or, equivalently, a residue calculation. We know that $\theta_n(sa/c)$ has simple zeros at $s = (c/a)\beta_{n,j}$, $j = 1,2,\ldots,n$, $n \geq 1$. Hence

$$\psi_n(r,t) = \frac{1}{2\pi\mathrm{i}}\int_{\mathrm{Br}} \Psi_n(r,s)\,\mathrm{e}^{st}\,\mathrm{d}s = \frac{c}{a}\sum_{j=1}^{n} a_{n,j}(r)\exp\left(\beta_{n,j}\,ct/a\right),$$

where

$$a_{n,j}(r) = \frac{a^n\,\theta_n(\beta_{n,j}\,r/a)}{r^n\,\theta_n'(\beta_{n,j})} = \frac{r\,k_n(\beta_{n,j}r/a)}{a\,k_n'(\beta_{n,j})}\exp\{(r-a)\beta_{n,j}/a\} \tag{7.16}$$

and the second form comes by differentiating (7.9):

$$\theta_n'(\beta) = (2/\pi)\beta^{n+1}\mathrm{e}^{\beta}\,k_n'(\beta) \quad \text{when} \quad k_n(\beta) = 0.$$

We conclude that $\psi_n(r,t)$ is a linear combination of n exponential functions of t with coefficients that are rational functions of r. As the dependence on r and t is separated, the integral term in (7.15) becomes

$$\frac{c}{a}\sum_{j=1}^{n} a_{n,j}(r)\exp\left(\beta_{n,j}\,cT/a\right)\int_0^T d_n(t')\exp\left(-\beta_{n,j}\,ct'/a\right)\mathrm{d}t'.$$

When this is used in (7.15), we obtain a formula stated by Wilcox [885] in 1959; see also [386, eqn (31)] and [356, eqn (13)]. Computationally, this formula may not be useful 'due to catastrophic cancellation in carrying out the summation' [356, p. 193]: the coefficients $a_{n,j}(r)$ grow exponentially with n (see (7.19)). Greengard et al. [356] advocate using a recursive version of (7.15).

Second Approach: Close the Contour

Although the convolution form (7.15) is attractive, we could evaluate $w_n(r,t)$ directly from the contour integral (7.14) by closing the contour. Thus, if $T < 0$ ($r > a+ct$, ahead of the wavefront), we close the contour to the right with a large semicircular contour, whereas if $T > 0$ ($r < a+ct$, behind the wavefront), we close to the left.

We then use a rotated form of Jordan's lemma to justify discarding the contribution from the semicircular contour as its radius $\to \infty$. See, for example, [148, §31], [869, §3.3] or [216, §2.4]. There are no singularities to the right of Br, so $w_n(r,t) = 0$ when $T < 0$. When $T > 0$, there will be residue contributions from the poles to the left of Br; in some applications, there may be branch points.

The integrand in (7.14) has n simple poles coming from $\Psi_n(r,s)$ (at the zeros of $\theta_n(sa/c)$) and additional poles coming from $D_n(s)$. To examine the latter, let us consider the scattering of a plane sound pulse, defined by (7.1). The boundary condition for scattering by a soft sphere, $u + u_{\rm inc} = 0$ at $r = a$, gives $d = -u_{\rm inc}$ with

$$d(\theta,t) = -w_{\rm inc}(t - [1+\cos\theta]a/c)\,\mathrm{H}(t - [1+\cos\theta]a/c).$$

Then $D = \mathscr{L}\{d\}$ is given by

$$\begin{aligned} D(\theta,s) &= -\int_{(1+\cos\theta)a/c}^{\infty} w_{\rm inc}(t - [1+\cos\theta]a/c)\,\mathrm{e}^{-st}\,\mathrm{d}t \\ &= -\mathrm{e}^{-sa/c}\mathrm{e}^{-(sa/c)\cos\theta}\,W_{\rm inc}(s) \\ &= \mathrm{e}^{-sa/c}W_{\rm inc}(s)\sum_{n=0}^{\infty}(-1)^{n+1}(2n+1)\,i_n(sa/c)P_n(\cos\theta) \end{aligned}$$

where $W_{\rm inc}(s) = \mathscr{L}\{w_{\rm inc}\}$ and we have used [661, eqn 10.60.9]

$$\mathrm{e}^{-w\cos\theta} = \sum_{n=0}^{\infty}(-1)^n(2n+1)\,i_n(w)P_n(\cos\theta).$$

Comparison with (7.6) gives

$$D_n(s) = (2n+1)(-1)^{n+1}\mathrm{e}^{-sa/c}\,i_n(sa/c)W_{\rm inc}(s), \tag{7.17}$$

which shows that D_n inherits its singularities from those of $W_{\rm inc}$. For example, if $w_{\rm inc}(t) = \sin\omega_0 t$, $W_{\rm inc}(s) = \omega_0/(s^2+\omega_0^2)$, which has simple poles at $s = \pm\mathrm{i}\omega_0$.

From (2.31), $2(-x)^{n+1}i_n(x) = \mathrm{e}^{-x}\theta_n(x) - \mathrm{e}^{x}\theta_n(-x)$. Hence, if β is a zero of θ_n, (7.17) gives

$$2D_n(\beta c/a) = -(2n+1)\beta^{-n-1}\theta_n(-\beta)W_{\rm inc}(\beta c/a).$$

This is useful when evaluating D_n at the poles of Ψ_n.

For another example, consider a spherical step pulse (4.15) generated by a simple source located at a point $(0,0,z')$ inside the sphere ($0 \le z' < a$), so that we are solving a radiation problem. We have

$$d(\theta,t) = R^{-1}w_{\rm inc}(t - [R - R_0]/c)\,\mathrm{H}(t - [R - R_0]/c),$$

where $R(\theta) = (a^2 + z'^2 - 2az'\cos\theta)^{1/2}$ and $R_0 = a - z'$. Then $D = \mathscr{L}\{d\}$ is given by

$$D(\theta,s) = \frac{1}{R}\int_{(R-R_0)/c}^{\infty} w_{\mathrm{inc}}(t-[R-R_0]/c)\,\mathrm{e}^{-st}\,\mathrm{d}t = \frac{\mathrm{e}^{-sR/c}}{R}\,\mathrm{e}^{sR_0/c}\,W_{\mathrm{inc}}(s)$$
$$= \frac{2s}{\pi c}\,\mathrm{e}^{sR_0/c}\,W_{\mathrm{inc}}(s)\sum_{n=0}^{\infty}(2n+1)\,i_n(sz'/c)\,k_n(sa/c)\,P_n(\cos\theta)$$

for $0 \le z' < a$, where we have used [661, eqn 10.60.3]. We can then write down $D_n(s)$ after comparison with (7.6).

Asymptotic Behaviour of $a_{n,j}(r)$ for Large n

The coefficients $a_{n,j}(r)$ are defined by (7.16). Their growth with n has been studied [356, 296, 600]. To investigate, let β denote any zero of Θ_n: $\Theta_n(\beta) = k_n(\beta) = 0$. As $k_n(z) = \sqrt{\pi/(2z)}\,K_\nu(z)$, where K_ν is a modified Bessel function and $\nu = n+\frac{1}{2}$, $k_n'(\beta) = \sqrt{\pi/(2\beta)}\,K_\nu'(\beta)$. It is known that $\beta \sim n\zeta\,\mathrm{e}^{\mathrm{i}\pi}$ as $n \to \infty$ for some ζ with $0.66 < |\zeta| \le 1$ and $|\arg\zeta| < \frac{1}{2}\pi$; see the discussion below (7.9). Hence we need the asymptotics of $K_\nu(\nu z)$ for large ν. These are given by [661, eqn 10.41.4] or [3, eqn 9.7.8] but only when $|\arg z| < \frac{1}{2}\pi$, so we first continue K_ν analytically, using [661, eqn 10.34.2] or [3, eqn 9.6.31]:

$$K_\nu(\nu\zeta\mathrm{e}^{\mathrm{i}\pi}) = \mathrm{e}^{-\nu\pi\mathrm{i}}K_\nu(\nu\zeta) - \pi\mathrm{i}\,I_\nu(\nu\zeta). \tag{7.18}$$

Thus we also need the asymptotics of the other modified Bessel function $I_\nu(\nu z)$, as given in [661, eqn 10.41.3] or [3, eqn 9.7.7]. Hence

$$K_\nu(\nu\zeta\mathrm{e}^{\mathrm{i}\pi}) \sim \sqrt{\frac{\pi}{2\nu}}\,\frac{1}{(1+\zeta^2)^{1/4}}\left(\mathrm{e}^{-\nu\pi\mathrm{i}}\mathrm{e}^{-\nu\eta} - \mathrm{i}\,\mathrm{e}^{\nu\eta}\right) \quad \text{as } \nu \to \infty,$$

where

$$\eta(\zeta) = (1+\zeta^2)^{1/2} + \log\{\zeta/[1+(1+\zeta^2)^{1/2}]\}.$$

We have $\eta(1) \simeq 0.534$ and $\eta'(\zeta) = \zeta^{-1}(1+\zeta^2)^{1/2}$.

Similarly, differentiating (7.18) with respect to ζ gives

$$K_\nu'(\nu\zeta\mathrm{e}^{\mathrm{i}\pi}) = -\mathrm{e}^{-\nu\pi\mathrm{i}}K_\nu'(\nu\zeta) + \pi\mathrm{i}\,I_\nu'(\nu\zeta)$$
$$\sim \sqrt{\frac{\pi}{2\nu}}\,\frac{(1+\zeta^2)^{1/4}}{\zeta}\left(\mathrm{e}^{-\nu\pi\mathrm{i}}\mathrm{e}^{-\nu\eta} + \mathrm{i}\,\mathrm{e}^{\nu\eta}\right) \quad \text{as } \nu \to \infty,$$

after use of [661, eqns 10.41.5 and 10.41.6] or [3, eqns 9.7.9 and 9.7.10].

For the simplest case, consider $\zeta = 1$ ($\beta \sim -n$). Then, writing $\rho = r/a$,

$$a_{n,j}(r) = \frac{\rho\,k_n(\beta\rho)}{\mathrm{e}^{(1-\rho)\beta}\,k_n'(\beta)} \sim \frac{-\sqrt{\rho}\,\mathrm{e}^{n\phi}}{2^{1/4}(1+\rho^2)^{1/4}} \quad \text{as } n \to \infty, \tag{7.19}$$

where the exponent

$$\phi(\rho) = \eta(\rho) - \eta(1) + 1 - \rho.$$

We have $\phi(1) = 0$ and $\phi'(\rho) > 0$ for $\rho \ge 1$. This gives the expected exponential growth with n for any $r > a$; the growth rate increases with r.

Similar results can be derived for other values of ζ, including complex values.

7.2.2 Pressure Boundary Condition

For the axisymmetric pressure condition, $-\rho\,\partial u/\partial t = p_a(\theta,t)$ on $r=a$ for $0\le\theta\le\pi$ and $t>0$. If we write

$$p_a(\theta,t) = -\rho\sum_{n=0}^{\infty} q_n(t)P_n(\cos\theta), \tag{7.20}$$

we find that B_n in (7.4) is given by $sB_n(s)\,k_n(sa/c) = Q_n(s)$, where $Q_n(s)=\mathscr{L}\{q_n\}$. Hence, as in Section 7.2.1, (7.3) and (7.8) give $u(r,\theta,t)$ with

$$u_n(r,t) = \frac{1}{2\pi\mathrm{i}}\int_{\mathrm{Br}} \frac{Q_n(s)}{s}\,\frac{k_n(sr/c)}{k_n(sa/c)}\,\mathrm{e}^{st}\,\mathrm{d}s. \tag{7.21}$$

As $s^{-1}Q_n(s) = \mathscr{L}\int_0^t q_n(\eta)\,\mathrm{d}\eta$, we can write (7.21) as (7.15) with $d_n(t)$ replaced by $\int_0^t q_n(\eta)\,\mathrm{d}\eta$,

$$\begin{aligned} u_n(r,t) &= \frac{a}{r}\left\{\int_0^T q_n(\eta)\,\mathrm{d}\eta + \int_0^T \psi_n(r,T-t')\int_0^{t'} q_n(\eta)\,\mathrm{d}\eta\,\mathrm{d}t'\right\}\mathrm{H}(T) \\ &= \frac{a}{r}\int_0^T q_n(\eta)\left\{1+\int_\eta^T \psi_n(r,T-t')\,\mathrm{d}t'\right\}\mathrm{d}\eta\,\mathrm{H}(T). \end{aligned} \tag{7.22}$$

This agrees with (6.26) when $n=0$.

7.2.3 Neumann Boundary Condition

For the axisymmetric Neumann problem, the boundary condition is $\partial u/\partial r = v(\theta,t)$ on $r=a$ for $0\le\theta\le\pi$ and $t>0$. Expanding $v(\theta,t)=\sum_{n=0}^{\infty} v_n(t)P_n(\cos\theta)$ and differentiating (7.4) with respect to r, we obtain

$$(s/c)\,k_n'(sa/c)B_n(s) = V_n(s) = \mathscr{L}\{v_n\}.$$

Hence (7.3) and (7.4) give $u(r,\theta,t)$ with

$$u_n(r,t) = \frac{1}{2\pi\mathrm{i}}\int_{\mathrm{Br}} V_n(s)\,\frac{k_n(sr/c)}{(s/c)\,k_n'(sa/c)}\,\mathrm{e}^{st}\,\mathrm{d}s. \tag{7.23}$$

From [661, eqn 10.51.5], $xk_n'(x) = nk_n(x) - xk_{n+1}(x)$. Combining this with (7.9) gives

$$(2/\pi)xk_n'(x) = x^{-n-1}\mathrm{e}^{-x}\phi_n(x) \quad \text{with} \quad \phi_n(x) = n\theta_n(x) - \theta_{n+1}(x),$$

which shows that the denominator in (7.23) has $n+1$ zeros. Hence (7.23) becomes

$$u_n(r,t) = \frac{a}{2\pi\mathrm{i}}\left(\frac{a}{r}\right)^{n+1}\int_{\mathrm{Br}} V_n(s)\,\frac{\theta_n(sr/c)}{\phi_n(sa/c)}\,\mathrm{e}^{sT}\,\mathrm{d}s, \tag{7.24}$$

with $T=t-(r-a)/c$, as usual.

Denote the zeros of $\phi_n(x)$ by $\beta_{n,m}'$: $\phi_n(\beta_{n,m}')=0$ for $m=1,2,\dots,n+1$. They are the zeros of $k_n'(x)$ and they have the same qualitative properties as the zeros of $\theta_n(x)$. They are tabulated in [382, Table 1] for $n \leq 25$.

As $\theta_0 = 1$ and $\phi_0(x) = -\theta_1(x) = -(1+x)$, we can confirm that the spherically symmetric component, u_0, agrees with (6.29) and (6.30). More generally, and motivated by (6.30), define

$$\Lambda_n(r,s) = \frac{a^{n+1}\,\theta_n(sr/c)}{c\,r^n\,\phi_n(sa/c)}.$$

As $\Lambda_n(r,s) \to 0$ as $s \to \infty$, there is a function $\lambda_n(r,t)$ with $\mathscr{L}\{\lambda_n\} = \Lambda_n$. Hence

$$u_n(r,t) = \frac{ac}{r}\,\mathrm{H}(T) \int_0^T v_n(\eta)\lambda_n(r,T-\eta)\,\mathrm{d}\eta. \tag{7.25}$$

We could now calculate λ_n in the same way as we calculated ψ_n in Section 7.2.1, making use of the zeros of $\phi_n(sa/c)$.

7.2.4 Literature

There are many papers where Laplace transforms are used to solve the wave equation exterior to a sphere. For problems that are not spherically symmetric (see Section 6.2), the earliest work is by Brillouin [129]; his long two-part French paper was given a detailed exposition by Hanish [385, §§2.1 and 7.4]. A variety of Neumann radiation problems are solved. Longhorn [569] also solved a Neumann problem, with $\partial u/\partial r = v_1(t)\cos\theta$ on $r = a$.

Friedlander [310, pp. 166–174] constructed Green's function for a hard sphere (Neumann boundary condition); the incident field is generated by a simple source at a point on the z-axis outside the sphere. At about the same time, Wilcox [885] published a 'preliminary report' on Problem DI_0; his short note is discussed in [356].

In 1960, Barakat [49] discussed the scattering of a plane pulse, $u_{\mathrm{inc}} = \mathrm{H}(z-ct)\mathrm{e}^{\mathrm{i}k(z-ct)}$, by both soft and hard spheres. For other incident plane pulses, see [186, 724, 725, 331]. Problem NI_0, with various forcings, has been studied in other papers from the 1960s [496, 458, 837]. For example, Tupholme [837] gave a detailed study when only a cap of the sphere moves; see also [21]. For scattering by a sphere with a hole (a Helmholtz resonator [696, §7-4]), see [510].

Applications to scattering by spherical shells and membranes were also made, starting with a 1957 paper by Mann-Nachbar [588]; later papers include [624, 431, 812, 433, 96, 330, 456, 434]. See [389] for many additional references and [843] for a review. For acoustic scattering by a solid sphere, see [855].

Huang & Gaunaurd [432] consider acoustic scattering of a plane step pulse in pressure by a hard sphere with emphasis on calculating p at $r = a$. The series expansion of this quantity, using spherical harmonics, is not uniformly convergent: this is an example of Gibbs' phenomenon [351]. To compensate for this phenomenon, the authors use Cesàro summation, extending previous work by others [96, 906]. Better remedies are available [335, 351], but these do not seem to have been used for transient scattering problems.

Hamilton & Astley [382] solved Problem NI_0 with various forcings. Greengard et al. [356] gave results for Problem DI_0 and for the analogous problem with a Robin condition on the sphere. For the latter problem, see [827].

Wu et al. [893] were interested in solving an inverse problem, reconstructing surface pressures from transient near-field measurements; see also [892, Chapter 9].

There is an extensive literature on electromagnetic scattering by a sphere. Twentieth-century papers include [878, 475, 476, 715, 306, 606, 17, 770]. More recent papers usually employ the Fourier transform (Section 5.6.1) combined with efficient implementations of Mie's solution for time-harmonic scattering by a dielectric sphere: representative papers are [352, 611, 82, 450, 523, 567, 568, 898]. For electromagnetic scattering by a spherical shell, see [409, 246, 800].

For elastodynamic scattering or radiation by a sphere, see [268, 647, 668, 838, 884], [681, §VI.4], [269, §8.15] and [623, §III.C].

7.3 Residual Potential Methods, Nonreflecting Boundary Kernels and Dirichlet-to-Neumann Mappings

The outgoing spherically symmetric wavefunction $u_0(r,t) = r^{-1}f(r-ct)$ satisfies

$$\frac{\partial u_0}{\partial r} + \frac{1}{r}u_0 + \frac{1}{c}\frac{\partial u_0}{\partial t} = 0$$

exactly, for any smooth f. More generally, write

$$\frac{\partial u_n}{\partial r} + \frac{1}{r}u_n + \frac{1}{c}\frac{\partial u_n}{\partial t} = \frac{1}{r}u_n^{\mathrm{R}}, \tag{7.26}$$

where u_n is defined by (7.4) and satisfies (2.21). Equation (7.26) can be viewed as defining $u_n^{\mathrm{R}}(r,t)$ in terms of $u_n(r,t)$. Geers [332] introduced the analogue of u_n^{R} for the two-dimensional wave equation in 1969; he called it the *residual potential.* Equation (7.26) can be found in [11, eqn (8)] and [334, eqn (9)].

What is u_n^{R} and why is it useful? Inserting the Laplace-transform representation (7.4) in (7.26) gives

$$u_n^{\mathrm{R}}(r,t) = \frac{1}{2\pi \mathrm{i}}\int_{\mathrm{Br}} R_n(sr/c)\,B_n(s)\,k_n(sr/c)\,\mathrm{e}^{st}\,\mathrm{d}s$$

where [906, eqn (5)]

$$R_n(\sigma) = \frac{\sigma k_n'(\sigma)}{k_n(\sigma)} + \sigma + 1. \tag{7.27}$$

(It is easy to check, using (2.28), that $R_0 = 0$, as expected.) Making use of (2.29), we find that $R_n(\sigma) \sim -\frac{1}{2}n(n+1)\sigma^{-1}$ as $\sigma \to \infty$, implying that there is a function $\mathscr{R}_n(t)$ such that $\mathscr{L}\{\mathscr{R}_n(t)\} = R_n(s)$. Then (1.75) gives

$$R_n(sr/c) = (c/r)\,\mathscr{L}\{\mathscr{R}_n(ct/r)\}$$

and the convolution theorem (1.76) gives

$$u_n^{\mathrm{R}}(r,t) = \frac{c}{r}\int_0^t \mathscr{R}_n(c[t-\tau]/r)\, u_n(r,\tau)\,\mathrm{d}\tau. \tag{7.28}$$

When this formula is substituted on the right-hand side of (7.26), we obtain a relation between $u_n(r,t)$, $\partial u_n/\partial r$ and $\partial u_n/\partial t$ on the sphere of radius r:

$$\frac{\partial u_n}{\partial r} + \frac{1}{r}u_n(r,t) + \frac{1}{c}\frac{\partial u_n}{\partial t} = \frac{c}{r^2}\int_0^t \mathscr{R}_n(c[t-\tau]/r)\, u_n(r,\tau)\,\mathrm{d}\tau. \tag{7.29}$$

As $\partial u_n/\partial t$ can be derived from $u_n(r,t)$, we obtain a connection between u_n and $\partial u_n/\partial r$: it is an exact formula for the *Dirichlet-to-Neumann mapping* on a sphere.

We can invert (7.27) to obtain

$$\mathscr{R}_n(\xi) = \frac{1}{2\pi\mathrm{i}}\int_{\mathrm{Br}} R_n(\sigma)\,\mathrm{e}^{\xi\sigma}\,\mathrm{d}\sigma = \sum_{m=1}^{n} \beta_{n,m}\,\mathrm{e}^{\xi\beta_{n,m}}, \quad n = 1,2,\ldots,$$

where $k_n(\beta_{n,m}) = 0$, $m = 1,2,\ldots,n$; see below (7.9) for information on the zeros $\beta_{n,m}$.

We note that $\mathscr{R}_n(\xi)$ is an explicit dimensionless function of a dimensionless variable; it appears above in the form $\mathscr{R}_n(ct/r)$. In early applications [332, 333, 11, 12, 906], (7.29) was applied on the surface of a structure at $r = a$. Thus 'once the functions $\mathscr{R}_n(\xi)$ have been tabulated, we can rigorously account for the effects of fluid loading on the motion of the structure without having to invert a complicated Laplace transform' [333, p. 1506]. For extensions to elastodynamic problems, see [13, 14].

However, (7.29) can be used on a fictitious spherical surface surrounding a scatterer, thus forming the basis for a radiation boundary condition that can be used to truncate computational domains [379, 380]. It is in this context that the utility of $\mathscr{R}_n(\xi)$ was rediscovered by Alpert et al. [15, 16]; they call $\mathscr{R}_n$ a *nonreflecting boundary kernel*. Specifically, for (7.27), see [15, eqn (2.17)] or [16, eqn (13)]. The connection between residual potential methods and nonreflecting boundary kernels was noted by Geers & Sprague [334, p. 678]. Efficient schemes for implementing these methods continue to be developed.

7.4 Application of the Similarity Representation

Let us use the similarity representation (2.36) for the wave field generated in the exterior of a sphere of radius a. This approach is described in [600, §4]. We take $t_0 = a/c$ and $t_1 = 0$ in (2.36), giving

$$u_n(r,t) = \frac{a}{r}\int_0^{t-(r-a)/c} f_n(\tau)P_n\left(\frac{c}{r}\left[t-\tau+\frac{a}{c}\right]\right)\mathrm{d}\tau. \tag{7.30}$$

Zero initial conditions, $u_n = 0$ and $\partial u_n/\partial t = 0$ at $t = 0$ for $r > a$, are enforced by requiring that $f_n(\tau) = 0$ for $\tau < 0$.

Before using (7.30), we confirm that it is equivalent to the more familiar representation obtained by combining a Laplace transform with separation of variables

(Section 7.2). Let $T = t - (r-a)/c$ so that (7.30) becomes

$$u_n(r,t) = \frac{a}{r}\int_0^T f_n(\tau)P_n\Big(\frac{c}{r}\Big[T - \tau + \frac{r}{c}\Big]\Big)\,\mathrm{d}\tau, \quad T > 0. \tag{7.31}$$

The right-hand side is a Laplace convolution, so we take the Laplace transform with respect to T,

$$\int_0^\infty u_n(r,t)\,\mathrm{e}^{-sT}\,\mathrm{d}T = \frac{a}{r}F_n(s)K(s;r), \tag{7.32}$$

where $F_n(s) = \mathscr{L}\{f_n\}$ is the Laplace transform of f_n,

$$K(s;r) = \int_0^\infty \mathrm{e}^{-st}P_n\Big(\frac{c}{r}\Big[t + \frac{r}{c}\Big]\Big)\,\mathrm{d}t = \frac{2r}{\pi c}\mathrm{e}^{sr/c}k_n(sr/c) \tag{7.33}$$

and we have used [353, 7.143.1]. Once F_n has been determined using the boundary condition (see below), we invert the Laplace transform, which means we multiply (7.32) by e^{sT} and integrate over the Bromwich contour, Br. This gives

$$u_n(r,t) = \frac{a}{\pi^2 c\,\mathrm{i}}\int_{\mathrm{Br}} F_n(s)k_n(sr/c)\,\mathrm{e}^{s(t+a/c)}\,\mathrm{d}s, \tag{7.34}$$

which has the same form as (7.4). Alternatively, we could invert $F_n(s)$, giving $f_n(t)$, and then (7.30) provides a different representation for $u_n(r,t)$, one whose properties have not been fully investigated.

When (7.30) is combined with a boundary condition at $r = a$, the result will be a Volterra integral equation; see below for details. Such integral equations have a large literature [133]. In our applications, it is not surprising that they can be solved exactly. However, direct treatments may be worth further investigation. For example, Volterra integral equations of the second kind can usually be solved by iteration. There is also the possibility of using the 'convolution quadrature method' (Section 10.4); an application of this method to (7.35) is outlined in Section 10.4.3.

7.4.1 Dirichlet Boundary Condition

For the Dirichlet boundary condition, use of $u_n(a,t) = d_n(t)$ in (7.30) gives

$$\int_0^t f_n(\tau)P_n\Big(\frac{c}{a}\Big[t - \tau + \frac{a}{c}\Big]\Big)\,\mathrm{d}\tau = d_n(t), \quad t > 0, \tag{7.35}$$

a Volterra integral equation of the first kind for f_n. Implicit in this equation is the constraint

$$d_n(0) = 0. \tag{7.36}$$

Taking the Laplace transform of (7.35) gives

$$F_n(s)K(s;a) = D_n(s), \tag{7.37}$$

where $D_n = \mathscr{L}\{d_n\}$ and K is defined by (7.33). Solving for F_n followed by substitution in (7.34) gives precisely (7.8).

Let us go further. From (7.9), (7.33) and (7.37), we have

$$F_n(s) = \frac{D_n(s)}{K(s;a)} = \frac{(sa/c)^n}{\theta_n(sa/c)}\, sD_n(s) = (1+X_n(s))\, sD_n(s), \tag{7.38}$$

say, where

$$X_n(s) = \frac{(sa/c)^n - \theta_n(sa/c)}{\theta_n(sa/c)} = \mathscr{L}\{\chi_n\}$$

for some function $\chi_n(t)$. This function can be found by inverting X_n:

$$\chi_n(t) = \frac{c}{a}\sum_{j=1}^{n} \frac{(\beta_{n,j})^n}{\theta_n'(\beta_{n,j})} \exp(\beta_{n,j}ct/a).$$

We have $sD_n(s) = \mathscr{L}\{d_n'\}$ (recall (7.36)) and $X_n(s) = \mathscr{L}\{\chi_n\}$. Hence, inverting (7.38),

$$f_n(t) = d_n'(t) + \int_0^t d_n'(\eta)\, \chi_n(t-\eta)\, \mathrm{d}\eta.$$

Substituting in (7.31) gives

$$\begin{aligned}
u_n(r,t) &= \frac{a}{r}\int_0^T f_n(\tau)\mathscr{Q}_n(T-\tau)\,\mathrm{d}\tau \\
&= \frac{a}{r}\int_0^T d_n'(\eta)\mathscr{Q}_n(T-\eta)\,\mathrm{d}\eta + \frac{a}{r}\int_0^T\int_0^\tau d_n'(\eta)\,\chi_n(\tau-\eta)\,\mathscr{Q}_n(T-\tau)\,\mathrm{d}\eta\,\mathrm{d}\tau \\
&= \frac{a}{r}\int_0^T d_n'(\eta)\left\{\mathscr{Q}_n(T-\eta) + \int_\eta^T \chi_n(\tau-\eta)\,\mathscr{Q}_n(T-\tau)\,\mathrm{d}\tau\right\}\mathrm{d}\eta \\
&= \frac{a}{r}\int_0^T d_n'(\eta)L_n(r,T-\eta)\,\mathrm{d}\eta,
\end{aligned} \tag{7.39}$$

where $\mathscr{Q}_n(t) = P_n(1+ct/r)$ and

$$L_n(r,t) = \mathscr{Q}_n(t) + \int_0^t \chi_n(\sigma)\mathscr{Q}_n(t-\sigma)\,\mathrm{d}\sigma.$$

Notice that $\mathscr{Q}_n$ depends on r but χ_n does not. Equation (7.39) is an exact formula for the solution of Problem DI_0.

Let us make another observation concerning the Volterra integral equation of the first kind for f_n, (7.35). Make the substitution $f_n(t) = g_n'(t)$ with $g_n(0) = 0$. After an integration by parts, we find that g_n satisfies

$$g_n(t) + \frac{c}{a}\int_0^t g_n(\tau)\, P_n'(1+[t-\tau]c/a)\,\mathrm{d}\tau = d_n(t), \quad t>0, \tag{7.40}$$

a Volterra integral equation of the second kind. Similar equations (with different right-hand sides) can be found in [311, eqn (3.12)] and [465, eqn (A.6)].

7.4.2 Pressure Boundary Condition

For Problem PI_0, differentiate (7.30) with respect to t,

$$\frac{\partial u_n}{\partial t} = \frac{a}{r} f_n(T) + \frac{ac}{r^2} \int_0^T f_n(\tau) P_n'\left(\frac{c}{r}\left[t-\tau+\frac{a}{c}\right]\right) \mathrm{d}\tau,$$

using $P_n(1)=1$. The boundary condition, $\partial u_n/\partial t = q_n(t)$ at $r=a$ (see (7.20)), gives

$$f_n(t) + \frac{c}{a}\int_0^t f_n(\tau) P_n'(1+[t-\tau]/t_0)\,\mathrm{d}\tau = q_n(t), \quad t>0, \tag{7.41}$$

where $t_0 = a/c$. Equation (7.41) is a Volterra integral equation of the second kind for f_n. To solve it, we apply $\mathscr{L}$. From (7.33), we have

$$\mathscr{L}\{P_n(1+t/t_0)\} = (2/\pi) t_0 \mathrm{e}^{st_0} k_n(st_0) = K(s; ct_0), \tag{7.42}$$

whence

$$\mathscr{L}\{P_n'(1+t/t_0)\} = t_0(sK-1) \tag{7.43}$$

and (7.41) gives $sK(s;a)F_n(s) = Q_n(s)$. Substitution for F_n in (7.34) leads back to (7.21).

7.4.3 Neumann Boundary Condition

For Problem NI_0, the boundary condition is $\partial u_n/\partial r = v_n(t)$ on $r=a$. Differentiating (7.30) gives

$$\frac{\partial u_n}{\partial r} = -\frac{u_n}{r} - \frac{a}{rc} f_n(T) - \frac{ac}{r^3}\int_0^T f_n(\tau)\left(t-\tau+\frac{a}{c}\right) P_n'\left(\frac{c}{r}\left[t-\tau+\frac{a}{c}\right]\right)\mathrm{d}\tau.$$

Applying the boundary condition, we obtain

$$f_n(t) + \frac{c}{a}\int_0^t f_n(\tau)\mathscr{K}_n(t-\tau)\,\mathrm{d}\tau = -cv_n(t), \quad t>0, \tag{7.44}$$

where

$$\mathscr{K}_n(t) = P_n(1+t/t_0) + (1+t/t_0)P_n'(1+t/t_0).$$

Equation (7.44) is a Volterra integral equation of the second kind for f_n. Again, to solve it, we apply $\mathscr{L}$. Using (7.42), (7.43) and

$$\mathscr{L}\{tP_n'(1+t/t_0)\} = -t_0^2\frac{\partial K}{\partial t_0} = -t_0^2\left\{(s+t_0^{-1})K + (2/\pi)st_0\mathrm{e}^{st_0}k_n'(st_0)\right\},$$

we obtain

$$\begin{aligned}\mathscr{L}\{\mathscr{K}(t)\} &= K + t_0(sK-1) - t_0\left\{(s+t_0^{-1})K + (2/\pi)st_0\mathrm{e}^{st_0}k_n'(st_0)\right\}\\ &= -t_0 - (2/\pi)st_0^2\mathrm{e}^{st_0}k_n'(st_0),\end{aligned}$$

and then (7.44) gives

$$(2/\pi)st_0\mathrm{e}^{st_0}k_n'(st_0)F_n(s) = cV_n(s). \tag{7.45}$$

Solving for F_n and substitution in (7.34) gives precisely (7.23).

7.5 Moving Spheres

For time-dependent problems, it is natural to consider moving obstacles. The earliest work is in the context of electromagnetic waves, and it is well surveyed in the book by Van Bladel [848]. The prototype problem is the scattering of a plane wave by a sphere that is moving with a constant velocity $\boldsymbol{U}$. The fields are governed by Maxwell's equations and they are invariant under Lorentz transformations [848, §1.6]. Therefore, we can solve the problem using the 'frame-hopping method'. Introduce two inertial frames, K_{lab} and K_{mov}, where K_{lab} is the laboratory frame and K_{mov} is attached to the sphere. The two frames coincide initially and then K_{mov} moves away with velocity $\boldsymbol{U}$. The incident wave is specified relative to K_{lab}, and is then Lorentz-transformed to K_{mov}. The wave is then scattered by the sphere; this calculation is done using a standard method for scattering by a fixed sphere. Finally, the scattered field is then Lorentz-transformed back to the laboratory frame K_{lab}. This basic method is described in [848, §5.17] and [622], and it has been applied to several problems involving spheres [713, 772, 325] and spheroids [26]. For electromagnetic scattering by rotating objects, see, for example, [813] and [848, Chapter 10]. For problems arising in the context of micro-Doppler effects, see [172]. Electromagnetic scattering by several objects, moving independently, requires a time-stepping approach 'known as the "stop-go-stop" model, i.e., the target is motionless when it is interacting with a radar pulse and then moves to the next position to interact with the next pulse' [905, p. 2024].

The situation for acoustic problems is more complicated because a translating sphere, for example, will disrupt the basic flow around the sphere. This implies that it may not be appropriate to solve the wave equation exterior to the moving obstacle, $S(t)$; such an IBVP has been studied mathematically [194, 195, 791] but its relevance to the physical acoustics problem is unclear. This fact was noted by Oestreicher [671, p. 1224], and then in more detail by Myers & Hausmann [650, p. 2595]:

A necessary condition for $\square^2 u = 0$ to govern the total disturbance in the fluid is that the aerodynamic perturbations, which are introduced by the motion of the body, and the acoustic perturbations are both of sufficiently small magnitude. The order of magnitude of the former is established by the body normal Mach number $M_n = \boldsymbol{U} \cdot \boldsymbol{n}/c$ [where $\boldsymbol{U}$ is the velocity of the body and $\boldsymbol{n}$ is the unit normal to the boundary]. *In addition, however, ... nonlinear aerodynamic effects must be accounted for along with the acoustic perturbations unless the sound exceeds M_n^2 in order of magnitude. This is a severe limitation on M_n. It means that an analysis of scattering based on $\square^2 u = 0$ for bodies moving at finite Mach number is valid only if the thickness ratio of the body is very small and if it has sharp pointed leading and trailing edges.*

Evidently, then, such an analysis is not valid for a translating sphere!

When $\boldsymbol{U}$ is a constant vector, we may contemplate using the convected wave equation, (1.26), or an approximation of (1.26) obtained when the Mach number $M = |\boldsymbol{U}|/c \ll 1$ [816, 29]. For some numerical results, see [539, 427, 428].

At this point, we should recall Lighthill's acoustic analogy, as described in Section 4.7.3. It starts by assuming that the basic flow has been calculated first, perhaps by solving the incompressible Navier–Stokes equations numerically. The output of this calculation is then used to drive an acoustic problem: this two-step process is at the heart of *computational aeroacoustics*; see Section 9.3.2 for some discussion and references.

Returning to exact studies, Leppington & Levine [544, 548] have given a clear analysis of a specific problem for a pulsating and translating sphere. The sphere has radius $a(t)$ and it is moving along the z-axis so that its centre is at $(0,0,\zeta(t))$. There is no flow through the sphere, so the boundary condition is

$$\frac{\partial u_{\text{ex}}}{\partial r'} = \frac{\mathrm{d}a}{\mathrm{d}t} + \frac{\mathrm{d}\zeta}{\mathrm{d}t}\cos\theta' \quad \text{at } r' = a(t),$$

where u_{ex} is the exact velocity potential and (r', θ') are spherical polar coordinates relative to the centre of the sphere. (For derivations of the boundary condition when there is an ambient flow, see [649, 285].) Let a_0 be a typical value of $a(t)$. Let U_0 be a typical speed that characterises the larger of $|a'(t)|$ and $|\zeta'(t)|$. Then $t_0 = a_0/U_0$ is a typical time scale and $M_0 = U_0/c$ is a (dimensionless) Mach number. The authors start from the exact nonlinear PDE for u_{ex}, (1.44) [544, eqn (1.4)], and then develop matched asymptotic approximations, assuming that $M_0 \ll 1$. In the inner region (close to the sphere), the flow is nearly incompressible: the relevant length scale is a_0. In the outer region (far from the sphere), compressibility is important: the relevant length scale is a typical wavelength, $ct_0 = a_0/M_0$. Approximate solutions are constructed in both regions and then matched. One outcome is the following approximation in the outer region [544, eqn (3.29)],

$$u_{\text{ex}} \sim \frac{f_0(\tau_0)}{M(\tau_0)} + \frac{\partial}{\partial z}\left(\frac{f_1(\tau_0)}{M(\tau_0)}\right) + \frac{f_2(\tau_0)}{M(\tau_0)}, \tag{7.46}$$

where $f_0(t) = -a^2 a'(t)$, $f_1(t) = \frac{1}{2}a^3\zeta'(t)$,

$$M(\tau_0) = R\left(1 - \frac{\boldsymbol{v}\cdot\boldsymbol{R}}{cR}\right) \quad \text{evaluated at } \tau = \tau_0,$$

$\boldsymbol{v}(\tau) = (0,0,\zeta'(\tau))$, $\boldsymbol{R}(\tau) = (x,y,z-\zeta(\tau))$, $R = |\boldsymbol{R}|$ and $\tau_0(x,y,z,t)$ is the unique solution of $\tau_0 = t - c^{-1}R(\tau_0)$. The notation here is the same as in Section 2.6. Thus, the first term in (7.46) represents a point source, the second term is a z-directed dipole and the third term is another point source, all located at the centre of the sphere and translating with the sphere. In terms of the small Mach number M_0, the first term gives the leading approximation, but this term is absent if $a(t) \equiv a_0$: the sphere's pulsations give the dominant effect. The next term is smaller but it is absent if the sphere is not translating. The third term is smaller still; its strength $f_2(t)$ is given by a lengthy formula [544, eqn (3.25)].

In the absence of translational motion ($\zeta' = 0$), the leading term in (7.46) reduces to

$$u_{\text{ex}} \sim -(4\pi r)^{-1} V'(t - r/c), \tag{7.47}$$

a simple source (2.11) at the centre of the sphere (taken to be at the origin). Here

$V(t) = \frac{4}{3}\pi a^3$ is the volume of the sphere. The result (7.47) was obtained by Frost & Harper [317, eqn (52)]; they also used matched asymptotic expansions. The corresponding estimate for the pressure is $\rho(4\pi r)^{-1}V''(t-t/c)$, which is reminiscent of a formula given by Strasberg [798, eqn (8)]: he does not have the retarded argument and he assumes small-amplitude pulsations (although he uses his formula to estimate sound generation by large-amplitude pulsations).

For small-amplitude pulsations, we have $V'(t) \simeq 4\pi a_0^2 a'(t)$ and then (7.47) reduces to

$$u \sim -r^{-1}a_0^2 a'(t-r/c). \tag{7.48}$$

This is exactly what we would have obtained if we had sought the solution in the form of a simple source (see Section 6.1), $u(r,t) = r^{-1}f(t-r/c)$, and then applied the boundary condition $\partial u/\partial r = a'(t)$ on the fixed sphere, $r = a_0$. Thus, doing this, we find that

$$f(t-a_0/c) + (a_0/c)f'(t-a_0/c) = -a_0^2 a'(t),$$

which is [559, p. 66, eqn (168)]. The left-hand side gives the first two terms in the Taylor expansion of $f(t)$ about $t-a_0/c$; if we accept this approximation, we obtain $f(t) \simeq -a_0^2 a'(t)$, which leads back to (7.48).

The sphere problem with translational motion but without pulsations is considered further in [544, §4]. For related work on irrotational inviscid compressional flow around moving spheres, see [206, 243, 816], [245, §9.3], [244] and [421, §1.12].

Returning to the nonlinear approximation (7.47), suppose that we have monochromatic forcing at some frequency ω, so that

$$a(t) = a_0\,(1+2b\cos\omega t),$$

where b is a (dimensionless) constant. Then

$$f_0(t) = -a^2 a' = 2a_0^3 b\omega\left\{(1-b^2)\sin\omega t + 2b\sin 2\omega t + b^2\sin 3\omega t\right\}.$$

This shows that the nonlinearity in the problem causes two higher harmonics to be generated; of course, these effects are small when b is small.

More generally, suppose that we have forcing at two frequencies, ω_1 and ω_2,

$$a(t) = a_0\,(1+b_1\cos\omega_1 t + b_2\cos\omega_2 t).$$

If we suppose that the constants b_1 and b_2 are small and we discard cubic terms in b_1 and b_2, we obtain

$$\begin{aligned} a_0^{-3} f_0(t) \simeq{}& b_1\omega_1\sin\omega_1 t + b_2\omega_2\sin\omega_2 t + b_1^2\omega_1\sin 2\omega_1 t + b_2^2\omega_2\sin 2\omega_2 t \\ &+ b_1 b_2(\omega_1+\omega_2)\sin(\omega_1+\omega_2)t + b_1 b_2(\omega_1-\omega_2)\sin(\omega_1-\omega_2)t. \end{aligned}$$

Now we see sum and difference frequencies, $\omega_1 \pm \omega_2$. A related problem of this kind arises when an incident wave of frequency ω_1 is scattered by a sphere that is forced to vibrate at frequency ω_2. This problem was first considered by Censor [150]. He assumed that $\Box^2 u_{\text{ex}} = 0$ and then calculated nonlinear effects due to the motion of the spherical boundary. Rogers [719] pointed out that Censor had overlooked the nonlinear effects caused by the fluid itself. About ten years later, Censor [151] returned to the problem and a new analysis was made by Piquette & Van Buren [698]. Further discussion ensued; see [699] for references. At the time, experimental techniques were not sufficiently advanced so as to distinguish Censor's boundary (Doppler) effects from bulk fluid effects [698]. That had to wait for another twenty years [648, 895]: it turns out that there are situations when each effect can be dominant.

7.6 Scattering by a Spheroid

It is natural to consider using spheroidal coordinates and spheroidal wavefunctions (Section 2.5) for scattering by a spheroid. This approach has been used extensively for time-harmonic problems and, indeed, one could argue that solving such problems led to a better understanding of spheroidal wavefunctions and their properties. For time-harmonic acoustic problems, see, for example, [752, 495, 6, 348] and [599, §4.12.2]. For time-harmonic electromagnetic problems, see [752] and [553]. For time-harmonic elastodynamic scattering by a spheroidal inclusion, see [447].

The analogous approach for transient problems is less well developed, as we shall see. Before embarking on a study of acoustic scattering by a spheroid, let us revisit the same problem for scattering by a sphere. We use Laplace transforms with respect to t, as discussed in Section 7.2. For simplicity, consider the Dirichlet problem with zero initial conditions (Problem DI_0). Then $U = \mathscr{L}\{u\}$, the Laplace transform of u, can be written as

$$U(r,\theta,\phi,s) = \sum_{m=0}^{\infty}\sum_{n=m}^{\infty} k_n(sr/c)\,P_n^m(\cos\theta)\,\mathscr{B}_n^m(\phi,s), \quad r>a, \tag{7.49}$$

where

$$\mathscr{B}_n^m(\phi,s) = B_{mn}^{\mathrm{c}}(s)\cos m\phi + B_{mn}^{\mathrm{s}}(s)\sin m\phi. \tag{7.50}$$

Here r, θ and ϕ are spherical polar coordinates and the sphere has radius a; k_n is a modified spherical Bessel function, P_n^m is an associated Legendre function, and B_{mn}^{c} and B_{mn}^{s} are to be determined using the boundary condition at $r=a$. The axisymmetric form of (7.49) is (7.2). The form of (7.49) and (7.50) (with $\cos m\phi$, $\sin m\phi$ and outer summation over m) will be convenient later when we compare with similar formulas involving spheroidal wavefunctions.

We notice two things about the form of the expansion (7.49). First, the angular functions, $P_n^m(\cos\theta)\cos m\phi$ and $P_n^m(\cos\theta)\sin m\phi$, do not depend on s. Second, the radial function, $k_n(sr/c)$, does not depend on the mode number m. This structure will be lost when we consider scattering by a spheroid.

Returning to (7.49), let us apply the boundary condition $u(a,\theta,\phi,t) = d(\theta,\phi,t)$, where d is a specified function. Suppose that $D = \mathscr{L}\{d\}$ has the expansion

$$D(\theta,\phi,s) = \sum_{m=0}^{\infty}\sum_{n=m}^{\infty} P_n^m(\cos\theta)\,\mathscr{D}_n^m(\phi,s), \tag{7.51}$$

where

$$\mathscr{D}_n^m(\phi,s) = D_{mn}^{\mathrm{c}}(s)\cos m\phi + D_{mn}^{\mathrm{s}}(s)\sin m\phi, \tag{7.52}$$

with coefficients D_{mn}^{c} and D_{mn}^{s}. Then the boundary condition yields

$$U(r,\theta,\phi,s) = \sum_{m=0}^{\infty}\sum_{n=m}^{\infty} \frac{k_n(sr/c)}{k_n(sa/c)}\,P_n^m(\cos\theta)\,\mathscr{D}_n^m(\phi,s). \tag{7.53}$$

We then invert U, as in Section 7.2.1. The zeros of the denominator, $k_n(sa/c)$, imply that $U(r,\theta,\phi,s)$ has singularities (simple poles) in the complex s-plane; these singularities play a crucial role. We note that the zeros of $k_n(sa/c)$, sometimes known as *natural frequencies* (see Section 8.6), obviously do not depend on the azimuthal mode number m. Also, there may be additional singularities arising from the form of $D_{mn}^{\mathrm{c}}(s)$ and $D_{mn}^{\mathrm{s}}(s)$. Once all the singularities have been located, we can contemplate moving the Bromwich contour in the inversion formula (5.3) to the left, picking up residue contributions.

Let us try to mimic the sphere calculation for a prolate spheroid. We use prolate spheroidal coordinates ξ, η and ϕ, defined by (2.89). We expand $U = \mathscr{L}\{u\}$ using prolate spheroidal wavefunctions (PSWFs), giving

$$U(\xi,\eta,\phi,s) = \sum_{m=0}^{\infty}\sum_{n=m}^{\infty} R_{mn}^{(3)}(\mathrm{i}p,\xi)\,S_{mn}(\mathrm{i}p,\eta)\,\mathscr{B}_n^m(\phi,s), \quad \xi > \xi_0, \tag{7.54}$$

with $\mathscr{B}_n^m$ defined by (7.50). In this formula, $-1 \le \eta \le 1$, $-\pi \le \phi < \pi$ and $\xi = \xi_0$ is the boundary of the spheroid; the parameter

$$p = s\ell/c, \tag{7.55}$$

where the spheroid has its foci on the z-axis at $z = \pm\ell$. We have used the 'old' notation for PSWFs, see (2.106): $R_{mn}^{(3)}$ is a radial PSWF and S_{mn} is an angular PSWF.

Equation (7.54) should be compared with (7.49): the radial part $R_{mn}^{(3)}(\mathrm{i}p,\xi)$ depends on both m and n, and the angular part $S_{mn}(\mathrm{i}p,\eta)$ depends on s (see (7.55)).

For the Dirichlet boundary condition, $u(\xi_0,\eta,\phi,t) = d(\eta,\phi,t)$, we expand $D = \mathscr{L}\{d\}$ as

$$\begin{aligned} D(\eta,\phi,s) &= \sum_{m=0}^{\infty}\sum_{n=m}^{\infty} S_{mn}(\mathrm{i}p,\eta)\,\mathscr{D}_n^m(\phi,s) \\ &= \sum_{m=0}^{\infty} \mathscr{D}_m^{\mathrm{c}}(\eta,s)\cos m\phi + \sum_{m=1}^{\infty} \mathscr{D}_m^{\mathrm{s}}(\eta,s)\sin m\phi, \end{aligned} \tag{7.56}$$

with $\mathscr{D}_n^m(\phi,s)$ given by (7.52) and

$$\mathscr{D}_m^{\mathrm{c,s}}(\eta,s) = \sum_{n=m}^{\infty} D_{mn}^{\mathrm{c,s}}(s)\,S_{mn}(\mathrm{i}p,\eta).$$

Application of the boundary condition, $U = D$ at $\xi = \xi_0$, leads to

$$U(\xi,\eta,\phi,s) = \sum_{m=0}^{\infty}\sum_{n=m}^{\infty} \frac{R_{mn}^{(3)}(\mathrm{i}p,\xi)}{R_{mn}^{(3)}(\mathrm{i}p,\xi_0)} S_{mn}(\mathrm{i}p,\eta)\,\mathscr{D}_n^m(\phi,s) \tag{7.57}$$

$$= \sum_{m=0}^{\infty} \mathscr{U}_m^{\mathrm{c}}(\xi,\eta,s)\cos m\phi + \sum_{m=1}^{\infty} \mathscr{U}_m^{\mathrm{s}}(\xi,\eta,s)\sin m\phi,$$

with

$$\mathscr{U}_m^{\mathrm{c,s}}(\xi,\eta,s) = \sum_{n=m}^{\infty} \frac{R_{mn}^{(3)}(\mathrm{i}p,\xi)}{R_{mn}^{(3)}(\mathrm{i}p,\xi_0)} S_{mn}(\mathrm{i}p,\eta) D_{mn}^{\mathrm{c,s}}(s).$$

Prior to inverting $U = \mathscr{L}\{u\}$, we require some information on properties of the various functions of s occurring. In particular, as with (7.53), we are interested in zeros of the denominator in (7.57):

$$\text{solve} \quad R_{mn}^{(3)}(\gamma,\xi_0) = 0 \quad \text{for } s \text{ where} \quad \gamma = \mathrm{i}p = \mathrm{i}s\ell/c. \tag{7.58}$$

Such solutions for s are the natural frequencies.

The literature on the problem (7.58) is sparse; this probably explains why solving transient scattering problems for spheroids using PSWFs has not been pursued in detail. Lauvstad [522, p. 42] noted that 'the difficulties of finding the solutions of (7.58) preclude an explicit evaluation' of the inversion formula, whereas Sidman [774, p. 881] observed that solving (7.58) 'would involve a rather painful numerical analysis'.

The earliest work on calculating natural frequencies for spheroids is for electromagnetic problems. In 1938, Page & Adams [677] used PSWFs and attempted analytical estimates. Much later, when the Singularity Expansion Method (see Section 8.6) was topical, Marin [593, 595] gave numerical results obtained using integral equations. More extensive results were computed in [844] using PSWFs, and in [615] using T-matrix methods [599, Chapter 7].

For acoustic problems, numerical results for Dirichlet problems were given by Boström [120] in 1982 using a T-matrix method. At about the same time, Peterson et al. [692] gave results for fluid-loaded solid spheres and spheroids and Bollig & Langenberg [113] gave results for axisymmetric ($m = 0$) Neumann problems using PSWFs. Further numerical results were published in the 1980s [214, 844, 615]. For the limiting case of a flat circular disc, see [505].

There is also some work on calculating the natural frequencies of fluid-loaded spheroidal shells when $m = 0$ [393]. Numerical results for axisymmetric pulse scattering by such a shell have been published [86, 95, 870, 455].

Asymptotic estimates for the problem (7.58) are feasible (see [602] for some results in this direction) but the problem has to solved for each choice of $m \geq 0$, $n \geq m$ and $\xi_0 > 1$. It is unlikely that this problem will attract much more attention because the scattering problem itself can be solved using other methods (such as those based on integral equations; see Chapter 10).

8
Scattering Frequencies and the Singularity Expansion Method

A natural method for solving initial-boundary value problems is to use a Laplace (or Fourier) transform with respect to time (Chapter 5). Inversion then motivates a study of singularities in an appropriate complex plane as this information can be exploited to good effect. This basic idea has led to a wide range of research areas, ranging from very theoretical (such as Lax–Phillips scattering theory) to very formal (such as Baum's Singularity Expansion Method). We attempt a survey of these topics, covering scattering frequencies (also known as complex resonances or natural frequencies), the use of boundary integral equations, and the singular value decomposition (SVD). For another survey, mainly in the context of water waves, see [396]. For a third survey, mainly in the context of optics, see [512].

8.1 Fourier Transforms

We defined the Fourier transform and its inverse in Section 5.6.1; we recall the relevant formulas and notation here:

$$\widehat{u}(\boldsymbol{r},\omega) = \mathscr{F}\{u\} = \int_{-\infty}^{\infty} u(\boldsymbol{r},t)\,\mathrm{e}^{\mathrm{i}\omega t}\,\mathrm{d}t, \tag{8.1}$$

$$u(\boldsymbol{r},t) = \mathscr{F}^{-1}\{\widehat{u}\} = \frac{1}{2\pi}\int_{-\infty}^{\infty} \widehat{u}(\boldsymbol{r},\omega)\,\mathrm{e}^{-\mathrm{i}\omega t}\,\mathrm{d}\omega. \tag{8.2}$$

In these definitions, we have assumed that ω is real but this constraint will be relaxed soon.

As outlined in Section 5.6.1, a viable strategy for solving a time-domain problem is to solve a related frequency-domain problem followed by inverse Fourier transformation. To fix ideas, consider Problem DI_0, with $u = d$ on the boundary S, and then formulate a boundary-value problem for $\widehat{u}$. The governing PDE in the exterior domain B_e is the Helmholtz equation,

$$(\nabla^2 + (\omega/c)^2)\widehat{u} = 0. \tag{8.3}$$

The boundary condition gives $\widehat{u} = \widehat{d} = \mathscr{F}\{d\}$ on S. In addition, $\widehat{u}$ must satisfy the

Sommerfeld radiation condition (5.18), implying that $\widehat{u}$ behaves as $r^{-1}e^{i\omega r/c}$ (multiplied by the far-field pattern, a function of direction $\boldsymbol{r}/r$) as $r = |\boldsymbol{r}| \to \infty$. The problem for $\widehat{u}(\boldsymbol{r},\omega)$ is then solved assuming that ω is real and positive. Formally, we have

$$\widehat{u} = \mathscr{R}\widehat{d}, \tag{8.4}$$

where $\mathscr{R}$ could be called a resolvent operator; we note that both $\mathscr{R}$ and $\widehat{d}$ depend on ω. Finally, $\widehat{u}(\boldsymbol{r},\omega)$ is inverted (using (5.16)) to obtain $u(\boldsymbol{r},t)$.

For an explicit example, suppose that S is a sphere of radius a. Then we can expand $d(\theta,\phi,t)$ on the sphere $r = a$ using spherical harmonics Y_n^m,

$$d(\theta,\phi,t) = \sum_{n,m} d_n^m(t)\, Y_n^m(\theta,\phi),$$

where (r,θ,ϕ) are spherical polar coordinates and $d_n^m(t)$ can be determined from d. It follows that

$$\widehat{u}(\boldsymbol{r},\omega) = \sum_{n,m} \widehat{d_n^m}(\omega)\, \frac{h_n(\omega r/c)}{h_n(\omega a/c)}\, Y_n^m(\theta,\phi), \tag{8.5}$$

where $h_n \equiv h_n^{(1)}$ is a spherical Hankel function of the first kind, chosen so that $\widehat{u}$ satisfies (8.3) and the radiation condition.

8.2 Scattering Frequencies

Now, suppose that, somehow, we can define $\widehat{u}(\boldsymbol{r},\omega)$ in the complex ω-plane. Then we may be able to evaluate the inversion formula (8.2) using residue calculus and a suitable contour of integration. This motivates an investigation of $\widehat{u}(\boldsymbol{r},\omega)$ regarded as a function of the complex variable ω. As u solves an IBVP, we can assume that $u(\boldsymbol{r},t) = 0$ for all $t < 0$, whence (8.1) simplifies to

$$\widehat{u}(\boldsymbol{r},\omega) = \int_0^\infty u(\boldsymbol{r},t)\, e^{i\omega t}\, dt. \tag{8.6}$$

This defines a function analytic in an upper half-plane (even if u grows as $t \to \infty$); if u is bounded as $t \to \infty$, which we henceforth assume, the half-plane is $\operatorname{Im}\omega > 0$.

The function $\widehat{u}(\boldsymbol{r},\omega)$ can be continued analytically into the lower half-plane. When this is done, singularities will be encountered; in general, these could be poles or branch points. The precise details of the singularity structure (for example, location and nature of poles) will depend on the specific problem being solved. Referring to the resolvent form (8.4), let us write $\widehat{u}(\omega) = \mathscr{R}(\omega)\widehat{d}(\omega)$, emphasising the dependence on ω. This shows that some singularities can come from the specific choice of the Dirichlet data, via $\widehat{d}$, whereas others come from the scattering process itself, via $\mathscr{R}$. Let us call these *data singularities* and *resolvent singularities*, respectively.

For our acoustic problems, governed by the three-dimensional wave equation, it is known ([527, Chapter V], [454, §7.8], [614, Chapter 5], [817, §9.7]) that the analytic continuation is meromorphic: all the resolvent singularities are poles. The values of

ω at these poles are known as *scattering frequencies*; this terminology was perhaps first introduced by Lax & Phillips [530, p. 85]. Scattering frequencies are also known as *complex resonances*, thus indicating that they do not correspond to genuine (real) resonances. Cho [178, Chapter 10] calls them *exterior resonant frequencies*. They are known as *natural frequencies* in the context of the Singularity Expansion Method (Section 8.6).

According to Newton [658, p. 512], the 'first paper using the modern approach of analytic continuation ... off the real energy axis was by A. J. F. Siegert' in 1939 [775] in the context of nuclear scattering. Thus: 'We investigate the singularities of the cross section which occur at certain complex values of the energy. Those singularities which lie near enough to the real axis cause a sharp resonance maximum on the real axis' [775, p. 750]. See also [657, §5], [815, Chapter 13] and [779, Chapter 7]. For overviews, see [241, 910, 250].

Suppose that $\omega_* = \omega_\mathrm{r} + \mathrm{i}\omega_\mathrm{i}$ is a scattering frequency, where ω_r and ω_i are real with $\omega_\mathrm{i} < 0$. The poles are symmetric about the imaginary axis: $-\overline{\omega_*} = -\omega_\mathrm{r} + \mathrm{i}\omega_\mathrm{i}$ is also a scattering frequency. Numerical methods for locating scattering frequencies have been devised [454, §7.10], [505, 875, 597, 586, 652], [512, §3]; for spheroids, see the discussion near the end of Section 7.6; for water waves, see [543, 395, 619, 620]. See also [625] for a discussion of the perils of numerical analytic continuation in the presence of poles.

Inspection of the solution of Problem DI_0 for a sphere, (8.5), reveals that the data singularities come from singularities of $\widehat{d_n^m}(\omega)$, whereas the scattering frequencies are at $\omega_{n,p}$, where $z = \omega_{n,p}a/c$ is the pth zero of $h_n(z)$. Using (2.30) and [661, eqn 10.47.13], we obtain

$$\mathrm{i}z^{n+1}h_n(z) = (2/\pi)(-\mathrm{i}z)^{n+1}k_n(-\mathrm{i}z) = \mathrm{e}^{\mathrm{i}z}\theta_n(-\mathrm{i}z), \tag{8.7}$$

where k_n is a modified spherical Bessel function and $\theta_n(x)$ is a reverse Bessel polynomial, a polynomial in x of degree n; see discussion following (7.9) for properties of θ_n. Thus $h_n(\omega a/c)$ has n zeros, all simple, in the lower half of the ω-plane, located symmetrically about the imaginary axis.

Once the singularities have been found, we can contemplate moving the integration contour in (8.2); this will be discussed in Section 8.5.

8.3 Boundary Integral Equations

One way to effect analytic continuation into the lower half of the complex ω-plane is to use boundary integral equations. This is a standard method for solving frequency-domain boundary-value problems [189, 599]. For example, we may write $\widehat{u}$ as a single-layer potential,

$$\widehat{u}(\boldsymbol{r},\omega) = (\mathrm{S}\mu)(\boldsymbol{r}) \equiv \int_S \mu(\boldsymbol{r}')\,G(\boldsymbol{r},\boldsymbol{r}')\,\mathrm{d}s(\boldsymbol{r}') \tag{8.8}$$

where $G = -\exp(\mathrm{i}\omega R/c)/(2\pi R)$ and $R = |\boldsymbol{r} - \boldsymbol{r}'|$. (Time-dependent layer potentials for the wave equation will be introduced in Section 9.4.) The representation (8.8) ensures that $\widehat{u}$ satisfies (8.3) for any choice of μ; it also satisfies the Sommerfeld radiation condition when ω is real and positive. The boundary condition gives

$$\mathrm{S}\mu = \widehat{d}, \tag{8.9}$$

a Fredholm integral equation of the first kind for μ. The homogeneous integral equation $\mathrm{S}\mu = 0$ is known to have non-trivial solutions when ω/c is an eigenvalue of the corresponding interior Dirichlet problem (solve (8.3) inside S with $\widehat{u} = 0$ on S). These eigenvalues (known as *irregular frequencies*) are real and they are of no interest here: they are a consequence of using the representation (8.8). Of more interest are complex values of ω for which $\mathrm{S}\mu = 0$ has non-trivial solutions: these are the scattering frequencies.

Irregular frequencies can be removed by using a different representation [599, §6.8]. Actually, 'removed' should be 'moved': changing the representation typically moves the irregular frequencies off the real axis, and so they may then be confused with genuine scattering frequencies [627, 792].

For theoretical purposes, it is often better to use a Fredholm integral equation of the second kind. Instead of starting with (8.8), we try a double-layer potential

$$\widehat{u}(\boldsymbol{r},\omega) = \int_S \nu(\boldsymbol{r}')\,\frac{\partial G}{\partial n_q}(\boldsymbol{r},\boldsymbol{r}')\,\mathrm{d}s(\boldsymbol{r}'), \tag{8.10}$$

where $\partial/\partial n_q$ denotes normal differentiation at the point $q \in S$ with position vector $\boldsymbol{r}'$. Application of the boundary condition gives

$$(I - \overline{K^*})\nu = -\widehat{d} \tag{8.11}$$

where

$$(\overline{K^*}\nu)(\boldsymbol{r}) = \int_S \nu(\boldsymbol{r}')\,\frac{\partial G}{\partial n_q}(\boldsymbol{r},\boldsymbol{r}')\,\mathrm{d}s(\boldsymbol{r}'),$$

and $\boldsymbol{r}$ is the position vector of a point $p \in S$. The notation $\overline{K^*}$ (which is also used in [599, §5.3]) arises as follows. First, we introduce the standard inner product for $L^2(S)$,

$$(u,v) = \int_S u\bar{v}\,\mathrm{d}s,$$

where $\bar{v}$ is the complex conjugate of v. Then, for an operator A, its adjoint A^* is defined by $(Au,v) = (u,A^*v)$; if $A = A^*$, A is self-adjoint. Some calculation shows that $\mathrm{S}^* = \overline{\mathrm{S}}$ and $(\overline{K^*})^* = \overline{K}$ where

$$(K\nu)(\boldsymbol{r}) = \int_S \nu(\boldsymbol{r}')\,\frac{\partial G}{\partial n_p}(\boldsymbol{r},\boldsymbol{r}')\,\mathrm{d}s(\boldsymbol{r}'),$$

and $\partial/\partial n_p$ denotes normal differentiation at $p \in S$. We conclude that S and $\overline{K^*}$ are not self-adjoint operators.

Just like $S\mu = 0$, $(I - \overline{K^*})\nu = 0$ suffers from irregular frequencies: it has nontrivial solutions when ω/c is an eigenvalue of the corresponding interior Neumann problem. However, $I - \overline{K^*}$ is of the form 'identity plus compact' and so we can appeal to Steinberg's theorem [793] in order to prove that the analytic continuation into the lower half of the ω-plane is meromorphic; for details, see [712, Theorem VI.14], [241, §A.3], [817, §9.7] and [239, 543].

Boundary integral equations can also be solved numerically so as to locate scattering frequencies [853, 587, 627, 792]. After discretisation, the problem is reduced to a nonlinear eigenvalue problem that may be solved by a variety of methods [370]. A popular choice is the Sakurai–Sugiura method, which determines eigenvalues inside a given contour; for more information, see [735, 28, 846], [370, §5.3] and [627, §3.2].

8.4 Generalised Eigenfunctions; Quasinormal Modes

When $\omega = \omega_n = \omega_r + i\omega_i$ (with $\omega_i < 0$) is a scattering frequency, there will be a wavefunction $\widehat{u}_n(\boldsymbol{r})\,e^{-i\omega_n t}$ that satisfies a homogeneous boundary condition together with 'outgoing behaviour' inherited from the Sommerfeld radiation condition (5.18); the wavefunction behaves as

$$r^{-1}e^{-i\omega_n(t-r/c)} = r^{-1}e^{-i\omega_r(t-r/c)}\,e^{|\omega_i|r/c}\,e^{-t|\omega_i|}, \tag{8.12}$$

which means it grows exponentially with r but decays exponentially with t. Thus $\widehat{u}_n$ satisfies

$$\left(\nabla^2 + (\omega_n/c)^2\right)\widehat{u}_n = 0 \quad \text{in } B_e,$$

together with a homogeneous boundary condition and an outgoing condition at infinity. Evidently, $\widehat{u}_n$ is some kind of eigenfunction. We can call it a *generalised eigenfunction*; in the physics literature, the name *quasinormal mode* is used [512]. Such functions were first encountered in the nineteenth century by J. J. Thomson:

Perhaps the earliest application of the method of complex eigenfrequencies to linear systems with an infinite number of degrees of freedom was made by Thomson [823], *in his treatment of the free modes of oscillation of the electromagnetic field around a perfectly conducting sphere. He determined them by the requirement that they should contain only outgoing radiation*... [and he] *found them to be of the form* $\omega_n = \omega_n' - i\gamma_n$ ($\gamma_n > 0$), *corresponding to a time dependence* $\exp(-i\omega_n t)$, *and he accordingly interpreted* ω_n' *as the frequency, and* γ_n *as the damping constant, for the* nth *natural mode of oscillation.*

It was subsequently remarked by Lamb [514] *that Thomson's modes behave like*

$$r^{-1}\exp\{-i\omega_n[t-(r/c)]\} \quad \textit{as } r \to \infty,$$

so that they also give rise to an "exponential catastrophe." Lamb pointed out that the difficulty arises from the unphysical assumption that the modes have been in existence for an indefinitely long time, and that it may be overcome by taking into account the excitation conditions, of which he gave an example. Another example was treated by Love [574].

(Nussenzveig [670, p. 162])

The theory of generalised eigenfunctions is complicated because of the exponential growth (8.12) and because of the lack of self-adjointness; recall the discussion in Section 8.3 concerning the integral operators S and $\overline{K^*}$. In particular, we do not know how many different generalised eigenfunctions exist for each scattering frequency. Computationally, the exponential growth means that normalisation is an issue; for a review of this topic, see [512, §4].

For the sphere problem with a homogeneous Dirichlet boundary condition at $r = a$, the generalised eigenfunctions are given by

$$h_n(\omega_{n,p} r/c)\, Y_n^m(\theta,\phi)\, \mathrm{e}^{-\mathrm{i}\omega_{n,p} t},$$

where $h_n(\omega_{n,p} a/c) = 0$; see (8.5) and the discussion around (8.7).

Überall and his co-authors have characterised generalised eigenfunctions in terms of constructive interference of waves over the surface of a scatterer S. Thus [616, p. 1464]

propagating circumferential waves (surface waves) are generated during the excitation process, and the resonances $[\omega_\mathrm{r}]$ *are actually caused by the phase matching of such waves as they encircle the object one or more times. These waves are dispersive, and phase matching will occur at those (complex) frequencies* $[\omega_n = \omega_\mathrm{r} + \mathrm{i}\omega_\mathrm{i}]$ *where standing waves are formed around the circumference. ... The principle of phase matching may be* [applied to surfaces S], *where it can be formulated in the sense of Fermat's principle* [789, p. 356]. *The surface wave then propagates over* [S] *along a geodesic. The phase matching principle thus has to be stated in integral form, such that ... the path length element measured locally in units of the wavelength and then integrated over the entire closed path results in an integer.*

At first, this idea was used to interpret computations for simple scatterers (circular cylinders, spheres, spheroids) [841, 842, 616]. Later it was used as the basis of a numerical method for computing scattering frequencies [617], which requires the computation of the relevant geodesics. This is a problem of differential geometry. It is straightforward when S is axisymmetric (a spheroid is a good example): 'the geodesics of a surface of revolution can be found by quadratures' [802, p. 134]; see [802, p. 134, eqn (2-4)] or [617, eqn (4a)]. However, we are not aware of applications to non-axisymmetric surfaces S.

8.5 Moving the Inversion Contour

There are applications where the scattering frequency closest to the real axis, ω_1 say, is of most interest because it decays most slowly over time, as $\exp(-t\,|\mathrm{Im}\,\omega_1|)$ (see [676] and references therein). The main idea here is to move the integration contour in (8.2) down from the real axis to a parallel contour, picking up residue contributions as poles are crossed. Let us list the scattering frequencies ω_n with decreasing imaginary parts, $0 > \mathrm{Im}\,\omega_1 \geq \mathrm{Im}\,\omega_2 \geq \mathrm{Im}\,\omega_3 \geq \cdots$. Similarly, list the frequencies corresponding to the data singularities ϖ_n in the same way, $0 > \mathrm{Im}\,\varpi_1 \geq \mathrm{Im}\,\varpi_2 \geq$

$\operatorname{Im} \varpi_3 \geq \cdots$. If we assume that all poles are simple, we may expect to obtain an asymptotic expansion of the form

$$u(\boldsymbol{r},t) \sim \sum_{n=1}^{\infty} E_n(\boldsymbol{r})\,\mathrm{e}^{-\mathrm{i}\omega_n t} + \sum_{n=1}^{\infty} \widetilde{E}_n(\boldsymbol{r})\,\mathrm{e}^{-\mathrm{i}\varpi_n t} \quad \text{as } t \to \infty. \tag{8.13}$$

This gives the expected exponential decay with t. For E_n and $\widetilde{E}_n$, residue calculations give

$$E_n(\boldsymbol{r}) = -\mathrm{i} \lim_{\omega \to \omega_n} \{(\omega - \omega_n)\widehat{u}(\boldsymbol{r},\omega)\},$$

with a similar formula for $\widetilde{E}_n$. For a formula similar to (8.13), see [707, eqn (1.26)]. (Extension to higher-order poles is straightforward; for example, $E_n(\boldsymbol{r})$ in (8.13) would be replaced by $E_n(\boldsymbol{r},t)$ with a dependence on t.)

Some authors take another step and replace E_n in (8.13) by a multiple of a generalised eigenfunction $\widehat{u}_n$, giving

$$u(\boldsymbol{r},t) \sim \sum_{n=1}^{\infty} c_n \widehat{u}_n(\boldsymbol{r})\,\mathrm{e}^{-\mathrm{i}\omega_n t} + u_{\mathrm{data}}(\boldsymbol{r},t) \quad \text{as } t \to \infty, \tag{8.14}$$

with coefficients c_n, where u_{data} is the contribution from the data singularities. For a similar formula, see [50, eqn (28)]. We shall discuss an equivalent formula below; see (8.16).

8.6 The Singularity Expansion Method (SEM)

Let us rotate the picture, moving from a Fourier-transform view to an equivalent Laplace-transform view. Thus put $\omega = \mathrm{i}s$ in (8.6) giving the Laplace transform of $u(\boldsymbol{r},t)$,

$$U(\boldsymbol{r},s) \equiv \widehat{u}(\boldsymbol{r},\mathrm{i}s) = \int_0^{\infty} u(\boldsymbol{r},t)\,\mathrm{e}^{-st}\,\mathrm{d}t.$$

We see that $U(\boldsymbol{r},s)$ is analytic for $\operatorname{Re} s > 0$. The pole at $\omega = \omega_*$ becomes a pole at $s = s_* = -\mathrm{i}\omega_* = \omega_\mathrm{i} - \mathrm{i}\omega_\mathrm{r}$ in the left half of the complex s-plane, $\operatorname{Re} s < 0$. As there is also a pole at $\omega = -\overline{\omega_*}$, there is a corresponding pole at $s = \overline{s_*}$: poles appear in complex-conjugate pairs. The values of s at these poles are known as *natural frequencies*, especially in the context of the *Singularity Expansion Method* (SEM) as developed by Baum and his collaborators [596, 592, 594, 69, 70]:

> *The general SEM formalism began* [in 1971] *from experimental observations concerning the transient electromagnetic response of complicated scatterers such as missiles and aircraft. It was observed that damped sinusoids were dominant features of typical transient responses. Such damped sinusoids corresponded to pole pairs in the* [s-plane]. *Poles were one type of singularity in the s plane and this led to the concept of using all the s plane singularities to form a description of the transient and frequency-domain responses. It was also noted that the poles corresponded to the natural frequencies which had been discussed in some of the older literature* [799]. (Baum [70, p. 1603])

Later papers include [71, 74, 72, 73]. For elastic scatterers, see [840] and [669, §5].

To recover u from $U = \mathscr{L}\{u\}$, we use the inversion formula (5.3). Moving the Bromwich contour to the left, we pick up residue contributions. List the natural frequencies s_n with decreasing real parts, $0 > \mathrm{Re}\, s_1 \geq \mathrm{Re}\, s_2 \geq \mathrm{Re}\, s_3 \geq \cdots$, and do the same for the frequencies $\tilde{s}_n$ corresponding to the data singularities. If we assume that all poles are simple, we expect to obtain a rotated form of (8.13),

$$u(\boldsymbol{r},t) \sim \sum_{n=1}^{\infty} E_n(\boldsymbol{r})\, \mathrm{e}^{s_n t} + u_{\mathrm{data}}(\boldsymbol{r},t) \quad \text{as } t \to \infty, \tag{8.15}$$

where u_{data} comes from the data singularities and

$$E_n(\boldsymbol{r}) = \lim_{s \to s_n} \{(s - s_n) U(\boldsymbol{r}, s)\}.$$

Corresponding to a natural frequency s_n, there is a *natural mode*, $U_n(\boldsymbol{r})$ satisfying

$$\left(\nabla^2 - (s_n/c)^2\right) U_n = 0 \quad \text{in } B_{\mathrm{e}},$$

together with a homogeneous boundary condition and an outgoing condition at infinity. Again, instead of (8.15), we may hope to have a rotated form of (8.14),

$$u(\boldsymbol{r},t) \sim \sum_{n=1}^{\infty} c_n U_n(\boldsymbol{r})\, \mathrm{e}^{s_n t} + u_{\mathrm{data}}(\boldsymbol{r},t) \quad \text{as } t \to \infty. \tag{8.16}$$

Such an asymptotic expansion was given in 1969 by Lax & Phillips [528, eqn (1.7)]. It is not in their 1967 book [527]. It is in their 1971 paper [529, eqn (9.7)] with an additional remark [529, p. 176]: 'c_n is a constant; if U_n is a generalised eigenfunction, c_n is a power of t'. In the 1989 revision of their book [532, p. 277, eqn (10)], the term $c_n U_n$ in (8.16) is replaced by a 'projection onto the nth eigenspace'. This indicates a realisation that there may be more that one natural mode U_n corresponding to each natural frequency s_n; exactly the same criticism applies to (8.14).

Conventional SEM takes another step and replaces the asymptotic approximations in (8.14) and (8.16) by equalities:

$$u(\boldsymbol{r},t) - u_{\mathrm{data}}(\boldsymbol{r},t) = \sum_{n=1}^{\infty} c_n \widehat{u}_n(\boldsymbol{r})\, \mathrm{e}^{-\mathrm{i}\omega_n t} = \sum_{n=1}^{\infty} c_n U_n(\boldsymbol{r})\, \mathrm{e}^{s_n t}.$$

As might be expected, this 'formalism' has been the subject of some criticism [239, 240, 707, 686, 638, 687]. See also [118, §8.1]. We conclude this section with some remarks by Heyman & Felsen [406, p. 706]:

Transient fields scattered by an object may be synthesized in terms of progressing waves (wavefronts) or oscillatory waves (resonances) [291]. *In a progressing wave formulation, a causal wavefront is tracked from the source to the scatterer, where it undergoes an interactive process that is conveyed to the observer by successive wavefront arrivals corresponding to multiple passes around the object and/or multiple diffraction from scattering centers located on the object. This description, which is sensitive to* local *features encountered along the wavefront trajectory, becomes cumbersome at late observation times, . . .*

The alternative oscillatory representation, formalized by the SEM [69], *emphasizes* global *features of the scatterer by expressing the field in terms of its resonant modes which are damped in time. Therefore, this formulation is most convenient for late observation times.*

8.7 The Eigenmode Expansion Method (EEM)

In this section, we review some results from the functional analysis of operators and we connect them with boundary integral equations and Baum's EEM.

8.7.1 Self-Adjoint Operators and Normal Operators

Let $A : \mathscr{H} \to \mathscr{H}$ be a compact linear operator, where $\mathscr{H}$ is a Hilbert space. Let $A^* : \mathscr{H} \to \mathscr{H}$ be the adjoint operator; A^* is also compact. If $A\varphi = \lambda\varphi$ for some $\varphi \not\equiv 0$, λ is an eigenvalue and φ is an eigenfunction.

Suppose that A is compact and *self-adjoint*, $A = A^*$. This could be described as the textbook case, for which we have the Hilbert–Schmidt theorem [712, Theorem VI.16]. It gives the following results. The non-zero eigenvalues of A form a non-empty sequence $\lambda_1, \lambda_2, \ldots$, ordered so that $|\lambda_1| \geq |\lambda_2| \geq \cdots$ with each eigenvalue repeated in the sequence according to its finite multiplicity. All eigenvalues λ_n are real. When the sequence $\{\lambda_n\}$ is infinite, $\lambda_n \to 0$ as $n \to \infty$. For each λ_n there is an eigenfunction φ_n so that $A\varphi_n = \lambda_n\varphi_n$. The sequence $\{\varphi_n\}$ forms a complete orthonormal basis for $\mathscr{H}$.

To solve $Au = f$ for $u \in \mathscr{H}$ and $f \in \mathscr{H}$, we put $u = \sum_n a_n\varphi_n + u_0$ where $Au_0 = 0$. Then $Au = \sum_n a_n\lambda_n\varphi_n$. If we expand $f = \sum_n f_n\varphi_n$ with $f_n = (f, \varphi_n)$, we obtain $a_n = f_n/\lambda_n$ for non-zero λ_n; zero eigenvalues have been taken into account through u_0. Hence

$$u = u_0 + \sum_{\lambda_n \neq 0} \frac{(f, \varphi_n)}{\lambda_n}\varphi_n. \tag{8.17}$$

For this to give a solution in $\mathscr{H}$, we require $(u, u) < \infty$; this gives $\sum_n |(f, \varphi_n)/\lambda_n|^2 < \infty$, which can be viewed as a constraint on f. Also, as $(f, u_0) = (Au, u_0) = (u, Au_0)$, the given function f must be orthogonal to all solutions of the homogeneous equation $Au_0 = 0$, $(f, u_0) = 0$.

For more information on the self-adjoint problem, see [784, Chapter VII], [712, §VI.5], [403, §72], [193, Chapter II, §5] and [501, §15.3], for example. But, as an early warning, we quote from a 2002 survey article:

> *At present, the subject* [of non-self-adjoint operators] *has hardly moved beyond the foothills, but the clear message is that if one wishes to travel in this direction, one must give up any hope that theorems about self-adjoint operators will provide useful signposts: they regularly lead in quite the wrong direction.* (Davies [217, p. 516])

Let us enlarge the class of operators and suppose that A is a compact *normal* operator, which means $AA^* = A^*A$; evidently, all self-adjoint operators are normal but normal operators need not be self-adjoint. Nevertheless, it turns out that all the results given above for self-adjoint operators hold for normal operators, with two exceptions. First, eigenvalues need not be real. Second, the constraint on f for $Au = f$ to be solvable, $(f, u_0) = 0$, is replaced by $(f, v_0) = 0$ for every solution of the homogeneous adjoint equation, $A^* v_0 = 0$. For more details, see [403, §71] and [193, Chapter II, §7].

8.7.2 The Time-Harmonic Single-Layer Operator

As an example, suppose that A is the single-layer operator S, (8.8), with $\mathscr{H} = L^2(S)$. As already noted in Section 8.3, S is not self-adjoint. In particular, for a sphere of radius a, we can use the bilinear expansion of G [599, Theorem 6.4] to show that

$$\mathrm{S}Y_n^m = -2\mathrm{i}\,a^2 k j_n(ka)h_n(ka)\,Y_n^m \quad \text{and} \quad \mathrm{S}^*Y_n^m = 2\mathrm{i}\,a^2\,\overline{k j_n(ka)h_n(ka)}\,Y_n^m,$$

where $k = \omega/c$ and Y_n^m is a spherical harmonic. This identifies the eigenvalues and eigenfunctions. Moreover, for S, we see that to the eigenvalue $-2\mathrm{i}\,a^2 k j_n(ka)h_n(ka)$ there are $2n+1$ eigenfunctions, Y_n^m, $m = 0, \pm 1, \pm 2, \ldots, \pm n$. Further calculation gives

$$\mathrm{S}\mathrm{S}^*Y_n^m = \mathrm{S}^*\mathrm{S}Y_n^m = 4a^2\,|ka\,j_n(ka)h_n(ka)|^2\,Y_n^m. \tag{8.18}$$

Thus S is a normal operator when S is a sphere. This is a known result [707, p. 579]. Dolph [238, p. 381] states that S is normal for a sphere but not for an ellipsoid. In fact, Betcke et al. [98] have proved that, for real k, S is normal only when S is a sphere.

8.7.3 Singular Value Decomposition (SVD)

Returning to the general case, suppose now that $A : \mathscr{H} \to \mathscr{H}$ is a non-normal compact operator. The analysis of such operators is more complicated. There may not be any eigenvalues. When they do exist, they may not yield a complete set of eigenfunctions. In general, eigenfunctions will not be orthogonal, and so-called root vectors may be required; see [707, p. 564] and the book by Gohberg & Kreĭn [346]. Some would argue that it is a mistake to contemplate searching for eigenfunctions:

Eigenvalues and eigenvectors are an imperfect tool for analyzing non-normal matrices and operators, a tool that has often been abused. Physically, it is not always the eigenmodes that dominate what one observes in a highly non-normal system. (Trefethen [831, p. 234])

Everyone has seen a time-dependent problem where eigenvalues are misleading, and such effects are always intriguing. (Trefethen & Embree [832, p. 135])

Instead of investigating eigenvalues, another option is to introduce singular values, as first done by Schmidt [746]; see also [784, Chapter VIII], [794] and [501, §15.4]. These are defined as follows, starting with the observation that the operators $A^*A : \mathscr{H} \to \mathscr{H}$ and $AA^* : \mathscr{H} \to \mathscr{H}$ are compact and self-adjoint. If we suppose that $A^*A\varphi = \lambda\varphi$ with $\varphi \not\equiv 0$, then $\lambda(\varphi, \varphi) = (A^*A\varphi, \varphi) = (A\varphi, A\varphi)$. This shows that the eigenvalue λ is real and non-negative. Moreover, $AA^*A\varphi = \lambda A\varphi$ shows that λ is also an eigenvalue of AA^* with eigenfunction $A\varphi$. As an example, λ is given by (8.18) when $A = \mathrm{S}$ and S is a sphere.

The number $\sigma = +\sqrt{\lambda}$ is called a *singular value* of A. Let $\sigma_1, \sigma_2, \ldots$ be the non-zero singular values of the compact linear operator A ($\neq 0$), repeated according to their multiplicities. Then, we have the following results [501, Theorem 15.16]. There are two orthonormal sequences in $\mathscr{H}$, $\{\phi_n\}$ and $\{\psi_n\}$, such that

$$A\phi_n = \sigma_n \psi_n, \quad A^* \psi_n = \sigma_n \phi_n, \quad n = 1, 2, \ldots. \tag{8.19}$$

For each $u \in \mathscr{H}$, we have the *singular value decomposition* (SVD)

$$u = u_0 + \sum_{n=1}^{\infty} a_n \phi_n \quad \text{with } a_n = (u, \phi_n), \tag{8.20}$$

where $Au_0 = 0$, and

$$Au = \sum_{n=1}^{\infty} \sigma_n a_n \psi_n. \tag{8.21}$$

Then, to solve $Au = f$ for $u \in \mathscr{H}$ and $f \in \mathscr{H}$, we expand u as (8.20) whereas (8.21) implies that we should expand f as

$$f = \sum_{n=1}^{\infty} f_n \psi_n \quad \text{with } f_n = (f, \psi_n).$$

As $Au = f$, we obtain $\sigma_n a_n = f_n$, whence

$$u = u_0 + \sum_{n=1} \frac{(f, \psi_n)}{\sigma_n} \phi_n, \tag{8.22}$$

which may be compared with (8.17). For (8.22) to give a solution in $\mathscr{H}$, we require $(u, u) < \infty$; this gives $\sum_n |(f, \psi_n)/\sigma_n|^2 < \infty$, a constraint on f. Also, as $(f, v) = (u, A^* v)$, f must be orthogonal to all solutions of the homogeneous adjoint equation $A^* v = 0$.

8.7.4 The Eigenmode Expansion Method (EEM)

The EEM was introduced by Baum [70] 'for the purpose of further understanding SEM, extending its utility and tying it into some other perhaps related concepts' [71, p. 648]. It starts with (the electromagnetic equivalent of) the integral equation $S\mu = f$, but written in terms of the Laplace-transform variable $s = -\mathrm{i}\omega$,

$$\langle G(\boldsymbol{r}, \cdot\,; s)\, \mu(\cdot\,; s)\rangle \equiv \int_S G(\boldsymbol{r}, \boldsymbol{r}'; s)\, \mu(\boldsymbol{r}'; s)\, \mathrm{d}s(\boldsymbol{r}') = f(\boldsymbol{r}; s), \quad \boldsymbol{r} \in S, \tag{8.23}$$

where $G = -\exp(-sR/c)/(2\pi R)$, $R = |\boldsymbol{r} - \boldsymbol{r}'|$ and we have emphasised the dependence on s. Baum [70, p. 1609] then defines 'eigenvalues and eigenmodes of the simple variety', λ_n and g_n, respectively, by

$$\langle G(\boldsymbol{r}, \cdot\,; s)\, g_n(\cdot\,; s)\rangle = \lambda_n(s)\, g_n(\boldsymbol{r}; s),$$

he assumes that $\langle g_m(\cdot\,; s) g_n(\cdot\,; s)\rangle = \delta_{mn}$, and he states that the solution of the integral equation (8.23) 'takes the form' [70, eqn (4.10)]

$$\mu(\boldsymbol{r}; s) = \sum_n \frac{1}{\lambda_n(s)} \langle f(\cdot\,; s)\, g_n(\cdot\,; s)\rangle\, g_n(\boldsymbol{r}; s). \tag{8.24}$$

As we saw in Section 8.7.1, this formalism would be justified if the single-layer operator on the left-hand side of (8.23) were self-adjoint or normal. In general, it is not: 'even the formal expressions are often meaningless' [239, p. 893].

The special case of real s is worth noting. In this case, $G(\boldsymbol{r},\boldsymbol{r}';s)$ is real and symmetric, the single-layer operator is self-adjoint, and the formula (8.24) is justified. However, it is very clear that Baum wants s to be complex: his idea is to find those values of s and n for which $\lambda_n(s) = 0$, using a numerical discretisation of the integral equation (8.23): 'each natural frequency s_m "belongs" to a particular eigenvalue $\lambda_n(s)$' [70, p. 1610]. For a later, but essentially unchanged, description of both EEM and SEM, see [73].

An alternative approach is to exploit the singular value decomposition of Section 8.7.3. If we take $A = \mathrm{S}$, the quantities of interest become the singular values $\sigma_n(\omega)$; the goal is to find complex frequencies for which $\sigma_n = 0$. As far as we know, this approach has not been pursued, although Marks [597] did suggest using the singular value decomposition in the context of transient scattering.

9
Integral Representations

In potential theory (Laplace's equation) and for frequency-domain problems (Helmholtz equation), we can represent solutions using certain integral formulas. The integrands contain a fundamental solution (sometimes called Green's function). For the wave equation, the basic integral formula is due to Kirchhoff; we give several derivations in Sections 9.1–9.3. Then, in Section 9.4, we introduce layer potentials (single layers and double layers) and give their properties. The machinery of this chapter will be used in Chapter 10 so as to derive time-domain boundary integral equations for solving IBVPs.

Integral representations for electromagnetic problems are given in Section 10.5. For elastodynamic problems, see Section 10.6. For hydrodynamic problems, see Section 10.7.

9.1 Kirchhoff's Formula: Classical Derivations

There are three things that every beginner . . . should know. The first is that "Kirchhoff" has two h's in it. (Marsden & Hughes [598, p. xi])

Let us recall our basic notation. We suppose that B is a three-dimensional region bounded by a smooth closed (fixed) surface S. The unbounded exterior of S is denoted by B_e. The unit normal vector on S, $\boldsymbol{n}$, points from S into B_e. Points in $B \cup B_\mathrm{e}$ are denoted by upper-case letters, P, Q. Points on S are denoted by lower-case letters, p, q.

The point P has position vector $\boldsymbol{r} = (x,y,z) = (x_1,x_2,x_3)$. The point Q has position vector $\boldsymbol{r}' = (x',y',z') = (x_1',x_2',x_3')$.

We follow Baker & Copson [38, §5.2, pp. 38–40]. They give a direct proof of Kirchhoff's formula which they note is similar to one given by Gutzmer [371]. Alternative derivations are available; see Section 9.3, where generalised functions are deployed.

9.1.1 Kirchhoff's Formula for Bounded Domains

Assume that the function v is twice-differentiable in the bounded region B. (Later, in Section 9.1.6, we consider weakening this smoothness assumption.) Let $R = |\boldsymbol{r} - \boldsymbol{r}'|$ denote the distance between P and Q. Suppose first that P is fixed and located outside B (in B_e). Then Green's theorem gives

$$\int_S \left\{ v \frac{\partial}{\partial n_q}\left(\frac{1}{R}\right) - \frac{1}{R}\frac{\partial v}{\partial n} \right\} \mathrm{d}s(\boldsymbol{r}') + \int_B \frac{1}{R} \nabla_Q^2 v \, \mathrm{d}V(\boldsymbol{r}') = 0. \tag{9.1}$$

Next, we introduce what has become a standard notation. For any $\phi(x', y', z', t)$, we write

$$[\phi] = \phi(x', y', z', t - R/c) = \phi(\boldsymbol{r}', t - R/c) \tag{9.2}$$

and call $[\phi]$ the *retarded value* of ϕ.

Let u be a wavefunction that has no singularities in B or on S. Then, in (9.1), take $v = [u] = u(\boldsymbol{r}', t - R/c)$. We have

$$\begin{aligned} \frac{\partial v}{\partial x_i'} &= \left(\frac{\partial u}{\partial x_i'}(x_1', x_2', x_3', \tau) - \frac{1}{c}\frac{\partial R}{\partial x_i'}\frac{\partial u}{\partial \tau}(x_1', x_2', x_3', \tau) \right)\Bigg|_{\tau = t - R/c} \\ &= \left[\frac{\partial u}{\partial x_i'}\right] - \frac{1}{c}\frac{\partial R}{\partial x_i'}\left[\frac{\partial u}{\partial t}\right], \end{aligned} \tag{9.3}$$

and so

$$\frac{\partial v}{\partial n} = \left[\frac{\partial u}{\partial n}\right] - \frac{1}{c}\frac{\partial R}{\partial n_q}\left[\frac{\partial u}{\partial t}\right]. \tag{9.4}$$

Using (9.3), we obtain

$$\frac{\partial}{\partial x_i'}\left[\frac{\partial u}{\partial x_i'}\right] = [\nabla_Q^2 u] - \frac{1}{c}\frac{\partial R}{\partial x_i'}\left[\frac{\partial^2 u}{\partial x_i' \partial t}\right],$$

$$\frac{\partial}{\partial x_i'}\left[\frac{\partial u}{\partial t}\right] = \left[\frac{\partial^2 u}{\partial x_i' \partial t}\right] - \frac{1}{c}\frac{\partial R}{\partial x_i'}\left[\frac{\partial^2 u}{\partial t^2}\right],$$

and so

$$\begin{aligned} \nabla_Q^2 v &= \frac{\partial}{\partial x_i'}\left[\frac{\partial u}{\partial x_i'}\right] - \frac{1}{c}\frac{\partial R}{\partial x_i'}\frac{\partial}{\partial x_i'}\left[\frac{\partial u}{\partial t}\right] - \frac{1}{c}\nabla_Q^2 R\left[\frac{\partial u}{\partial t}\right] \\ &= [\nabla_Q^2 u] - \frac{1}{c}\frac{\partial R}{\partial x_i'}\left[\frac{\partial^2 u}{\partial x_i' \partial t}\right] - \frac{1}{c}\frac{\partial R}{\partial x_i'}\left(\left[\frac{\partial^2 u}{\partial x_i' \partial t}\right] - \frac{1}{c}\frac{\partial R}{\partial x_i'}\left[\frac{\partial^2 u}{\partial t^2}\right]\right) - \frac{2}{cR}\left[\frac{\partial u}{\partial t}\right] \\ &= [\nabla_Q^2 u] - \frac{2}{c}\frac{\partial R}{\partial x_i'}\left[\frac{\partial^2 u}{\partial x_i' \partial t}\right] + \frac{1}{c^2}|\mathrm{grad}_Q R|^2\left[\frac{\partial^2 u}{\partial t^2}\right] - \frac{2}{cR}\left[\frac{\partial u}{\partial t}\right] \\ &= \frac{2}{c^2}\left[\frac{\partial^2 u}{\partial t^2}\right] - \frac{2}{cR}\left[\frac{\partial u}{\partial t}\right] - \frac{2}{c}\frac{\partial R}{\partial x_i'}\left[\frac{\partial^2 u}{\partial x_i' \partial t}\right], \end{aligned} \tag{9.5}$$

using the wave equation satisfied by u, (1.46),

$$\frac{\partial R}{\partial x_i'} = \frac{x_i' - x_i}{R}, \quad |\mathrm{grad}_Q R| = 1, \quad \nabla_Q^2 R = \frac{2}{R}.$$

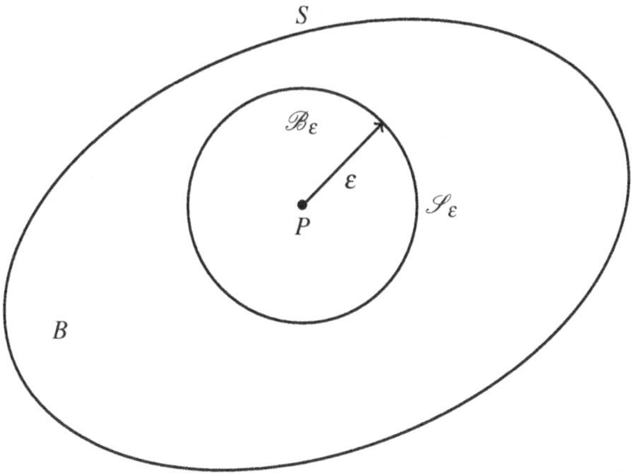

Figure 9.1 The small ball $\mathscr{B}_\varepsilon$ is centred at $P \in B$. It has radius ε and surface $\mathscr{S}_\varepsilon$.

Next, we have

$$\frac{\partial}{\partial x_i'}\left(\frac{1}{R}\frac{\partial R}{\partial x_i'}\left[\frac{\partial u}{\partial t}\right]\right) = \left(\frac{\nabla_Q^2 R}{R} - \frac{1}{R^2}|\mathrm{grad}_Q R|^2\right)\left[\frac{\partial u}{\partial t}\right] + \frac{1}{R}\frac{\partial R}{\partial x_i'}\left(\left[\frac{\partial^2 u}{\partial x_i' \partial t}\right] - \frac{1}{c}\frac{\partial R}{\partial x_i'}\left[\frac{\partial^2 u}{\partial t^2}\right]\right)$$
$$= \frac{1}{R^2}\left[\frac{\partial u}{\partial t}\right] + \frac{1}{R}\frac{\partial R}{\partial x_i'}\left[\frac{\partial^2 u}{\partial x_i' \partial t}\right] - \frac{1}{cR}\left[\frac{\partial^2 u}{\partial t^2}\right].$$

Hence comparison with (9.5) gives

$$\frac{1}{R}\nabla_Q^2 v = -\frac{2}{c}\frac{\partial}{\partial x_i'}\left(\frac{1}{R}\frac{\partial R}{\partial x_i'}\left[\frac{\partial u}{\partial t}\right]\right),$$

which means we can convert the volume integral in (9.1) to a surface integral using the divergence theorem,

$$\int_B \frac{1}{R}\nabla_Q^2 v \,\mathrm{d}V(\boldsymbol{r}') = -\frac{2}{c}\int_S \frac{1}{R}\frac{\partial R}{\partial n_q}\left[\frac{\partial u}{\partial t}\right]\mathrm{d}s(\boldsymbol{r}'). \tag{9.6}$$

Substituting in (9.1), together with (9.4), gives

$$0 = \int_S \left\{[u]\frac{\partial}{\partial n_q}\left(\frac{1}{R}\right) - \frac{1}{R}\left[\frac{\partial u}{\partial n}\right] - \frac{1}{cR}\frac{\partial R}{\partial n_q}\left[\frac{\partial u}{\partial t}\right]\right\}\mathrm{d}s_q, \quad P \in B_\mathrm{e}, \tag{9.7}$$

where we have written $\mathrm{d}s_q$ for $\mathrm{d}s(\boldsymbol{r}')$. Equation (9.7) is *Kirchhoff's formula* when the point P is outside the bounded volume B.

When P is in B, we excise from B a small ball $\mathscr{B}_\varepsilon$, centred at P, with surface $\mathscr{S}_\varepsilon$ and radius ε; see Fig. 9.1. Equation (9.1) is modified to

$$
\begin{aligned}
0 &= \int_{S\cup\mathscr{S}_\varepsilon} \left\{ v\frac{\partial}{\partial n_q}\left(\frac{1}{R}\right) - \frac{1}{R}\frac{\partial v}{\partial n} \right\} \mathrm{d}s_q + \int_{B\setminus\mathscr{B}_\varepsilon} \frac{1}{R}\nabla_Q^2 v \,\mathrm{d}V(\boldsymbol{r}') \\
&= \int_{S\cup\mathscr{S}_\varepsilon} \left\{ [u]\frac{\partial}{\partial n_q}\left(\frac{1}{R}\right) - \frac{1}{R}\left[\frac{\partial u}{\partial n}\right] - \frac{1}{cR}\frac{\partial R}{\partial n_q}\left[\frac{\partial u}{\partial t}\right] \right\} \mathrm{d}s_q,
\end{aligned} \tag{9.8}
$$

where the normal vector on $\mathscr{S}_\varepsilon$ points into $\mathscr{B}_\varepsilon$. The term in $[u]$ gives

$$
\lim_{\varepsilon\to 0} \int_{\mathscr{S}_\varepsilon} u\left(\boldsymbol{r}', t - \frac{R}{c}\right) \frac{\partial}{\partial n_q}\left(\frac{1}{R}\right) \mathrm{d}s_q = 4\pi u(\boldsymbol{r},t),
$$

as $R = |\boldsymbol{r} - \boldsymbol{r}'|$ and

$$
\frac{\partial}{\partial n_q}\left(\frac{1}{R}\right) = -\frac{\mathrm{d}}{\mathrm{d}R}\left(\frac{1}{R}\right)\bigg|_{R=\varepsilon} = \frac{1}{\varepsilon^2}.
$$

The other terms in the integral over $\mathscr{S}_\varepsilon$ in (9.8) vanish as $\varepsilon \to 0$. Hence

$$
-4\pi u(\boldsymbol{r},t) = \int_S \left\{ [u]\frac{\partial}{\partial n_q}\left(\frac{1}{R}\right) - \frac{1}{R}\left[\frac{\partial u}{\partial n}\right] - \frac{1}{cR}\frac{\partial R}{\partial n_q}\left[\frac{\partial u}{\partial t}\right] \right\} \mathrm{d}s_q, \quad P \in B. \tag{9.9}
$$

This is Kirchhoff's formula when the point P is inside the bounded volume B. Recall that $\boldsymbol{n}$ points out of B, which is the opposite of the normal vector used by Baker & Copson [38, p. 37]. Equation (9.9) is [871, p. 216, eqn (97)], [200, p. 141, eqn (27)], [799, p. 427, eqn (22)], and it is in [374, §152]. We note that zero initial conditions are easily incorporated by supposing that $u(\boldsymbol{r},t) \equiv 0$ for $t < 0$.

Many authors use $(\partial/\partial n_q)R^{-1} = -R^{-2}(\partial R/\partial n_q)$ and then (9.9) becomes

$$
4\pi u(\boldsymbol{r},t) = \int_S \left\{ \frac{1}{R}\left[\frac{\partial u}{\partial n}\right] + \left(\frac{[u]}{R^2} + \frac{1}{cR}\left[\frac{\partial u}{\partial t}\right] \right) \frac{\partial R}{\partial n_q} \right\} \mathrm{d}s_q, \quad P \in B. \tag{9.10}
$$

For example, this is the formula derived by Love [572, eqn (3)] and it is [425, eqn (1.1)]. An equivalent formula can be found in [513, p. 197], once the following formula is used:

$$
\frac{\partial}{\partial n_q}\left(\frac{u(\boldsymbol{r}'', t - R/c)}{R} \right)\bigg|_{\boldsymbol{r}''=\boldsymbol{r}'} = -\left(\frac{[u]}{R^2} + \frac{1}{cR}\left[\frac{\partial u}{\partial t}\right] \right) \frac{\partial R}{\partial n_q}.
$$

Some authors, including Webster [871, p. 215], use a special notation for the operation on the left, denoting it by $\delta/\delta n$.

9.1.2 Kirchhoff's Formula for Exterior Domains

For exterior domains B_e, we start by considering a region bounded internally by S and externally by a large sphere S_1. Then:

If we make the surface S_1 recede to infinity on all sides the surface integrals can in many cases be made to vanish. We may suppose, for instance, that in distant regions of space the function u has been zero until some definite instant t_0. The time $t - R/c$ then always falls below t_0 when R is sufficiently large and so all the quantities in square brackets vanish.

(Bateman [67, p. 185])

As noted by Baker & Copson [38, p. 38], the contribution from S_1 will vanish if u behaves as $f(t-r/c)/r$ for large r, where $f(t)$ and $f'(t)$ are bounded as $t \to -\infty$. Bateman's condition will be met when u represents an expanding wavefront with $u \equiv 0$ ahead of the wavefront: this is the case when we have an IBVP with zero initial conditions.

Assuming that there is no contribution from S_1, and recalling that we take $\boldsymbol{n}$ pointing into B_e, we obtain (cf. (9.9), (9.10))

$$4\pi u(\boldsymbol{r},t) = \int_S \left\{ [u] \frac{\partial}{\partial n_q}\left(\frac{1}{R}\right) - \frac{1}{R}\left[\frac{\partial u}{\partial n}\right] - \frac{1}{cR}\frac{\partial R}{\partial n_q}\left[\frac{\partial u}{\partial t}\right] \right\} \mathrm{d}s_q \tag{9.11}$$

$$= -\int_S \left\{ \frac{1}{R}\left[\frac{\partial u}{\partial n}\right] + \left(\frac{[u]}{R^2} + \frac{1}{cR}\left[\frac{\partial u}{\partial t}\right]\right)\frac{\partial R}{\partial n_q} \right\} \mathrm{d}s_q, \quad P \in B_e. \tag{9.12}$$

This is Kirchhoff's formula for an exterior domain.

The fact that Kirchhoff's formula involves integrals of retarded quantities makes the underlying space-time operators more difficult to handle: 'time and space variables are intertwined' [377, p. 307]. This intertwining can be unravelled using Laplace transforms, as we shall see in Section 9.1.5.

9.1.3 Special Cases of Kirchhoff's Formula

Let us make two simple checks. If u does not depend on t and satisfies $\nabla^2 u = 0$ (so that u is a wavefunction), (9.11) reduces to

$$4\pi u(\boldsymbol{r}) = \int_S \left\{ u \frac{\partial}{\partial n_q}\left(\frac{1}{R}\right) - \frac{1}{R}\frac{\partial u}{\partial n} \right\} \mathrm{d}s_q, \quad P \in B_e, \tag{9.13}$$

which is a standard integral representation for harmonic functions that are regular at infinity, $u = O(r^{-1})$ as $r = |\boldsymbol{r}| \to \infty$.

If we take $u(\boldsymbol{r},t) = U(\boldsymbol{r})\,\mathrm{e}^{-\mathrm{i}\omega t}$,

$$[u] = U\,\mathrm{e}^{-\mathrm{i}\omega(t-R/c)}, \quad \left[\frac{\partial u}{\partial t}\right] = -\mathrm{i}\omega[u], \quad \left[\frac{\partial u}{\partial n}\right] = \frac{\partial U}{\partial n}\,\mathrm{e}^{-\mathrm{i}\omega(t-R/c)}$$

and then (9.11) reduces to

$$4\pi U(\boldsymbol{r}) = \int_S \left\{ U\mathrm{e}^{\mathrm{i}kR} \frac{\partial}{\partial n_q}\left(\frac{1}{R}\right) - \frac{\mathrm{e}^{\mathrm{i}kR}}{R}\frac{\partial U}{\partial n} + \mathrm{i}k\frac{\mathrm{e}^{\mathrm{i}kR}}{R}\frac{\partial R}{\partial n_q}U \right\} \mathrm{d}s_q$$

$$= \int_S \left\{ U \frac{\partial}{\partial n_q}\left(\frac{\mathrm{e}^{\mathrm{i}kR}}{R}\right) - \frac{\mathrm{e}^{\mathrm{i}kR}}{R}\frac{\partial U}{\partial n} \right\} \mathrm{d}s_q, \quad P \in B_e, \tag{9.14}$$

where $k = \omega/c$. Equation (9.14) is Sommerfeld's formula for radiating solutions of the Helmholtz equation [599, eqn (5.36)].

For a third example, suppose that $u = r^{-1}f(t-r/c)$, with the origin inside B. This represents a simple source, (2.11). We have

$$[u] = \frac{f(T)}{r'}, \quad \left[\frac{\partial u}{\partial n}\right] = -\left(\frac{f(T)}{r'^2} + \frac{f'(T)}{r'c}\right)\frac{\partial r}{\partial n}, \quad \left[\frac{\partial u}{\partial t}\right] = \frac{f'(T)}{r'}$$

where $r' = |\boldsymbol{r}'|$ and $T = t - r'/c - R/c$. Hence (9.11) gives

$$\frac{4\pi}{r} f(t-r/c) = \int_S \left\{ \frac{f(T)}{r'} \frac{\partial}{\partial n_q}\left(\frac{1}{R}\right) + \frac{1}{r'R}\left(\frac{f(T)}{r'} + \frac{f'(T)}{c}\right)\frac{\partial r}{\partial n} - \frac{f'(T)}{cr'R}\frac{\partial R}{\partial n_q} \right\} \mathrm{d}s_q.$$

We do not have an independent verification of this formula but we can check a simple case, that with $f(\xi) = \xi$; we find

$$4\pi\left(\frac{t}{r} - \frac{1}{c}\right) = t \int_S \left\{ \frac{1}{r'}\frac{\partial}{\partial n_q}\left(\frac{1}{R}\right) - \frac{1}{R}\frac{\partial}{\partial n_q}\left(\frac{1}{r}\right) \right\} \mathrm{d}s_q$$
$$- \frac{1}{c}\int_S \frac{\partial}{\partial n_q}\left(\frac{1}{R}\right) \mathrm{d}s_q + \frac{1}{c}\int_S \frac{\partial}{\partial n_q}\left(\frac{1}{r}\right) \mathrm{d}s_q.$$

The first integral equals $4\pi/r$, using (9.13) (with $u = 1/r$ therein). The second integral vanishes because $1/R$ is a regular harmonic function inside S (recall that $P \in B_e$). The third integral equals -4π because $1/r$ is harmonic in B_e so that the integral over S can be replaced by the same integral but over a large sphere at infinity, and this integral is readily evaluated.

9.1.4 A Formula for the Pressure

The main quantities of physical interest are the pressure p and the velocity $\boldsymbol{v}$. As p is a wavefunction, we may replace u by p in (9.12) or, alternatively, we may use $p = -\rho\, \partial u/\partial t$. Either way, the result can be written as

$$4\pi p(\boldsymbol{r},t) = \rho \frac{\partial \Phi}{\partial t} - \operatorname{div} \boldsymbol{\Psi}, \quad P \in B_e, \tag{9.15}$$

where

$$\Phi(\boldsymbol{r},t) = \int_S \frac{1}{R}\left[\frac{\partial u}{\partial n}\right] \mathrm{d}s_q, \quad \boldsymbol{\Psi}(\boldsymbol{r},t) = \int_S \frac{1}{R}[p]\, \boldsymbol{n}(q)\, \mathrm{d}s_q.$$

Here, we have used

$$\operatorname{div}\left(\boldsymbol{n}\frac{[p]}{R}\right) = n_i \frac{\partial}{\partial x_i}\left(\frac{[p]}{R}\right) = \left(\frac{[p]}{R^2} + \frac{1}{cR}\left[\frac{\partial p}{\partial t}\right]\right)\frac{\partial R}{\partial n_q},$$

$\partial R/\partial x_i = -\partial R/\partial x_i'$ and a formula for $\partial v/\partial x_i$ similar to (9.3). The formula (9.15) represents the pressure in the exterior fluid in terms of the pressure and the normal velocity on the boundary S. It can be found in a paper by de Hoop [226, eqn (18)]; he also gives a formula for $\boldsymbol{v}$. See also [227, §5.4] and [387, §2.2.7].

9.1.5 Use of Laplace Transforms

Suppose that $u(\boldsymbol{r},t) = 0$ for all $t \le 0$, which is the case for zero initial conditions. Then take the Laplace transform of (9.12) with $\mathscr{L}\{u\}$ defined by (5.1),

$$\mathscr{L}\{u\} = U(\boldsymbol{r},s) = \int_0^\infty u(\boldsymbol{r},t)\mathrm{e}^{-st}\, \mathrm{d}t.$$

We have

$$\mathscr{L}\{[u]\} = \int_0^\infty u(\boldsymbol{r}, t - R/c)\mathrm{e}^{-st}\,\mathrm{d}t = \mathrm{e}^{-sR/c}U(\boldsymbol{r}, s), \quad \mathscr{L}\{[\partial u/\partial t]\} = s\mathrm{e}^{-sR/c}U(\boldsymbol{r}, s). \tag{9.16}$$

Hence the Laplace transform of (9.12) is

$$\begin{aligned} 4\pi U(\boldsymbol{r}, s) &= -\int_S \left\{ \frac{\mathrm{e}^{-sR/c}}{R}\frac{\partial U}{\partial n} + U\mathrm{e}^{-sR/c}\left(\frac{1}{R^2} + \frac{s}{cR}\right)\frac{\partial R}{\partial n_q} \right\} \mathrm{d}s_q \\ &= \int_S \left\{ U\frac{\partial}{\partial n_q}\left(\frac{\mathrm{e}^{-sR/c}}{R}\right) - \frac{\mathrm{e}^{-sR/c}}{R}\frac{\partial U}{\partial n} \right\} \mathrm{d}s_q, \quad P \in B_{\mathrm{e}}. \end{aligned} \tag{9.17}$$

This is a known integral representation for solutions of the modified Helmholtz equation, $\nabla^2 U - (s/c)^2 U = 0$, in B_{e}, assuming those solutions do not grow exponentially at infinity.

9.1.6 Weakening the Smoothness Assumptions

Love [572] was interested in weakening the smoothness (twice differentiable) assumptions in the known derivations of Kirchhoff's formula. We paraphrase his summary [572, §22]:

There are three types of waves with wavefronts, viz., (i) *waves in which neither u, nor any of its partial derivatives of the first order, is discontinuous at the wavefronts;* (ii) *waves in which u is continuous at the wavefronts, but its partial derivatives are not;* (iii) *waves in which u is discontinuous at the wavefronts.*

In case (i) *the conditions which are necessary to secure the validity of known proofs of Kirchhoff's and Poisson's integral formulæ . . . hold good.*

In case (ii) *it is shown that . . . the validity of Kirchhoff's and Poisson's integral formulæ is . . . unimpaired by the existence of a wavefront involving discontinuity in the temporal and spatial gradients of the disturbance.*

In the class (iii), *in which u is discontinuous at a wavefront, Kirchhoff's and Poisson's integral formulæ cannot be applied to the calculation of u.*

Case (i) is noted explicitly by Bateman [67, §2.61]. Love was especially interested in case (ii), and he argued that it is sufficient to have continuous u (so that the jump $[\![u]\!] = 0$) but discontinuous first partial derivatives. In fact, for zero initial conditions, we can see that his result is true by using the Laplace transform $\mathscr{L}$. Thus, as $[\![u]\!] = 0$, we know (Section 5.5) that application of $\mathscr{L}$ to the wave equation gives the modified Helmholtz equation for which we have a known integral representation, namely (9.17); inverting $\mathscr{L}$ then gives the Kirchhoff formula.

For a weak version of Kirchhoff's formula (assuming zero initial conditions), see [513].

9.1.7 Literature

The formula (9.9) was first published by Gustav Kirchhoff (1824–1887) in 1882 [482, p. 646] and reprinted one year later [483, p. 669]. He gave a similar proof in his book [484, pp. 22–27], and it is this proof that is given by Stratton [799, §8.1]. Kirchhoff's paper [482] has been translated [485] and analysed [138].

We have already noted that Baker & Copson [38] modelled their proof on a paper by Gutzmer [371] published in 1895. Love [572, p. 40] credits another paper from 1895 for the essentials of his proof: he cites Beltrami's short note [88].

The exterior formula (9.12) has been used in order to truncate computational domains. Thus enclose the scatterer S by a surface $\mathscr{S}$; this surface bounds the computational domain. The problem is to devise a boundary condition on $\mathscr{S}$ so that waves can pass through $\mathscr{S}$ without inducing spurious reflections. Ting & Miksis [826] proposed introducing a second surface $\mathscr{S}_2$ inside $\mathscr{S}$ but still enclosing S, and then expressed u outside $\mathscr{S}_2$ using Kirchhoff's formula (9.12). In particular, this formula can be used for u on $\mathscr{S}$ and, moreover, this formula is explicit if $\mathscr{S}_2$ is close to $\mathscr{S}$; this exploits the retarded arguments appearing on the right-hand side of (9.12). The resulting *non-reflecting boundary condition* was implemented by Givoli & Cohen [344]. They found that the basic scheme exhibits a long-time instability. Such instabilities are also commonly observed when time-domain boundary integral equations are discretised, as we shall see later (Section 10.3.2).

Kirchhoff's formula is applicable to the homogeneous wave equation, $\Box^2 u = 0$. It can be generalised to the inhomogeneous (forced) wave equation, $\Box^2 u = -q$, where q is a given function. In effect, one adds a particular solution, constructed using a retarded volume potential (4.34). The resulting formula can be found in Bateman's first book [62, p. 147, example 3]; he attributes it to Kirchhoff, but it is not in Kirchhoff's 1882 paper [482]. It can be found in Lorentz's book [571, p. 238, eqn (10)]. For later occurrences, see [67, p. 185], [799, p. 427], [212, eqn (2.4)], and [425]. For the convected wave equation (1.26), see [24, 539, 427].

It is possible to derive a version of Kirchhoff's formula when the surface S is moving, but, inevitably, it is much more complicated; see Section 9.3.2.

9.2 Kirchhoff's Formula: a Space-Time Derivation

It is possible to derive Kirchhoff's formula by direct computation in space-time using the causal fundamental solution (4.39), $\mathscr{G}(\boldsymbol{r},t) = (4\pi r)^{-1}\,\delta(t - r/c)$, together with the four-dimensional version of Green's formula (3.33),

$$\int_\Sigma \left(v\Box^2 u - u\Box^2 v\right) \mathrm{d}\Sigma = \int_{\mathscr{S}} \left(v\frac{\partial u}{\partial T} - u\frac{\partial v}{\partial T}\right) \mathrm{d}\mathscr{S}. \tag{9.18}$$

In this formula, Σ is a bounded region of space-time bounded by a hypersurface $\mathscr{S}$. The outward unit normal 4-vector to $\mathscr{S}$ is $\mathbf{N} = (N_1, N_2, N_3, N_4)$ and, from (3.34),

$$\frac{\partial u}{\partial T} = N_i \frac{\partial u}{\partial x_i} - \frac{N_4}{c}\frac{\partial u}{\partial t}.$$

As in Section 4.1, choose $\Sigma = B_\mathrm{e} \times \mathbb{T}$, where $\mathbb{T}$ is the time interval, $0 < t \le T$, with

$T > 0$. Thus Σ is a hypercylinder in space-time. The 'base' of Σ is $B_e \times \{0\}$; this is where the two initial conditions are given. The 'top' of Σ is $B_e \times \{T\}$. The lateral boundary of Σ is $S \times \mathbb{T}$; this is where the boundary condition is given.

Suppose that $u(\boldsymbol{r},t)$ is a wavefunction, $\Box^2 u = 0$. Take $v(\boldsymbol{r},t) = \mathscr{G}(\boldsymbol{r}' - \boldsymbol{r}, t' - t)$, where $(\boldsymbol{r}', t')$ locates a point in Σ. By construction, $\Box^2 v(\boldsymbol{r},t) = -\delta(\boldsymbol{r} - \boldsymbol{r}')\,\delta(t - t')$; see Section 4.7.1. Hence the left-hand side of (9.18) reduces to $u(\boldsymbol{r}', t')$.

Now, consider the right-hand side of (9.18). Assume that u solves an IBVP with zero initial conditions. Then there is no contribution from the base of Σ on which $\mathbf{N} = (0,0,0,-1)$. There is no contribution from the top of Σ (on which $\mathbf{N} = (0,0,0,1)$) because $\mathscr{G}(\boldsymbol{r}' - \boldsymbol{r}, t' - t) \equiv 0$ for $t > t'$. On the lateral boundary of Σ, the outward normal is $\mathbf{N} = (-n_1, -n_2, -n_3, 0)$ where $\boldsymbol{n}$ is the usual unit normal 3-vector on S pointing into B_e, whence $\partial u/\partial T = -\partial u/\partial n$. Thus, switching $(\boldsymbol{r},t)$ and $(\boldsymbol{r}',t')$, (9.18) becomes

$$u(\boldsymbol{r},t) = \int_S \int_0^T \left(u \frac{\partial \mathscr{G}}{\partial n_q} - \mathscr{G} \frac{\partial u}{\partial n} \right) \mathrm{d}t' \, \mathrm{d}s_q, \tag{9.19}$$

where $q \in S$ is at $\boldsymbol{r}'$ and

$$\mathscr{G}(\boldsymbol{r} - \boldsymbol{r}', t - t') = (4\pi R)^{-1}\,\delta(t - t' - R/c) = (4\pi R)^{-1}\,\delta(t' - t + R/c)$$

with $R = |\boldsymbol{r} - \boldsymbol{r}'|$. We have

$$4\pi \frac{\partial \mathscr{G}}{\partial n_q} = \left(\frac{1}{cR}\,\delta'(t' - t + R/c) - \frac{1}{R^2}\,\delta(t' - t + R/c) \right) \frac{\partial R}{\partial n_q}$$

and $f(t)\,\delta'(t - t_0) = -f'(t)\,\delta(t - t_0)$. Hence (9.19) becomes

$$4\pi u(\boldsymbol{r},t) = -\int_S \int_0^T \left\{ \frac{1}{R}\frac{\partial u}{\partial n} + \left(\frac{u}{R^2} + \frac{1}{cR}\frac{\partial u}{\partial t} \right) \frac{\partial R}{\partial n_q} \right\} \delta(t' - t + R/c)\,\mathrm{d}t' \, \mathrm{d}s_q.$$

Evaluating the integral over t' using the sifting property of the delta function, we obtain Kirchhoff's formula for an exterior domain (9.12).

The derivation given here is essentially that given by Morse & Feshbach [643, §7.3]. See also [451, §1.17], [360], [198, §2.2.1] and [164]. The formula (9.19), involving integrals over $S \times \mathbb{T}$, has been used as the starting point for the development of a numerical method for solving IBVPs [693, 694].

9.3 Kirchhoff's Formula: Use of Generalised Functions

Another way of deriving Kirchhoff's formula uses generalised functions and their properties. (See Section 4.8.1 for some discussion and references on this topic.) We outline this approach, giving a derivation of Kirchhoff's formula for an exterior domain, (9.12), in Section 9.3.1. We then use the same approach to obtain a version of Kirchhoff's formula for a moving surface in Section 9.3.2.

We define such a surface as follows. Given a function $\varphi(\boldsymbol{r},t)$, $\varphi =$ constant defines a moving surface in space. In particular, $\varphi = 0$ defines a surface $S(t)$: S is a *level set* of φ. Assume that S is closed, that the exterior of S is unbounded (denoted, as usual, by B_e), and that $\varphi > 0$ for $\boldsymbol{r} \in B_e$, with $\varphi < 0$ inside S (in B). A normal to S is $\operatorname{grad}\varphi$ evaluated at $\varphi = 0$, so that the unit normal to S, pointing into B_e, is $\boldsymbol{n} = \operatorname{grad}\varphi/|\operatorname{grad}\varphi|$ evaluated at $\varphi = 0$.

When S is moving, we have

$$0 = \frac{\mathrm{d}\varphi}{\mathrm{d}t} = \frac{\partial\varphi}{\partial t} + \frac{\partial\varphi}{\partial x_i}\frac{\mathrm{d}x_i}{\mathrm{d}t} = \frac{\partial\varphi}{\partial t} + |\operatorname{grad}\varphi|\,\bar{v}_n, \tag{9.20}$$

where $\bar{v}_n$ is the normal velocity at $\boldsymbol{r}$ on the surface S.

Next, let $\mathrm{H}(x)$ denote the Heaviside unit function, defined by $\mathrm{H}(x) = 1$ for $x > 0$ and $\mathrm{H}(x) = 0$ for $x < 0$. We shall regard H as a generalised function (distribution), so that $\mathrm{H}'(x) = \delta(x)$ [750, eqn (II,2;16)].

Given that u satisfies $\Box^2 u = 0$ in B_e, let us find an equation for $u(\boldsymbol{r},t)\,\mathrm{H}(\varphi(\boldsymbol{r},t))$ that is valid everywhere in space, including in $B \cup S$. We will see that $u\mathrm{H}$ satisfies a forced wave equation, an equation that we can solve using the methods of Section 4.7.

In what follows, derivatives are to be interpreted in the sense of generalised functions when necessary. We start with

$$\frac{\partial}{\partial x_i}(u\,\mathrm{H}(\varphi)) = \frac{\partial u}{\partial x_i}\mathrm{H}(\varphi) + u\frac{\partial \mathrm{H}(\varphi)}{\partial x_i}. \tag{9.21}$$

Note that

$$\frac{\partial \mathrm{H}(\varphi)}{\partial x_i} = \mathrm{H}'(\varphi)\frac{\partial\varphi}{\partial x_i} = \frac{\partial\varphi}{\partial x_i}\,\delta(\varphi), \tag{9.22}$$

which vanishes everywhere except on S. Differentiating (9.21) gives

$$\frac{\partial^2}{\partial x_i\,\partial x_j}(u\,\mathrm{H}(\varphi)) = \frac{\partial^2 u}{\partial x_i\,\partial x_j}\mathrm{H}(\varphi) + \frac{\partial u}{\partial x_i}\frac{\partial \mathrm{H}(\varphi)}{\partial x_j} + \frac{\partial}{\partial x_j}\left(u\frac{\partial \mathrm{H}(\varphi)}{\partial x_i}\right),$$

whence

$$\nabla^2(u\mathrm{H}(\varphi)) = \mathrm{H}(\varphi)\nabla^2 u + \frac{\partial u}{\partial x_i}\frac{\partial \mathrm{H}(\varphi)}{\partial x_i} + \frac{\partial}{\partial x_i}\left(u\frac{\partial \mathrm{H}(\varphi)}{\partial x_i}\right). \tag{9.23}$$

Similarly,

$$\frac{\partial^2}{\partial t^2}(u\mathrm{H}(\varphi)) = \mathrm{H}(\varphi)\frac{\partial^2 u}{\partial t^2} + \frac{\partial u}{\partial t}\frac{\partial \mathrm{H}(\varphi)}{\partial t} + \frac{\partial}{\partial t}\left(u\frac{\partial \mathrm{H}(\varphi)}{\partial t}\right). \tag{9.24}$$

Also, making use of (9.20) and (9.22),

$$\frac{\partial \mathrm{H}(\varphi)}{\partial t} = \mathrm{H}'(\varphi)\frac{\partial\varphi}{\partial t} = \frac{\partial\varphi}{\partial t}\,\delta(\varphi) = -\bar{v}_i\frac{\partial\varphi}{\partial x_i}\,\delta(\varphi) = -\bar{v}_i\frac{\partial \mathrm{H}(\varphi)}{\partial x_i}, \tag{9.25}$$

where $\bar{\boldsymbol{v}}(\boldsymbol{r},t)$ is the velocity of a point at $\boldsymbol{r} \in S(t)$. Here, the third equality is valid because we only need $\partial\varphi/\partial t$ when $\varphi = 0$, and that is given by (9.20).

Then, multiplying $\Box^2 u = 0$ by $\mathrm{H}(\varphi)$, making use of (9.23) and (9.24), we obtain

$$\Box^2(u\,\mathrm{H}(\varphi)) = -f \tag{9.26}$$

where

$$f(\boldsymbol{r},t) = -v_i \frac{\partial \mathrm{H}(\varphi)}{\partial x_i} - \frac{\partial}{\partial x_i}\left(u \frac{\partial \mathrm{H}(\varphi)}{\partial x_i}\right) + \frac{1}{c^2}\frac{\partial u}{\partial t}\frac{\partial \mathrm{H}(\varphi)}{\partial t} + \frac{1}{c^2}\frac{\partial}{\partial t}\left(u \frac{\partial \mathrm{H}(\varphi)}{\partial t}\right)$$
$$= -\left(v_i + \frac{\bar{v}_i}{c^2}\frac{\partial u}{\partial t}\right)\frac{\partial \mathrm{H}(\varphi)}{\partial x_i} - \frac{\partial}{\partial x_i}\left(u \frac{\partial \mathrm{H}(\varphi)}{\partial x_i}\right) - \frac{1}{c^2}\frac{\partial}{\partial t}\left(u\bar{v}_i \frac{\partial \mathrm{H}(\varphi)}{\partial x_i}\right) \tag{9.27}$$

and $v_i(\boldsymbol{r},t) = \partial u/\partial x_i$. Equations (9.26) and (9.27) were derived by Ffowcs Williams & Hawkings [295, eqn (2.10)].

9.3.1 Fixed Surface S

Suppose that S is a *fixed* surface so that φ does not depend on t and $\bar{\boldsymbol{v}} = \boldsymbol{0}$. (Moving surfaces will be considered in Section 9.3.2.) The forcing function in (9.26) reduces to

$$f(\boldsymbol{r},t) = -v_i \frac{\partial \mathrm{H}(\varphi)}{\partial x_i} - \frac{\partial}{\partial x_i}\left(u \frac{\partial \mathrm{H}(\varphi)}{\partial x_i}\right).$$

Solving (9.26) for $u\mathrm{H}$, using (4.34), we obtain

$$u(\boldsymbol{r},t)\,\mathrm{H}(\varphi(\boldsymbol{r})) = -\int v_i(\boldsymbol{r}',t-R/c)\frac{\partial}{\partial x_i'}\mathrm{H}(\varphi(\boldsymbol{r}'))\,\frac{\mathrm{d}\boldsymbol{r}'}{4\pi R}$$
$$- \frac{\partial}{\partial x_i}\int u(\boldsymbol{r}',t-R/c)\frac{\partial}{\partial x_i'}\mathrm{H}(\varphi(\boldsymbol{r}'))\,\frac{\mathrm{d}\boldsymbol{r}'}{4\pi R}$$

with $R = |\boldsymbol{r}-\boldsymbol{r}'|$. As a consequence of (9.22), the volume integrals can be reduced to surface integrals over S ($\varphi = 0$). The relevant formula is

$$\int F(\boldsymbol{r})\,\frac{\partial}{\partial x_i}\mathrm{H}(\varphi(\boldsymbol{r}))\,\mathrm{d}\boldsymbol{r} = \int F(\boldsymbol{r})\,\frac{\partial \varphi}{\partial x_i}\,\delta(\varphi(\boldsymbol{r}))\,\mathrm{d}\boldsymbol{r} = \int_S F(\boldsymbol{r})\,n_i(\boldsymbol{r})\,\mathrm{d}s(\boldsymbol{r}), \tag{9.28}$$

where F is a smooth function; see [452, p. 263], [422, §2.3.1] or [420, §2.1.1]. We obtain

$$u(\boldsymbol{r},t)\,\mathrm{H}(\varphi(\boldsymbol{r})) = -\int_S v_i(\boldsymbol{r}',t-R/c)\,n_i(\boldsymbol{r}')\,\frac{\mathrm{d}s(\boldsymbol{r}')}{4\pi R}$$
$$- \frac{\partial}{\partial x_i}\int_S u(\boldsymbol{r}',t-R/c)\,n_i(\boldsymbol{r}')\,\frac{\mathrm{d}s(\boldsymbol{r}')}{4\pi R}.$$

Evaluating the derivative in the second term, we obtain

$$4\pi u(\boldsymbol{r},t)\,\mathrm{H}(\varphi(\boldsymbol{r})) = -\int_S \left[\frac{\partial u}{\partial n}\right]\frac{\mathrm{d}s_q}{R} - \int_S \left(\frac{[u]}{R^2} + \frac{1}{cR}\left[\frac{\partial u}{\partial t}\right]\right)\frac{\partial R}{\partial n_q}\,\mathrm{d}s_q \tag{9.29}$$

where, as usual, the point $q \in S$ has position vector $\boldsymbol{r}'$ and the square brackets denote retarded values, as defined in (9.2). When restricted to points outside S (where $\mathrm{H}(\varphi) = 1$), (9.29) reduces to Kirchhoff's formula for an exterior domain, (9.12).

The method used above is probably not to every reader's taste but it often leads to quick derivations of formulas that can be derived by other methods. It is used extensively in Howe's books, such as [421, 422, 420], although Howe is aware of the discomfort felt by some readers [420, p. 36]. Its use for acoustic problems began with the 1969 paper by Ffowcs Williams & Hawkings [295] cited above. For (9.29) itself, see Farassat & Myers [286, §2]. In the next section, we shall use the same basic method so as to extend Kirchhoff's formula to moving surfaces.

9.3.2 Moving Surface $S(t)$

Suppose now that $S(t)$ is a moving surface, defined by $\varphi(\boldsymbol{r},t)=0$. The velocity at any point $\boldsymbol{r}\in S(t)$ is $\bar{\boldsymbol{v}}(\boldsymbol{r},t)$. If we proceed as in Section 9.3.1, using (9.26) and (9.27), we arrive at

$$\begin{aligned}4\pi u(\boldsymbol{r},t)\,\mathrm{H}(\varphi(\boldsymbol{r},t)) = &-\int_{S(\tau)}\left[\frac{\partial u}{\partial n}+\frac{M_n}{c}\frac{\partial u}{\partial t}\right]\frac{\mathrm{d}s_q}{R}\\ &-\frac{\partial}{\partial x_i}\int_{S(\tau)}[u]\,n_i\frac{\mathrm{d}s_q}{R}-\frac{1}{c}\frac{\partial}{\partial t}\int_{S(\tau)}[uM_n]\,\frac{\mathrm{d}s_q}{R}.\end{aligned} \tag{9.30}$$

In this formula, $\mathrm{d}s_q=\mathrm{d}s(\boldsymbol{r}')$, $R=|\boldsymbol{r}-\boldsymbol{r}'|$, $M_n=c^{-1}\bar{\boldsymbol{v}}\cdot\boldsymbol{n}$ is the local normal Mach number, $\boldsymbol{n}$ is the unit normal vector on the retarded surface $S(\tau)$, and the square brackets denote evaluation at the retarded time $\tau=t-R/c$: given $g(\boldsymbol{r}',t)$, $[g]=g(\boldsymbol{r}',\tau)$. The retarded surface $S(\tau)$ is defined as follows: given a point $\boldsymbol{r}$ and a time t, $S(\tau)$ is found by solving

$$\varphi(\boldsymbol{r}',\tau)=\varphi(\boldsymbol{r}',t-|\boldsymbol{r}-\boldsymbol{r}'|/c)=0 \quad \text{for all points } \boldsymbol{r}'.$$

One version of the *Ffowcs Williams–Hawkings equation* has been derived in a similar manner to (9.30); see [421, §2.2.1], [651, eqn (2.1)] and [345, Chapter 5].

The formula (9.30) is based on a fixed coordinate system associated with the position vector $\boldsymbol{r}$. However, as $S(t)$ is moving, it is often more convenient to introduce a coordinate system $\boldsymbol{\eta}$ that moves with $S(t)$. To fix ideas (here, we are following [295, §2] and [421, §2.2.2]), start by considering $\Box^2\psi(\boldsymbol{r},t)=-f(\boldsymbol{r},t)$ with solution

$$\begin{aligned}\psi(\boldsymbol{r},t)&=\int f(\boldsymbol{r}',t-R/c)\,\frac{\mathrm{d}\boldsymbol{r}'}{4\pi R}\\ &=\int_{-\infty}^{t}\int f(\boldsymbol{r}',\tau)\,\delta(\tau-t+R/c)\,\frac{\mathrm{d}\boldsymbol{r}'}{4\pi R}\,\mathrm{d}\tau.\end{aligned} \tag{9.31}$$

The source element at $\boldsymbol{r}'$ has a certain velocity with respect to the fixed system $\boldsymbol{r}$; denote this velocity by $c\boldsymbol{M}(\boldsymbol{\eta},\tau)$. Thus

$$\frac{\partial\boldsymbol{r}'}{\partial\tau}=c\boldsymbol{M}, \quad \boldsymbol{r}'(\boldsymbol{\eta},\tau)=\boldsymbol{\eta}+\int_{\tau_1}^{\tau}c\boldsymbol{M}(\boldsymbol{\eta},\tau')\,\mathrm{d}\tau',$$

using $\boldsymbol{r}'(\boldsymbol{\eta},\tau_1)=\boldsymbol{\eta}$ for some τ_1. (We also have $\boldsymbol{r}'=\boldsymbol{\eta}$ when $\boldsymbol{M}=\boldsymbol{0}$.) Denoting the Jacobian of the change of variables from $\boldsymbol{r}'$ to $\boldsymbol{\eta}$ by $J(\boldsymbol{\eta},\tau)$, we obtain

$$\psi(\boldsymbol{r},t)=\int_{-\infty}^{t}\int \tilde{f}(\boldsymbol{\eta},\tau)\,\delta(\tau-t+R/c)\,\frac{J\,\mathrm{d}\boldsymbol{\eta}}{4\pi R}\,\mathrm{d}\tau, \tag{9.32}$$

where $\tilde{f}(\boldsymbol{\eta},\tau) = f(\boldsymbol{r}'(\boldsymbol{\eta},\tau),\tau)$. Notice that R has become a function of τ: $R = |\boldsymbol{R}|$ with

$$\boldsymbol{R}(\boldsymbol{r};\boldsymbol{\eta},\tau) = \boldsymbol{r} - \boldsymbol{r}' = \boldsymbol{r} - \boldsymbol{\eta} - \int_{\tau_1}^{\tau} c\boldsymbol{M}(\boldsymbol{\eta},\tau')\,\mathrm{d}\tau'. \tag{9.33}$$

This is relevant when we evaluate the integral with respect to τ in (9.32); we do this next.

Suppressing the dependence on $\boldsymbol{r}$ and $\boldsymbol{\eta}$, denote the argument of the delta function in (9.32) by $g(\tau) = \tau - t + c^{-1}R(\tau)$. For subsonic motions, with $|\boldsymbol{M}| < 1$, $g(\tau)$ is an increasing function of τ with exactly one zero. To see this, we need $R'(\tau)$. From (9.33), $\boldsymbol{R}'(\tau) = -c\boldsymbol{M}$, and then

$$2RR' = \left(R^2\right)' = (\boldsymbol{R}\cdot\boldsymbol{R})' = 2\boldsymbol{R}\cdot\boldsymbol{R}' = -2c\boldsymbol{M}\cdot\boldsymbol{R}.$$

Hence $R' = -cM_r$ where $M_r = \boldsymbol{M}\cdot\boldsymbol{R}/R$ is the component of $\boldsymbol{M}$ in the direction of $\boldsymbol{R}$. Thus $g'(\tau) = 1 + R'/c = 1 - M_r > 0$ and so $g(\tau)$ has exactly one zero, at $\tau = \tau_{\mathrm{e}}$, say, known as the *emission time*. Using the property $\delta(g(\tau)) = \delta(\tau - \tau_{\mathrm{e}})/g'(\tau_{\mathrm{e}})$ [452, §8.2], (9.32) becomes

$$\psi(\boldsymbol{r},t) = \int \tilde{f}(\boldsymbol{\eta},\tau_{\mathrm{e}})\,\frac{J(\boldsymbol{\eta},\tau_{\mathrm{e}})\,\mathrm{d}\boldsymbol{\eta}}{4\pi R(1-M_r)}, \tag{9.34}$$

where R and M_r are evaulated at τ_{e}. This formula agrees with [295, eqn (3.21)] and [421, eqn (2.2.8)]; similar formulas were obtained by Lowson [575].

In more detail, τ_{e} is defined by solving

$$0 = g(\tau_{\mathrm{e}}) = \tau_{\mathrm{e}} - t - c^{-1}\,|\boldsymbol{R}(\boldsymbol{r};\boldsymbol{\eta},\tau_{\mathrm{e}})| \tag{9.35}$$

so that $\tau_{\mathrm{e}}(\boldsymbol{r},t;\boldsymbol{\eta})$ depends on $\boldsymbol{r}$, t and $\boldsymbol{\eta}$, as do J, R and M_r in (9.34).

Let us apply these results to evaluate (9.31) when f is given by (9.27). Evidently, we require

$$\varphi(\boldsymbol{r}'(\boldsymbol{\eta},\tau_{\mathrm{e}}),\tau_{\mathrm{e}}) = \tilde{\varphi}(\boldsymbol{\eta}) \equiv \tilde{\varphi}(\boldsymbol{r},t;\boldsymbol{\eta}),$$

say, after substituting $\tau_{\mathrm{e}}(\boldsymbol{r},t;\boldsymbol{\eta})$. So, given $\boldsymbol{r}$ and t, $\tilde{\varphi}(\boldsymbol{\eta}) = 0$ defines a certain fixed surface in $\boldsymbol{\eta}$-space, denoted by $S_0(\boldsymbol{r},t)$. A typical point $\boldsymbol{\eta}$ on this surface corresponds to a point $\boldsymbol{r}' \in S(\tau)$ at time $\tau = \tau_{\mathrm{e}}(\boldsymbol{r},t;\boldsymbol{\eta})$.

Next, consider one piece of $f(\boldsymbol{r},t)$ in (9.27), namely $f_1(\boldsymbol{r},t) = u\,\partial\mathrm{H}(\varphi)/\partial x_i$. When substituted in (9.34), the integrand is seen to contain

$$\tilde{f}_1(\boldsymbol{\eta},\tau_{\mathrm{e}}) = \tilde{u}(\boldsymbol{\eta},\tau_{\mathrm{e}})\,\frac{\partial}{\partial x_i'}\mathrm{H}(\tilde{\varphi}(\boldsymbol{\eta})) = \tilde{u}(\boldsymbol{\eta},\tau_{\mathrm{e}})\,\frac{\partial \eta_j}{\partial x_i'}\,\frac{\partial}{\partial \eta_j}\mathrm{H}(\tilde{\varphi}(\boldsymbol{\eta})),$$

whence (9.28) gives

$$\int \tilde{f}_1(\boldsymbol{\eta},\tau_{\mathrm{e}})\,\frac{J(\boldsymbol{\eta},\tau_{\mathrm{e}})\,\mathrm{d}\boldsymbol{\eta}}{4\pi R(1-M_r)} = \int_{S_0} \tilde{u}(\boldsymbol{\eta},\tau_{\mathrm{e}})\,\frac{\partial \eta_j}{\partial x_i'}\,\tilde{n}_j\,\frac{J(\boldsymbol{\eta},\tau_{\mathrm{e}})\,\mathrm{d}s(\boldsymbol{\eta})}{4\pi R(1-M_r)}, \tag{9.36}$$

where $\tilde{\boldsymbol{n}}$ is the unit normal vector on S_0. We have

$$|\mathrm{grad}_{\boldsymbol{\eta}}\,\tilde{\varphi}|\,\tilde{n}_j\,\frac{\partial \eta_j}{\partial x_i'} = \frac{\partial \tilde{\varphi}}{\partial \eta_j}\,\frac{\partial \eta_j}{\partial x_i'} = \frac{\partial \tilde{\varphi}}{\partial x_i'} = \left.\frac{\partial \varphi}{\partial x_i'}\right|_{\tau=\tau_{\mathrm{e}}} = |\mathrm{grad}_{\boldsymbol{r}'}\,\varphi|\,n_i,$$

where $\boldsymbol{n}$ is the unit normal at $\boldsymbol{r}' \in S(\tau_e)$. This reduces the right-hand side of (9.36) to

$$\int_{S_0} \tilde{u}(\boldsymbol{\eta}, \tau_e) \frac{n_i A \, \mathrm{d}s(\boldsymbol{\eta})}{4\pi R(1-M_r)} \quad \text{with} \quad A = J(\boldsymbol{\eta}, \tau) \left. \frac{|\mathrm{grad}_{r'} \varphi(\boldsymbol{r}', \tau)|}{|\mathrm{grad}_{\boldsymbol{\eta}} \tilde{\varphi}(\boldsymbol{\eta})|} \right|_{\tau=\tau_e};$$

for some discussion of the quantity A, see [295, p. 330], [453, p. 217] or [421, p. 114]. Using this result for each term in (9.27), we obtain

$$\begin{aligned} 4\pi u(\boldsymbol{r},t)\mathrm{H}(\varphi(\boldsymbol{r},t)) = &-\frac{\partial}{\partial x_i} \int_{S_0(\boldsymbol{r},t)} \left[\frac{u n_i A}{R(1-M_r)} \right] \mathrm{d}s(\boldsymbol{\eta}) \\ &- \frac{1}{c^2} \frac{\partial}{\partial t} \int_{S_0(\boldsymbol{r},t)} \left[\frac{u \bar{v}_n A}{R(1-M_r)} \right] \mathrm{d}s(\boldsymbol{\eta}) \\ &- \int_{S_0} \left[\left(\frac{\partial u}{\partial n} + \frac{\bar{v}_n}{c^2} \frac{\partial u}{\partial \tau} \right) \frac{A}{R(1-M_r)} \right] \mathrm{d}s(\boldsymbol{\eta}), \end{aligned} \tag{9.37}$$

where the square brackets denote evaluation at $\tau = \tau_e$ and $\bar{v}_n$ is the surface normal velocity. Equation (9.37) agrees with [295, eqn (5.3)]; see also [286, §3] and [583, §3].

It is tempting to try to evaluate the derivatives outside the surface integrals in (9.37), but this is not straightforward because of the dependence of S_0 on $\boldsymbol{r}$ and t. For some indications of the difficulties, see [152].

Early attempts to extend Kirchhoff's formula to moving surfaces were made by Morgans [639] in 1930 and by Khromov [477] in 1963, although their formulas are imperfect [295, 286]. Applications had to wait until the late 1960s, when noise generation by jet aircraft became a subject of intense study leading to what is now known as *computational aeroacoustics*. We mention a few more papers [640, 894, 24, 704] and five reviews, [453, §6] and [583, 284, 128, 865].

9.4 Layer Potentials

As in classical potential theory (for Laplace's equation), we introduce layer potentials for the wave equation. Apparently, this was first done by Larmor [519], although the relevant integrals occur in Kirchhoff's formula. Larmor also gave the jump conditions; see also Baker & Copson [38, §5.4]. For more modern treatments, see [377, 198, 742, 199].

9.4.1 Single-Layer Potential

We define a *single-layer* potential by

$$(\mathrm{S}\mu)(\boldsymbol{r},t) = \int_S \mu(\boldsymbol{r}', t - R/c)\, G_0(R)\, \mathrm{d}s(\boldsymbol{r}') = -\frac{1}{2\pi} \int_S \frac{[\mu]}{R}\, \mathrm{d}s_q, \tag{9.38}$$

where $R = |\boldsymbol{r} - \boldsymbol{r}'|$ and

$$G_0(R) = -(2\pi R)^{-1}$$

is a fundamental solution for Laplace's equation.

With $\boldsymbol{r}'$ as the position vector of a point $q \in S$, we will also write $\mu(q,t)$ for $\mu(\boldsymbol{r}',t)$. Similarly, with $\boldsymbol{r}$ as the position vector of a point P, we will also write $(\mathrm{S}\mu)(P,t)$ for $(\mathrm{S}\mu)(\boldsymbol{r},t)$.

Given $\mu(q,t)$ for $q \in S$, $(\mathrm{S}\mu)(P,t)$ is a wavefunction for $P \notin S$; we can think of $\mathrm{S}\mu$ as being generated by a distribution of simple sources (2.12) over the surface S. Note that the integrand in (9.38) is a wavefunction when considered as a function of x, y, z and t.

In the far field, where $r = |\boldsymbol{r}|$ is much larger than the diameter of S, we have

$$(\mathrm{S}\mu)(P,t) \sim r^{-1} f(t - r/c) \quad \text{as } r \to \infty, \text{ with} \quad f(\xi) = -\frac{1}{2\pi}\int_S \mu(q,\xi)\,\mathrm{d}s_q. \tag{9.39}$$

Thus $\mathrm{S}\mu$ behaves as a simple source in the far field, with no dependence on direction $\widehat{\boldsymbol{r}} = \boldsymbol{r}/r$. This result holds for fixed t. However, the corresponding *far-field pattern* does contain directional information; its definition, (2.18), requires that both r and t are large:

$$f_0(\widehat{\boldsymbol{r}},t) = \lim_{r\to\infty}\{r(\mathrm{S}\mu)(\boldsymbol{r},t+r/c)\} = -\frac{1}{2\pi}\int_S \mu(\boldsymbol{r}', t+\widehat{\boldsymbol{r}}\cdot\boldsymbol{r}'/c)\,\mathrm{d}s(\boldsymbol{r}'), \tag{9.40}$$

where we have used $r - R \sim \widehat{\boldsymbol{r}}\cdot\boldsymbol{r}'$ as $r \to \infty$.

The presence of G_0 means that the jump behaviour of acoustic layer potentials is the same as with classical layer potentials [474]. In particular, $(\mathrm{S}\mu)(P,t)$ is continuous in P as P crosses S but the normal derivative of $\mathrm{S}\mu$ is discontinuous across S with values on S given by

$$\frac{\partial}{\partial n_p}\mathrm{S}\mu = (\pm I + K)\mu, \tag{9.41}$$

where the upper (lower) sign corresponds to $P \to p \in S$ from the exterior (interior) of S, that is, from B_e (B). The normal derivative is in the outward direction ($\boldsymbol{n}$ points into B_e), I is the identity operator and K is defined by

$$\begin{aligned}(K\mu)(p,t) &= \int_S \frac{\partial}{\partial n_p}\{\mu(\boldsymbol{r}', t-R/c)\,G_0(R)\}\,\mathrm{d}s(\boldsymbol{r}')\\ &= -\frac{1}{2\pi}\int_S n_i(p)\frac{\partial}{\partial x_i}\{\mu(q,t-R/c)R^{-1}\}\,\mathrm{d}s_q\\ &= \frac{1}{2\pi}\int_S\left\{\frac{1}{cR_0}\dot{\mu}(q,t-R_0/c) + \frac{1}{R_0^2}\mu(q,t-R_0/c)\right\}\frac{\partial R}{\partial n_p}\,\mathrm{d}s_q \end{aligned}\tag{9.42}$$

where

$$\dot{\mu}(q,t) = \frac{\partial \mu}{\partial t}, \tag{9.43}$$

$$\frac{\partial R}{\partial n_p} = n_i(p)\frac{\partial R}{\partial x_i} = \frac{1}{R_0}(x_i - x_i')n_i(p) = \frac{(\boldsymbol{r}-\boldsymbol{r}')\cdot\boldsymbol{n}(\boldsymbol{r})}{|\boldsymbol{r}-\boldsymbol{r}'|}$$

and $R_0 = |p - q| = |\boldsymbol{r} - \boldsymbol{r}'|$ is the distance between $p \in S$ and $q \in S$.

Some authors prefer to write layer potentials in space-time using the causal fundamental solution (4.39), $\mathscr{G}(\boldsymbol{r},t) = (4\pi r)^{-1}\,\delta(t - r/c)$. For example, we can write the single-layer potential (9.38) as

$$(\mathrm{S}\mu)(\boldsymbol{r},t) = -2\int_S \int_0^t \mu(\boldsymbol{r}',t')\,\mathscr{G}(\boldsymbol{r}-\boldsymbol{r}',t-t')\,\mathrm{d}t'\,\mathrm{d}s(\boldsymbol{r}'); \tag{9.44}$$

see, for example, [372, eqn (2.2)]. In the formula (9.44), 'it is important to keep in mind the fact that the [upper] limit [in the t'-integral] is t^+ rather than just t' where $t^+ = t + \varepsilon$ and ε is small and positive, thus ensuring that the endpoint is not 'exactly at the peak of a delta function' [643, p. 836].

9.4.2 Double-Layer Potential

In classical potential theory, the double-layer potential is defined by

$$\int_S \nu(q)\,\frac{\partial G_0}{\partial n_q}\,\mathrm{d}s_q = \frac{1}{2\pi}\int_S \frac{\nu(q)}{R^2}\,\frac{\partial R}{\partial n_q}\,\mathrm{d}s_q, \tag{9.45}$$

where

$$\frac{\partial R}{\partial n_q} = n_i(q)\frac{\partial R}{\partial x_i'} = \frac{1}{R}(x_i' - x_i)n_i(q) = -\frac{(\boldsymbol{r}-\boldsymbol{r}')\cdot\boldsymbol{n}(\boldsymbol{r}')}{|\boldsymbol{r}-\boldsymbol{r}'|}.$$

Then, comparing (2.14), (9.42) and (9.45) suggests defining an acoustic double-layer potential by

$$(\mathrm{D}\nu)(P,t) = \frac{1}{2\pi}\int_S \left\{\frac{1}{cR}\dot{\nu}(q,t-R/c) + \frac{1}{R^2}\nu(q,t-R/c)\right\}\frac{\partial R}{\partial n_q}\,\mathrm{d}s_q \tag{9.46}$$

$$= \frac{1}{2\pi}\int_S \left\{\frac{1}{cR}\left[\frac{\partial \nu}{\partial t}\right] + \frac{[\nu]}{R^2}\right\}\frac{\partial R}{\partial n_q}\,\mathrm{d}s_q, \tag{9.47}$$

where $\nu(q,t)$ is given for $q \in S$ and we have used the notation (9.2) in (9.47).

By construction, $(\mathrm{D}\nu)(P,t)$ is a wavefunction for $P \notin S$. In the far field,

$$(\mathrm{D}\nu)(P,t) \sim r^{-1} g(\widehat{\boldsymbol{r}},t-r/c) \quad \text{as } r\to\infty, \text{ with} \tag{9.48}$$

$$g(\widehat{\boldsymbol{r}},\xi) = -\frac{1}{2\pi c}\int_S \widehat{\boldsymbol{r}}\cdot\boldsymbol{n}(q)\,\dot{\nu}(q,\xi)\,\mathrm{d}s_q \quad \text{and} \quad \widehat{\boldsymbol{r}} = \boldsymbol{r}/r.$$

We see that $\mathrm{D}\nu$ is an outgoing wavefunction. The corresponding far-field pattern (defined by (2.18)) is

$$f_0(\widehat{\boldsymbol{r}},t) = \lim_{r\to\infty}\{r(\mathrm{D}\mu)(\boldsymbol{r},t+r/c)\}$$

$$= -\frac{1}{2\pi c}\int_S \widehat{\boldsymbol{r}}\cdot\boldsymbol{n}(\boldsymbol{r}')\,\dot{\nu}(\boldsymbol{r}',t+\widehat{\boldsymbol{r}}\cdot\boldsymbol{r}'/c)\,\mathrm{d}s(\boldsymbol{r}'). \tag{9.49}$$

The far field of $\mathrm{D}\nu$ is determined by the first term in the integrand in (9.46). On the other hand, the jump behaviour is determined by the second term in the integrand

Table 9.1 *Notations and definitions*

This book	Ha-Duong [377]	Costabel [198]	Sayas [742]
$\mathrm{S}\mu$	$-2L\mu$	$-2\mathscr{S}(\mu)$	$-2\mathscr{S}*\mu$
$\mathrm{D}\nu$	$-2M\nu$	$-2\mathscr{D}(\nu)$	$-2\mathscr{D}*\nu$
$K\mu$	$-2K\mu$	$-2K'\mu$	$-2\mathscr{K}^t*\mu$
$K'\nu$	$-2K'\nu$	$-2K\nu$	$-2\mathscr{K}*\nu$
$N\nu$	$-2D\nu$	$-2W\nu$	$2\mathscr{W}*\nu$

(the more singular of the two), and this term is exactly the same as in the classical double layer (9.45). Hence

$$\mathrm{D}\nu = (\mp I + K')\nu, \tag{9.50}$$

where the upper (lower) sign corresponds to $P \to p \in S$ from B_{e} (B) (as in (9.41)) and

$$(K'\nu)(p,t) = \frac{1}{2\pi}\int_S \left\{ \frac{1}{cR_0}\dot{\nu}(q,t-R_0/c) + \frac{1}{R_0^2}\nu(q,t-R_0/c) \right\} \frac{\partial R}{\partial n_q}\,\mathrm{d}s_q, \tag{9.51}$$

which is almost the same as (9.42); the latter contains $\partial R/\partial n_p$.

We can define the normal derivative of $\mathrm{D}\nu$ at $p \in S$ if the density ν is sufficiently smooth,

$$(N\nu)(p,t) = \frac{\partial}{\partial n_p}\mathrm{D}\nu = n_i(p)\frac{\partial}{\partial x_i}(\mathrm{D}\nu)(\boldsymbol{r},t); \tag{9.52}$$

as in classical potential theory, the normal derivative of $\mathrm{D}\nu$ is continuous across S.

9.4.3 Notation and Definitions

The notations and definitions for layer potentials vary. There are three excellent sources, the review papers by Ha-Duong [377] and by Costabel [198] (updated as [199]) and the book by Sayas [742]; see Table 9.1 for comparisons. Note that the factors of -2 are there because we prefer to have $\pm I$ in (9.41). (Note also that there is a sign error between the two formulas for Kp in [377, eqn (12)].) There are detailed proofs of the jump relations in [742, §1.3].

9.4.4 Kirchhoff's Formula

Let us use the single-layer and double-layer notations to summarise the various forms of Kirchhoff's formula. From (9.7) and (9.9), we obtain

$$(\mathrm{S}(\partial u/\partial n))(P,t) - (\mathrm{D}u)(P,t) = \begin{cases} 0, & P \in B_{\mathrm{e}}, \\ -2u(P,t), & P \in B, \end{cases} \tag{9.53}$$

when u is a wavefunction in B. Similarly, (9.12) gives

$$(\mathrm{S}(\partial u_\mathrm{e}/\partial n))(P,t) - (\mathrm{D}u_\mathrm{e})(P,t) = \begin{cases} 2u_\mathrm{e}(P,t), & P \in B_\mathrm{e}, \\ 0, & P \in B, \end{cases} \tag{9.54}$$

when u_e is a wavefunction in B_e. In the far field, we can use (9.39) and (9.48) in (9.54); doing this gives

$$u_\mathrm{e}(P,t) \sim r^{-1} g(\widehat{\boldsymbol{r}}, t - r/c) \quad \text{as } r \to \infty, \text{ with} \tag{9.55}$$

$$g(\widehat{\boldsymbol{r}}, \xi) = \frac{1}{4\pi} \int_S \left\{ \frac{1}{c} \widehat{\boldsymbol{r}} \cdot \boldsymbol{n}(q)\, \dot{u}_\mathrm{e}(q,\xi) - \frac{\partial u_\mathrm{e}}{\partial n}(q,\xi) \right\} \mathrm{d}s_q. \tag{9.56}$$

This result holds for fixed t. Similarly, using (9.40) and (9.49), we find that the far-field pattern is given by

$$\begin{aligned} f_0(\widehat{\boldsymbol{r}}, t) &= \lim_{r\to\infty} \{ r u_\mathrm{e}(\boldsymbol{r}, t + r/c) \} \qquad (9.57) \\ &= \frac{1}{4\pi} \int_S \left\{ \frac{1}{c} \widehat{\boldsymbol{r}} \cdot \boldsymbol{n}(\boldsymbol{r}')\, \dot{u}_\mathrm{e}(\boldsymbol{r}', t + \widehat{\boldsymbol{r}} \cdot \boldsymbol{r}'/c) - \frac{\partial u_\mathrm{e}}{\partial n}(\boldsymbol{r}', t + \widehat{\boldsymbol{r}} \cdot \boldsymbol{r}'/c) \right\} \mathrm{d}s(\boldsymbol{r}'). \end{aligned}$$

If we let $P \to p \in S$ in (9.53), using (9.50), we obtain

$$(I - K')u + \mathrm{S}(\partial u/\partial n) = 0. \tag{9.58}$$

Similarly, (9.54) gives

$$(I + K')u_\mathrm{e} - \mathrm{S}(\partial u_\mathrm{e}/\partial n) = 0. \tag{9.59}$$

We can also compute the normal derivatives of (9.53) and (9.54) on S, using (9.41) and (9.52); the results are

$$(I + K)(\partial u/\partial n) - Nu = 0 \tag{9.60}$$

and

$$(I - K)(\partial u_\mathrm{e}/\partial n) + Nu_\mathrm{e} = 0. \tag{9.61}$$

10
Integral Equations

Problems governed by elliptic PDEs (including frequency-domain problems) can be reduced to boundary integral equations. There are two main methods for doing this, usually called the indirect method and the direct method; we shall describe time-domain versions of these methods. In the indirect time-domain method (Section 10.1), we start by seeking the solution in the form of a layer potential (Section 9.4). In the direct method (Section 10.2), we start from Kirchhoff's formula (Section 9.1). With either method, we obtain a *time-domain boundary integral equation*. It turns out that solving such an equation numerically is not always straightforward; for example, the simplest algorithms are often unstable. Numerical aspects, difficulties and remedies are discussed in Section 10.3. One attractive approach uses the *convolution quadrature method*; it is developed in Section 10.4. All these methods can be adapted for electromagnetic (Section 10.5), elastodynamic (Section 10.6) and hydrodynamic (Section 10.7) problems. Another approach (often used for elliptic boundary-value problems), the *method of fundamental solutions* (sometimes called the *equivalent source method*) is described in Section 10.8. In Section 10.9, we revisit the use of Fourier transforms so as to exploit what is known about solving frequency-domain problems. Finally, in Section 10.10, we consider problems in which the scatterer is thin; this includes screens and cracks. Such scatterers are common in applications but their treatment introduces new challenges.

10.1 Integral Equations: Indirect Method

As with time-harmonic problems [599, §5.9], this method starts by seeking solutions in the form of single-layer or double-layer potentials, or some combination of both. The specific choice made will determine the kind of integral equation obtained. In what follows, we shall focus on IBVPs with zero initial conditions (see Section 4.1); other kinds of problems can be treated using similar methods.

For the Dirichlet problem, Problem DI_0, let us look for a solution in the form of a single-layer potential (9.38). Thus we write

$$u(P,t) = (\mathrm{S}\mu)(P,t), \quad P \in B_\mathrm{e}, \tag{10.1}$$

where the density $\mu(p,t)$ is to be found. Zero initial conditions are enforced by requiring that $\mu(p,t) \equiv 0$ for $t \leq 0$ and $p \in S$. Imposing the boundary condition, $u(p,t) = d(p,t)$ for $t > 0$ and $p \in S$, leads to

$$\mathrm{S}\mu = d, \tag{10.2}$$

an integral equation for μ. Similarly, if (10.1) is used for the Neumann problem, Problem NI_0, the boundary condition $\partial u/\partial n = v$ and (9.41) give

$$(I + K)\mu = v. \tag{10.3}$$

Alternatively, consider using a double-layer potential (9.47)

$$u(P,t) = (\mathrm{D}\nu)(P,t), \quad P \in B_\mathrm{e}, \tag{10.4}$$

where the density $\nu(p,t)$ is to be found. Then, for Problem DI_0, we obtain

$$(I - K')\nu = -d, \tag{10.5}$$

whereas for Problem NI_0, we obtain

$$N\nu = v. \tag{10.6}$$

The integral equations (10.2), (10.3), (10.5) and (10.6) are examples of *indirect time-domain boundary integral equations*, where 'indirect' indicates that the unknown density functions do not have clear physical significance.

Equations (10.3) and (10.5) are usually referred to as equations of the second kind (because of the presence of I). However, as emphasised by Ha-Duong, 'one cannot consider the operator $I - K'$ as a second kind integral operator in the sense of Fredholm' [377, p. 318]. Nevertheless, some solvability results (in appropriate function spaces) are available: for example, (10.2) is uniquely solvable [377, Theorem 3] and (10.5) is uniquely solvable [377, Theorem 2].

It can be advantageous to use combinations of layer potentials; for an example, see (10.21).

10.2 Integral Equations: Direct Method

Direct time-domain boundary integral equations involve unknown quantities with physical significance. They are derived from the Kirchhoff formula.

For Problem DI_0, with $u = d$ on S, the basic unknown is the normal velocity on S, $\partial u/\partial n = v$. Then (9.54) gives the integral representation

$$2u(P,t) = (\mathrm{S}v)(P,t) - (\mathrm{D}d)(P,t), \quad P \in B_\mathrm{e}, \tag{10.7}$$

where v can be found by solving $\mathrm{S}v = (I+K')d$ or $(I-K)v = -Nd$, these equations coming from (9.59) and (9.61).

Similarly, for Problem NI_0, with a Neumann boundary condition $\partial u/\partial n = v$ on S, the basic unknown is u on S. The integral representation is

$$2u(P,t) = (\mathrm{S}v)(P,t) - (\mathrm{D}u)(P,t), \quad P \in B_\mathrm{e}, \tag{10.8}$$

where $u(p,t)$ can be found by solving $(I+K')u = \mathrm{S}v$ or $Nu = (K-I)v$.

For scattering problems, we can write the total velocity potential u as $u = u_{\mathrm{inc}} + u_{\mathrm{sc}}$, where u_{inc} is the given incident wave and u_{sc} is the unknown scattered field. For an incident plane step pulse (see Section 4.3), with $u_{\mathrm{inc}} \equiv 0$ ahead of the propagating wavefront, we can arrange that u_{sc} satisfies zero initial conditions. Suppose further that u_{inc} is continuous across the wavefront: this is the usual situation (see Section 3.5), and we henceforth make this assumption. Then we can use (9.53) and (9.54) to obtain

$$(\mathrm{S}(\partial u_{\mathrm{inc}}/\partial n))(P,t) - (\mathrm{D}u_{\mathrm{inc}})(P,t) = \begin{cases} 0, & P \in B_\mathrm{e}, \\ -2u_{\mathrm{inc}}(P,t), & P \in B, \end{cases} \tag{10.9}$$

$$(\mathrm{S}(\partial u_{\mathrm{sc}}/\partial n))(P,t) - (\mathrm{D}u_{\mathrm{sc}})(P,t) = \begin{cases} 2u_{\mathrm{sc}}(P,t), & P \in B_\mathrm{e}, \\ 0, & P \in B. \end{cases} \tag{10.10}$$

Adding these gives an integral representation for the total velocity potential,

$$2u(P,t) = 2u_{\mathrm{inc}}(P,t) + (\mathrm{S}(\partial u/\partial n)(P,t) - (\mathrm{D}u)(P,t), \quad P \in B_\mathrm{e}. \tag{10.11}$$

This is especially convenient when the scatterer is rigid (sound-hard), for then we have $\partial u/\partial n = 0$ on S, (10.11) reduces to $u_{\mathrm{sc}}(P,t) = -\frac{1}{2}(\mathrm{D}u)(P,t)$ and then (9.50) gives the integral equation

$$(I+K')u = 2u_{\mathrm{inc}}. \tag{10.12}$$

This is an equation for the boundary values of u but, as inspection of (9.51) reveals, (10.12) also involves $\partial u/\partial t$. Indeed, we may consider the time derivative of (10.12), giving

$$\frac{\partial u}{\partial t} + \frac{\partial}{\partial t}\left(K'u\right) = 2\frac{\partial u_{\mathrm{inc}}}{\partial t}. \tag{10.13}$$

This can be found in [264, eqn (4)], for example.

As the pressure $p(P,t)$ is also a wavefunction, it can be found using similar integral representations and time-domain boundary integral equations. However, p may not be continuous, even if the potential u is continuous, so care may be warranted.

Most of the computations reported in the literature have used some form of direct method; we shall give references later.

10.3 Integral Equations: Numerical Methods

Time-domain boundary integral equations are usually solved numerically. We shall describe some methods for doing this, framing our discussion around the simplest integral equation, $(\mathrm{S}\mu)(p,t) = d(p,t)$. This equation arises when the solution of Problem DI_0 is sought in the form of a single-layer potential; see (10.2). It is to be solved for $\mu(p,t)$ with $p \in S$ and $t > 0$. The function d is given; both $d(p,t)$ and $\mu(p,t)$ vanish for $t \le 0$ and $p \in S$. Written out explicitly,

$$\frac{1}{2\pi}\int_S \mu\left(\boldsymbol{r}', t - \frac{|\boldsymbol{r}-\boldsymbol{r}'|}{c}\right)\frac{\mathrm{d}s(\boldsymbol{r}')}{|\boldsymbol{r}-\boldsymbol{r}'|} = -d(\boldsymbol{r},t), \quad \boldsymbol{r} \in S. \tag{10.14}$$

A variety of time-stepping and Galerkin methods will be described in the remainder of this section. A newer class of methods, known as convolution quadrature methods, will be described in Section 10.4.

10.3.1 Basic Time-Stepping Method

Following [220, §2.1], break the surface S into N disjoint boundary elements, $S = \cup_{j=1}^N S_j$, and then use a piecewise-constant spatial approximation, $\mu(\boldsymbol{r},t) = \mu_j(t)$ for all $\boldsymbol{r} \in S_j$. This reduces (10.14) to

$$\frac{1}{2\pi}\sum_{j=1}^N \int_{S_j} \mu_j\left(t - \frac{|\boldsymbol{r}-\boldsymbol{r}'|}{c}\right)\frac{\mathrm{d}s(\boldsymbol{r}')}{|\boldsymbol{r}-\boldsymbol{r}'|} = -d(\boldsymbol{r},t). \tag{10.15}$$

We are going to march forward in time, in steps of size Δ. Choose a central point $\boldsymbol{r}_j \in S_j$ and collocate (10.15) at each of these points. This gives

$$\frac{1}{2\pi}\sum_{j=1}^N \int_{S_j} \mu_j\left(t_n - \frac{|\boldsymbol{r}_i-\boldsymbol{r}'|}{c}\right)\frac{\mathrm{d}s(\boldsymbol{r}')}{|\boldsymbol{r}_i-\boldsymbol{r}'|} = -d(\boldsymbol{r}_i,t_n), \quad i = 1,2,\ldots,N, \tag{10.16}$$

where we have evaluated at time level $t_n = n\Delta$. The 'self-patch' integral over S_i has a weak singularity and so must be treated separately from all the others over S_j with $j \ne i$. For non-singular integrands, we use a simple midpoint rule, $\int_{S_j} g(\boldsymbol{r}')\,\mathrm{d}s(\boldsymbol{r}') \simeq g(\boldsymbol{r}_j)\,|S_j|$, where $|S_j|$ is the surface area of the jth boundary element. Hence (10.16) is approximated by

$$C_{ii}\mu_i(t_n) + \sum_{\substack{j=1\\ j\ne i}}^N C_{ij}\,\mu_j\left(t_n - \frac{|\boldsymbol{r}_i-\boldsymbol{r}_j|}{c}\right) = -d(\boldsymbol{r}_i,t_n), \quad i = 1,2,\ldots,N, \tag{10.17}$$

where

$$C_{ii} = \frac{1}{2\pi}\int_{S_i}\frac{\mathrm{d}s(\boldsymbol{r}')}{|\boldsymbol{r}_i-\boldsymbol{r}'|}, \quad C_{ij} = \frac{|S_j|}{2\pi|\boldsymbol{r}_i-\boldsymbol{r}_j|} \quad \text{for } i \ne j.$$

We shall use a piecewise-linear temporal approximation for $\mu_j(t)$, so that

$$\mu_j(t) = \Delta^{-1}\left\{\mu_j^\ell(t-t_{\ell-1}) + \mu_j^{\ell-1}(t_\ell - t)\right\}, \quad t_{\ell-1} \le t \le t_\ell, \tag{10.18}$$

where $\mu_j^\ell = \mu_j(t_\ell)$. Define an integer m_{ij} by

$$t_{m_{ij}} \le |\boldsymbol{r}_i - \boldsymbol{r}_j|/c < t_{m_{ij}+1} \tag{10.19}$$

whence $t_{n-1-m_{ij}} < t_n - |\boldsymbol{r}_i - \boldsymbol{r}_j|/c \le t_{n-m_{ij}}$ and then $\mu_j(t_n - |\boldsymbol{r}_i - \boldsymbol{r}_j|/c)$ can be evaluated using (10.18) with $\ell = n - m_{ij}$. Equation (10.17) becomes

$$C_{ii}\,\mu_i^n = -d(\boldsymbol{r}_i, t_n) - \sum_{\substack{j=1\\ j\neq i}}^{N} C_{ij}\left\{(1-\varepsilon_{ij})\,\mu_j^{n-m_{ij}} + \varepsilon_{ij}\,\mu_j^{n-1-m_{ij}}\right\}, \quad i = 1,2,\ldots,N, \tag{10.20}$$

where $\varepsilon_{ij} = |\boldsymbol{r}_i - \boldsymbol{r}_j|/(c\Delta) - m_{ij}$; from (10.19), we have $0 \le \varepsilon_{ij} < 1$.

Equation (10.20) is a linear system for μ_i^n in terms of μ_j^k with $1 \le k < n$: it enables stepping forward in time. Moreover, if the time step Δ can be chosen so that $c\Delta$ is smaller than the minimum distance between $\boldsymbol{r}_i$ and $\boldsymbol{r}_j$ $(i \neq j)$, then $m_{ij} \ge 1$ and (10.20) becomes an explicit formula for μ_i^n in terms of quantities computed from previous time steps. This property made methods of this kind very popular: there are no linear algebraic systems to solve. It comes about when a signal from position $\boldsymbol{r}_j$ at time t_{n-1} travelling at speed c cannot reach $\boldsymbol{r}_i$ by time t_n: this is another example of causality (Section 1.5).

Most time-domain integral equations encountered are more complicated than the single-layer equation (10.14) because of the presence of the normal derivative of a single-layer potential, or double-layer potentials; see the definitions of the operators K and K', (9.42) and (9.51), respectively. These involve time derivatives of the unknown function within the integrands. Nevertheless, similar approximation schemes can be developed.

The basic time-stepping method described above can be refined (using better approximations for the spatial and time dependencies) and applied to other time-domain boundary integral equations. It was first used by Friedman & Shaw [316] in a paper published in 1962. The junior author, Richard Shaw, has described its genesis [765], starting as a PhD student

> *in the Guggenheim Institute for Flight Structures at Columbia University in the mid-1950's. The director, Prof. H. H. Bleich, suggested to a new faculty member, Prof. M. B. Friedman, that he and I (his first PhD student) look at acoustic shock scattering by underwater structures by replacing the body by some saltus problem, i.e. surfaces of sources and sinks. ...* [The] *first half of my own PhD thesis in 1960 on transient acoustic wave scattering* [316] *was carried out on a Monroe desk calculator* [whereas] *the second half* [767] *was done on an IBM 704.*

The two cited papers consider three-dimensional scattering by a long cylinder of square cross-section; a derivation of the direct integral equation for the total potential u, (10.12), is given and then an equation for the pressure p is obtained. Shaw used similar methods for other acoustic problems [760, 763, 766, 764]; one of these [766] contains results for scattering by a sound-soft (Dirichlet condition) sphere. Shaw was also aware of complications arising from discontinuous pressure pulses; see his comments [761] on a paper by Mitzner [629]. Mitzner applied the explicit method described above to (10.12) and gave results for scattering of a Gaussian pulse (5.21) by a sound-hard (Neumann condition) sphere; see also [251]. For a pulse of finite width, (4.14), see [810]. Farn & Huang [288] used an indirect method for Problem NI_0; they solved (10.3) for a radiating sphere. For scattering by a penetrable sphere, see [90, 399]. For scattering by axisymmetric objects, see [654]. More sophisticated schemes, using boundary element methods, have been developed; see, for example, [361, 237, 109].

10.3.2 Instabilities and Remedies

Unfortunately, the explicit schemes are prone to instability: errors can accumulate with each time step, leading to errors that grow exponentially. Apparently, this was first noted by Bennett & Mieras [90], Herman [399, Fig. 8], Groenenboom [360, p. 39] and, for two-dimensional problems, Cole et al. [187, Appendix C]. Analysis of the phenomenon was made by Rynne [727, 728], Davies [218] and Davies & Duncan [220, 221], and various methods for restoring stability and convergence have been devised; see [221] and [825, Chapter 3] for further discussion.

The precise cause of the instability is unclear. One candidate is *irregular frequencies*. If we pass to the frequency domain, we obtain an integral equation (such as (8.9) or (8.11)) that is not uniquely solvable at certain frequencies [599, Chapter 5]. The presence of these irregular frequencies may pollute the numerical solution in the time domain. This observation was made by Smith [783]; see also [264, 864, 768]. Frequency-domain integral equations without irregular frequencies are known [599, §6.8], and corresponding time-domain integral equations have been investigated [264, 164, 163, 441, 276, 318]. For example, a combined-layer method [599, §6.10.1] has been used for Problem DI_0, in which the solution is sought in the form

$$u(P,t) = (\mathrm{D}\mu)(P,t) + (\mathrm{S}\{a\dot{\mu} + b\mu\})(P,t), \tag{10.21}$$

where a and b are constants [259, 54].

Another candidate is the explicit nature of the original algorithms. Modern algorithms are usually implicit, so that a fairly small linear system has to solved at each time step; for an early implementation, see [362]. Indeed, with a reasonable choice of time step Δ, so that $c\Delta$ is comparable to the spatial discretisation, 'it has been demonstrated that implicit formulations, adopting such a time step, are for practical purposes immune from instability' [236, p. 301].

Of course, stability is not the only requirement of an effective numerical scheme; we also have to consider accuracy and efficiency.

For an accurate spatial approximation on the boundary S, much is known from the (static and frequency-domain) literature on boundary element methods, and this can be exploited. Instead of collocation methods, Galerkin (or discontinuous Galerkin) methods have been used for the spatial part of the approximation scheme [900, 1].

For the temporal approximation, several options have been tried, including continuous piecewise quadratics [109], B-splines [717, 223] and Knab's [489] approximate prolate spheroidal wavefunctions [265, 267]. Laguerre polynomials $L_n \equiv L_n^{(0)}$ [661, §18.3] and Laguerre transforms (Section 5.7) have also been used [566]. Davies & Duncan [222] used a set of basis functions built using the error function [661, eqn 7.2.1].

Another way to achieve a provably convergent and stable scheme is to use a space-time Galerkin method. This method was described in two papers by Bamberger & Ha Duong [42, 43], and has been developed by Ha Duong and his colleagues; see [377, 449] for reviews. For some applications of space-time Galerkin methods to acoustic problems, see [232, 378, 739, 740, 854]. These papers contain results for fairly simple problems, but there is potential for application to difficult technological problems. Of these, one is the prediction of noise generated by tyres on vehicles [556]; for applications of space-time Galerkin methods to this problem, see [47, 338, 341].

Pölz & Schanz [702] have solved (10.14) using a direct space-time discretisation, in which time is treated as another spatial variable.

We also mention the approach of Aimi et al. [10, 9], where the total energy in a time interval $0 \le t \le T$, $\mathscr{E}(T,u)$, is used to obtain a weak formulation; here, $\mathscr{E}(T,u) = \int_0^T E(t)\,\mathrm{d}t$, where $E(t) \equiv E(t;u)$ is defined by (4.5). An earlier technique [401] proceeds as follows: write (10.14) concisely as $F[\mu](\boldsymbol{r},t) = 0$ and then compute μ^* so as to minimise

$$\int_0^T \int_S \{F[\mu^*](\boldsymbol{r},t)\}^2 \,\mathrm{d}s(\boldsymbol{r})\,\mathrm{d}t;$$

this may be done using an iterative scheme [401].

Let us now turn to efficiency. If we examine the basic numerical scheme leading to (10.20), we quickly see that evaluating the right-hand side could be expensive, especially for large-scale scattering problems [235, 266]. This has motivated the development of various acceleration strategies. One of these, known as the *plane-wave time-domain* algorithm [263, 265, 266, 267, 810], makes use of Whittaker's plane-wave representation, (2.70).

Adaptive methods are also beginning to emerge: both spatial adaptivity [899, 34, 338] and temporal adaptivity [741] have been investigated to some extent, but these are areas of current research.

10.4 Convolution Quadrature Methods

Convolution quadrature methods (CQMs) were introduced by Lubich in 1988 [577] and then applied by him to time-domain boundary integral equations in 1994 [578]. Since then, CQMs have been developed extensively. Lubich [578, §5] considered the simplest integral equation, $(\mathrm{S}\mu)(p,t) = d(p,t)$; we shall do the same, starting with

(10.14), which we write as

$$\int_S \mu\left(\boldsymbol{r}', t - \frac{R}{c}\right) \frac{\mathrm{d}s(\boldsymbol{r}')}{2\pi R} = -d(\boldsymbol{r}, t), \quad \boldsymbol{r} \in S, \tag{10.22}$$

where $R = |\boldsymbol{r} - \boldsymbol{r}'|$. As before, we want to solve (10.22) for $\mu(\boldsymbol{r}, t) \equiv \mu(p, t)$ with $p \in S$ and $t > 0$. The given function $d(p, t)$ and the unknown function $\mu(p, t)$ both vanish for $t \le 0$ and $p \in S$; further conditions will be given later.

As $\mu(\boldsymbol{r}, t) = 0$ for $t \le 0$, the Laplace transform of (10.22) with respect to t reduces to

$$\int_S \mathrm{e}^{-sR/c} M(\boldsymbol{r}', s) \frac{\mathrm{d}s(\boldsymbol{r}')}{2\pi R} = -D(\boldsymbol{r}, s),$$

where $M(\boldsymbol{r}, s) = \mathscr{L}\{\mu(\boldsymbol{r}, t)\}$ and $D(\boldsymbol{r}, s) = \mathscr{L}\{d(\boldsymbol{r}, t)\}$. (There should be no confusion between the transform variable s and the element of surface area $\mathrm{d}s(\boldsymbol{r}')$.) The left-hand side contains a Laplace convolution; inverting gives

$$\int_S \int_0^t \mathscr{K}(R, \tau)\, \mu(\boldsymbol{r}', t - \tau)\, \mathrm{d}\tau \frac{\mathrm{d}s(\boldsymbol{r}')}{2\pi R} = -d(\boldsymbol{r}, t), \tag{10.23}$$

where $\mathscr{L}\{\mathscr{K}(R, t)\} = K(R, s) = \mathrm{e}^{-sR/c}$. At this point, we may interpret $\mathscr{K}$ as a Dirac delta function, $\mathscr{K}(R, t) = \delta(t - R/c)$, but we do not need this interpretation here.

10.4.1 A Volterra Integral Equation

CQMs give effective methods for approximating the time integration in (10.23). In order to focus on this aspect, following [577, 744, 630, 223], [743, Chapter 2] and [742, Chapter 4], we suppress the spatial dependence and consider the Volterra integral equation for $\mu(t)$,

$$\int_0^t \mathscr{K}(\tau)\, \mu(t - \tau)\, \mathrm{d}\tau = d(t), \tag{10.24}$$

where $d(t)$ is given and $\mathscr{K} = \mathscr{L}^{-1}\{K\}$, that is

$$\mathscr{K}(t) = \frac{1}{2\pi \mathrm{i}} \int_{\mathrm{Br}} K(s)\, \mathrm{e}^{st}\, \mathrm{d}s. \tag{10.25}$$

Here, Br is the Bromwich contour in the complex s-plane; see (5.3). Substituting (10.25) in (10.24) gives

$$d(t) = \frac{1}{2\pi \mathrm{i}} \int_{\mathrm{Br}} K(s)\, y(t; s)\, \mathrm{d}s, \tag{10.26}$$

where

$$y(t; s) = \int_0^t \mathrm{e}^{s\tau} \mu(t - \tau)\, \mathrm{d}\tau. \tag{10.27}$$

Suppressing the dependence on s, differentiating $y(t) \equiv y(t; s)$ and then integrating by parts, we obtain a simple first-order ordinary differential equation (ODE),

$$y'(t) = sy(t) + \mu(t) \quad \text{with} \quad y(0) = 0. \tag{10.28}$$

Returning to (10.24), we extend $\mu(t)$ by zero for $t < 0$. Suppose that the extension satisfies $\mu(0) = \mu'(0) = \cdots = \mu^{(m)}(0) = 0$ for some positive integer m. Repeated differentiations of (10.26) and (10.28) then give $y^{(p)} = 0$ and $d^{(p)}(0) = 0$ for $p = 0, 1, 2, \ldots, m+1$; we henceforth impose this (strong) consistency condition on the given function d.

Returning to the ODE (10.28), there are several methods for finding approximate solutions; we use the k-step linear multistep method with time step Δ [381, Chapter III],

$$\sum_{j=0}^{k} \alpha_j y_{n-j} = \Delta \sum_{j=0}^{k} \beta_j f_{n-j}, \quad n = 0, 1, 2, \ldots, \tag{10.29}$$

where $y_n(s) \simeq y(t_n; s)$, $t_n = n\Delta$ and $f_n = s y_n + \mu(t_n)$. Specific methods are obtained by making choices for k (with $k \leq m$), α_j and β_j, $j = 0, 1, \ldots, k$. For starting values, we take $y_{-j} = 0$, $j = 1, 2, \ldots, k$, which is consistent with wanting a causal solution that vanishes identically for $t \leq 0$.

Multiply (10.29) by ζ^n and sum over all $n \geq 0$, giving

$$\sigma(\zeta)\mathscr{Y}(\zeta; s) = \Delta\{s\mathscr{Y}(\zeta; s) + \mathscr{M}(\zeta)\}, \tag{10.30}$$

where ζ is a complex variable,

$$\mathscr{M}(\zeta) = \sum_{n=0}^{\infty} \mu(t_n)\zeta^n, \quad \mathscr{Y}(\zeta; s) = \sum_{n=0}^{\infty} y_n(s)\zeta^n, \tag{10.31}$$

$$\sigma(\zeta) = \frac{\alpha_0 + \alpha_1\zeta + \cdots + \alpha_k\zeta^k}{\beta_0 + \beta_1\zeta + \cdots + \beta_k\zeta^k}. \tag{10.32}$$

(Note that the definition of σ varies widely in the literature.)

The function $\mathscr{Y}$ is called the *generating function* [262, Chapter XIX] or the *z-transform* [459] of the sequence $\{y_n\}$. Assuming that $\mathscr{Y}(\zeta; s)$ is analytic in a neighbourhood of $\zeta = 0$, we can integrate along a contour in this neighbourhood around the origin to give

$$y_n(s) = \frac{1}{2\pi i} \oint \mathscr{Y}(\zeta; s)\,\zeta^{-n-1}\,d\zeta. \tag{10.33}$$

From (10.30), $\mathscr{Y} = \left(\Delta^{-1}\sigma - s\right)^{-1}\mathscr{M}$. Hence $\mathscr{Y}(\zeta; s)$ has a simple pole in the complex s-plane at $s = \Delta^{-1}\sigma(\zeta)$. If we choose the Bromwich contour in (10.26) to the right of this pole, we can close the contour to the left; Cauchy's integral formula then gives

$$\begin{aligned} d(t_n) &\simeq \frac{1}{2\pi i} \int_{\mathrm{Br}} K(s)\,y_n(s)\,ds = \frac{1}{(2\pi i)^2} \oint \frac{1}{\zeta^{n+1}} \int_{\mathrm{Br}} K(s)\,\mathscr{Y}(\zeta; s)\,ds\,d\zeta \\ &= \frac{1}{(2\pi i)^2} \oint \frac{\mathscr{M}(\zeta)}{\zeta^{n+1}} \int_{\mathrm{Br}} \frac{K(s)\,ds}{\Delta^{-1}\sigma(\zeta) - s}\,d\zeta \\ &= \frac{1}{2\pi i} \oint \frac{\mathscr{M}(\zeta)}{\zeta^{n+1}} K(\Delta^{-1}\sigma(\zeta))\,d\zeta. \end{aligned} \tag{10.34}$$

The right-hand side of this equation is equal to the coefficient of ζ^n in the power series expansion of $\mathscr{M}(\zeta)K(\Delta^{-1}\sigma(\zeta))$ about $\zeta = 0$. If we expand,

$$K(\Delta^{-1}\sigma(\zeta)) = \sum_{j=0}^{\infty} \omega_j(\Delta)\,\zeta^j, \tag{10.35}$$

and combine with the expansion of $\mathscr{M}(\zeta)$, (10.31), we obtain

$$\mathscr{M}(\zeta)\,K(\Delta^{-1}\sigma(\zeta)) = \sum_{\ell=0}^{\infty} \zeta^\ell \sum_{j=0}^{\ell} \omega_j(\Delta)\,\mu_{\ell-j}$$

where $\mu_n = \mu(t_n)$. Hence (10.34) gives the approximation

$$d(t_n) = \sum_{j=0}^{n} \omega_j(\Delta)\,\mu_{n-j}. \tag{10.36}$$

As $\mu_0 = \mu(0) = 0$, the term with $j = n$ is absent, whence

$$\mu_1 = \frac{d(\Delta)}{\omega_0(\Delta)} \quad \text{and} \quad \mu_n = \frac{1}{\omega_0(\Delta)}\left(d(t_n) - \sum_{j=1}^{n-1} \omega_j(\Delta)\,\mu_{n-j}\right) \tag{10.37}$$

for $n = 2, 3, \ldots$. This gives a time-stepping approximation for μ_n.

The left-hand side of (10.24) is a Laplace convolution. This explains why (10.36) is called a *convolution quadrature*: it gives an approximation to the integral in (10.24) using discrete values of $\mu(t)$; the weights ω_j are computable from (10.35).

10.4.2 Application to Time-Domain Boundary Integral Equations

Let us apply a CQM to (10.23) in which $K(s) \equiv K(R,s) = \mathrm{e}^{-sR/c}$. Adapting (10.36), we obtain

$$\sum_{j=0}^{n} \int_S \omega_j\left(\frac{R}{c\Delta}\right) \mu_{n-j}(\boldsymbol{r}')\,\frac{\mathrm{d}s(\boldsymbol{r}')}{2\pi R} = -d(\boldsymbol{r},t_n), \quad n = 1, 2, \ldots, \tag{10.38}$$

where $R = |\boldsymbol{r} - \boldsymbol{r}'|$, $\mu_n(\boldsymbol{r}) = \mu(\boldsymbol{r},t_n)$ and the weights ω_j are defined by the expansion

$$\mathrm{e}^{-\rho\sigma(\zeta)} = \sum_{j=0}^{\infty} \omega_j(\rho)\,\zeta^j. \tag{10.39}$$

The integration over S can be approximated in a standard way using, for example, boundary elements, collocation, or Galerkin's method.

The method described here goes back to Lubich's 1994 paper [578, §5]. There have been numerous extensions and applications since then. For overviews, see [579, 46, 390] and [742, Chapter 4]. For some criticism, see [19, §2.2.1].

The earliest implementations use *backward differentiation formulas*, denoted by BDF k. For these, $\sigma(\zeta) = \sum_{j=1}^{k}(1-\zeta)^j/j$ [381, p. 380], [630, p. 410]. In particular, for BDF 1, we have $\sigma(\zeta) = 1 - \zeta$ and then (10.39) gives $\omega_j(\rho) = \mathrm{e}^{-\rho}\rho^j/j!$.

For BDF 2, $\sigma(\zeta) = \frac{3}{2} - 2\zeta + \frac{1}{2}\zeta^2$. Then (10.39) and the generating function for

Hermite polynomials $H_n(x)$ [661, eqn 18.12.15] gives [866, eqn (26)], [372, eqn (3.6)], [630, p. 421],

$$\omega_j(\rho) = \frac{e^{-3\rho/2}}{j!}(\rho/2)^{j/2} H_j(\sqrt{2\rho}).$$

CQMs with BDF 2 have been applied to Dirichlet problems [44, 278, 168, 280], Neumann problems [162, 607], the Lippmann–Schwinger equation [535] and transmission problems for a penetrable scatterer [706]. For applications of BDF 1 and BDF 2 to two-dimensional scattering problems, see [4]. For a combination of a CQM and the fast multipole method, see [732].

There are other numerical methods for treating ODEs such as (10.28), including Runge–Kutta methods [381, Chapter II], and these have led to associated CQMs. For some applications of Runge–Kutta CQMs, see [44, 45, 391]. CQMs with variable time steps have also been developed [570, 738].

Another option is to start with (10.13), the time derivative of a standard direct time-domain boundary integral equation. After spatial discretisation, one obtains a system of first-order ODEs which can then be solved using a linear multistep method. This approach has been used for acoustic scattering by rigid obstacles (including a sphere) [171], following earlier work on electromagnetic problems based on the time derivative of (10.60) [845].

10.4.3 Application to Scattering by a Sphere

In Section 7.4.1, the problem of scattering by a sound-soft sphere was reduced to a sequence of uncoupled Volterra integral equations of the first kind (7.35), which we write as

$$\int_0^t P_N(1+c\tau/a)\, f_N(t-\tau)\, d\tau = d_N(t), \quad N = 0,1,2,\ldots, \tag{10.40}$$

where P_N is a Legendre polynomial, d_N is given and f_N is to be found. CQMs are applicable to such an equation. The Laplace transform of the kernel is given by (7.33),

$$K_N(s) = \int_0^\infty e^{-st} P_N(1+ct/a)\, dt = \frac{2a}{\pi c} e^{sa/c} k_N(sa/c), \tag{10.41}$$

where k_n is a modified spherical Bessel function. Thus $K_N(s)$ is a polynomial in s^{-1} of degree $N+1$. In detail, from (2.29),

$$e^x k_N(x) = \frac{\pi}{2} \sum_{p=0}^{N} \frac{a_p^N}{x^{p+1}} \quad \text{with} \quad a_p^N = \frac{(N+p)!}{2^p p!(N-p)!}. \tag{10.42}$$

Applying the CQM to (10.40), we obtain the approximation

$$d_n^N = \sum_{j=0}^{n} \omega_j^N f_{n-j}^N,$$

where $d_n^N = (c/a)d_N(t_n)$, $f_n^N = f_N(t_n)$, $t_n = n\Delta$ and the weights ω_j^N are defined by the expansion

$$\frac{c}{a} K_N(\Delta^{-1}\sigma(\zeta)) = \sum_{j=0}^{\infty} \omega_j^N\left(\frac{a}{c\Delta}\right)\zeta^j.$$

Thus, using (10.41) and (10.42),

$$\sum_{j=0}^{\infty} \omega_j^N(\rho)\,\zeta^j = \frac{2}{\pi}e^{\rho\sigma}k_N(\rho\sigma) = \sum_{p=0}^{N} \frac{a_p^N}{[\rho\,\sigma(\zeta)]^{p+1}}. \tag{10.43}$$

If we choose BDF 2 as the underlying scheme, we have

$$\sigma(\zeta) = \frac{3}{2} - 2\zeta + \frac{1}{2}\zeta^2 = \frac{3}{2}\left\{1 - 2\left(\frac{2}{\sqrt{3}}\right)\frac{\zeta}{\sqrt{3}} + \left(\frac{\zeta}{\sqrt{3}}\right)^2\right\}.$$

Then the generating function for Gegenbauer polynomials $C_n^\lambda(x)$ [661, eqn 18.12.4] gives

$$[\sigma(\zeta)]^{-p-1} = (2/3)^{p+1}\sum_{j=0}^{\infty} C_j^{p+1}(2/\sqrt{3})\,(\zeta/\sqrt{3})^j.$$

Substitution in (10.43) then defines each weight as a finite sum of Gegenbauer polynomials.

10.4.4 An Alternative View of CQMs

The derivation of CQMs in Section 10.4.1 makes use of an established numerical method for solving first-order ODEs together with z-transforms (generating functions). These ingredients can be used differently [99]. Thus start by writing the wave equation as a first-order system,

$$\dot{\mathbf{y}}(\boldsymbol{r},t) = E\,\mathbf{y}(\boldsymbol{r},t) \quad \text{with} \quad \mathbf{y}(\boldsymbol{r},0) = \mathbf{0}, \tag{10.44}$$

where $\dot{\mathbf{y}} = \partial\mathbf{y}/\partial t$,

$$\mathbf{y}(\boldsymbol{r},t) = \begin{pmatrix} u \\ \partial u/\partial t \end{pmatrix} \quad \text{and} \quad E = \begin{pmatrix} 0 & I \\ c^2\nabla^2 & 0 \end{pmatrix}.$$

To find an approximate solution of (10.44), use the k-step linear multistep method with time step Δ, as in Section 10.4.1. Then (see around (10.29) for details)

$$\sum_{j=0}^{k} \alpha_j \mathbf{y}_{n-j} = \Delta\sum_{j=0}^{k} \beta_j E\mathbf{y}_{n-j}, \quad n = 0,1,2,\ldots, \tag{10.45}$$

where $\mathbf{y}_n(\boldsymbol{r}) \simeq \mathbf{y}(\boldsymbol{r},t_n)$, $t_n = n\Delta$ and we take $\mathbf{y}_{-j} = \mathbf{0}$ for $j = 1,2,\ldots,k$.

Multiply (10.45) by ζ^n and sum over all $n \geq 0$, giving

$$\sigma(\zeta)\,\mathbf{Y}(\zeta;\boldsymbol{r}) = \Delta E\mathbf{Y}(\zeta,\boldsymbol{r}), \tag{10.46}$$

where ζ is a complex variable, $\sigma(\zeta)$ is defined by (10.32) and

$$\mathbf{Y}(\zeta;\boldsymbol{r}) = \sum_{n=0}^{\infty} \mathbf{y}_n(\boldsymbol{r})\,\zeta^n. \tag{10.47}$$

Solving (10.46) for the first component of $\mathbf{Y}$, Y_1, gives

$$\nabla^2 Y_1 + \kappa^2 Y_1 = 0, \tag{10.48}$$

where

$$\kappa(\zeta) = \mathrm{i}\,\frac{\sigma(\zeta)}{c\Delta} \quad \text{and} \quad Y_1(\zeta,\boldsymbol{r}) = \sum_{n=0}^{\infty} u(\boldsymbol{r},t_n)\,\zeta^n. \tag{10.49}$$

Thus Y_1 solves the Helmholtz equation (10.48) in B_e, but note that κ is complex, in general. To find $Y_1(\zeta,\boldsymbol{r})$, we impose a boundary condition on S and a far-field condition as $|\boldsymbol{r}| \to \infty$.

For the former, assume that we have a Dirichlet problem, with $u = d$ on S, and define

$$D(\zeta,\boldsymbol{r}) = \sum_{n=0}^{\infty} d(\boldsymbol{r},t_n)\,\zeta^n, \quad \boldsymbol{r} \in S.$$

The boundary condition on u then becomes $Y_1 = D$ on S.

In the far field, we may use the Sommerfeld radiation condition,

$$r\left(\frac{\partial Y_1}{\partial r} - \mathrm{i}\kappa Y_1\right) \to 0 \quad \text{as } r = |\boldsymbol{r}| \to \infty \tag{10.50}$$

uniformly in all directions. This will lead to a uniquely solvable problem for Y_1, *but only if* $\operatorname{Im}\kappa \geq 0$ [189, Chapter 3]; equivalently,

$$\operatorname{Re}\sigma(\zeta) \geq 0. \tag{10.51}$$

Note that (10.50) implies that Y_1 behaves as $r^{-1}\mathrm{e}^{\mathrm{i}\kappa r}$ as $r \to \infty$: $Y_1(\zeta,\boldsymbol{r})$ decays exponentially as $r \to \infty$ when $\operatorname{Im}\kappa > 0$.

Having solved for Y_1, we invert the z-transform in (10.49) (see (10.33)),

$$u(\boldsymbol{r},t_n) = \frac{1}{2\pi\mathrm{i}} \int_{C_\lambda} Y_1(\zeta,\boldsymbol{r})\,\zeta^{-n-1}\,\mathrm{d}\zeta.$$

The contour of integration C_λ is taken as a circle of radius λ centred at $\zeta = 0$. It must be chosen so that Y_1 has no singularities inside C_λ and so that (10.51) is satisfied for all $\zeta \in C_\lambda$. (This constraint was not noticed in [99].) Subject to these conditions, the value of the integral does not depend on λ. After parametrisation of C_λ followed by use of the trapezoidal rule with N integration points, we obtain

$$u(\boldsymbol{r},t_n) \simeq \frac{1}{N} \sum_{m=1}^{N} Y_1(\zeta_m^{(N)},\boldsymbol{r}) \left(\zeta_m^{(N)}\right)^{-n} \tag{10.52}$$

where $\zeta_m^{(N)} = \lambda\,\mathrm{e}^{2\pi\mathrm{i}m/N}$. This suggests the following algorithm. Choose N and then compute $\kappa_m = \kappa(\zeta_m^{(N)})$ from the first of (10.49). Next, solve the boundary-value problem for $Y_1(\zeta_m^{(N)}, \boldsymbol{r})$ with $\kappa = \kappa_m$ in (10.48); this may be done using a boundary integral equation method, for example. Finally, use (10.52) to compute the time-domain solution. Note that there is no requirement to computes the weights ω_j that appear in the basic CQM; see (10.35).

The description above is largely based on a paper by Betcke et al. [99]. They also show how a Runge–Kutta scheme can be used, and they point out that one strength of their general approach is that it provides a parallel-in-time method: the N problems to be solved for $Y_1(\zeta_m^{(N)}, \boldsymbol{r})$ can be solved in parallel. (In fact, as u is real, $\overline{Y_1(\zeta, \boldsymbol{r})} = Y_1(\overline{\zeta}, \boldsymbol{r})$, and this can be used to halve the number of Y_1-problems that have to be solved.) These problems are solved in [99] using frequency-domain software adapted to complex frequencies. Of course, one could argue that a direct use of a Fourier transform in t followed by discretisation of the inversion integral provides another way of parallelisation, this time parallel-in-frequency; see the discussion in Sections 5.6.1 and 10.9.

10.5 Electromagnetics

There are analogues of Kirchhoff's formulas for electromagnetic problems (Section 1.6.1). For scattering problems, we write

$$\boldsymbol{E}(\boldsymbol{r},t) = \boldsymbol{E}_{\text{inc}}(\boldsymbol{r},t) + \boldsymbol{E}_{\text{sc}}(\boldsymbol{r},t), \quad \boldsymbol{H}(\boldsymbol{r},t) = \boldsymbol{H}_{\text{inc}}(\boldsymbol{r},t) + \boldsymbol{H}_{\text{sc}}(\boldsymbol{r},t),$$

where $\{\boldsymbol{E}_{\text{inc}}, \boldsymbol{H}_{\text{inc}}\}$ is the given incident field and $\{\boldsymbol{E}_{\text{sc}}, \boldsymbol{H}_{\text{sc}}\}$ is the unknown scattered field. Then [451, p. 44, eqn (118)], [700, eqn (4.35)]

$$\begin{aligned}\boldsymbol{E}_{\text{sc}}(\boldsymbol{r},t) = -\frac{1}{4\pi}\int_S \Big\{ \mu\,[\boldsymbol{n}\times\dot{\boldsymbol{H}}] - [\boldsymbol{n}\times\boldsymbol{E}]\times\frac{\boldsymbol{R}}{R^2} - [\boldsymbol{n}\times\dot{\boldsymbol{E}}]\times\frac{\boldsymbol{R}}{cR} \\ - [\boldsymbol{n}\cdot\boldsymbol{E}]\frac{\boldsymbol{R}}{R^2} - [\boldsymbol{n}\cdot\dot{\boldsymbol{E}}]\frac{\boldsymbol{R}}{cR}\Big\}\frac{\mathrm{d}s_q}{R},\end{aligned} \tag{10.53}$$

where q is at $\boldsymbol{r}'$, $\boldsymbol{n} = \boldsymbol{n}_q$, $\boldsymbol{R} = \boldsymbol{r} - \boldsymbol{r}'$, $R = |\boldsymbol{R}|$ and we have used the notation defined in (9.2) and (9.43).

To verify (10.53), suppose that all fields vanish at $t = 0$ and then apply the Laplace transform, as in Section 9.1.5. Write $\mathscr{L}\{\boldsymbol{E}(\boldsymbol{r},t)\} = \widetilde{\boldsymbol{E}}(\boldsymbol{r},s)$, $\mathscr{L}\{\boldsymbol{E}_{\text{sc}}\} = \widetilde{\boldsymbol{E}_{\text{sc}}}$ and $\mathscr{L}\{\boldsymbol{H}\} = \widetilde{\boldsymbol{H}}$. Making use of (9.16), we find that the Laplace transform of (10.53) is

$$\begin{aligned}2\widetilde{\boldsymbol{E}_{\text{sc}}}(\boldsymbol{r},s) &= -\int_S \Big\{ \mu s(\boldsymbol{n}\times\widetilde{\boldsymbol{H}}) - (\boldsymbol{n}\times\widetilde{\boldsymbol{E}})\times\frac{\boldsymbol{R}}{R^2} - s(\boldsymbol{n}\times\widetilde{\boldsymbol{E}})\times\frac{\boldsymbol{R}}{cR} \\ &\qquad - (\boldsymbol{n}\cdot\widetilde{\boldsymbol{E}})\frac{\boldsymbol{R}}{R^2} - s(\boldsymbol{n}\cdot\widetilde{\boldsymbol{E}})\frac{\boldsymbol{R}}{cR}\Big\}\frac{\mathrm{e}^{-sR/c}}{2\pi R}\,\mathrm{d}s_q, \\ &= -\int_S \Big\{ -\mu s(\boldsymbol{n}\times\widetilde{\boldsymbol{H}})\mathscr{G} + (\boldsymbol{n}\times\widetilde{\boldsymbol{E}})\times\operatorname{grad}_q\mathscr{G} + (\boldsymbol{n}\cdot\widetilde{\boldsymbol{E}})\operatorname{grad}_q\mathscr{G}\Big\}\,\mathrm{d}s_q,\end{aligned} \tag{10.54}$$

where $\mathscr{G} = \mathrm{e}^{-sR/c}/(-2\pi R)$. Equation (10.54) holds in B_{e} and it is recognised as the Stratton–Chu formula: put $\omega = \mathrm{i}s$ in [599, eqn (6.38)] (a formula that holds in B).

There is a formula for $\boldsymbol{H}_{\text{sc}}$ in B_{e} corresponding to (10.53): it is [700, eqn (4.37)]

$$\boldsymbol{H}_{\text{sc}}(\boldsymbol{r},t) = \frac{1}{4\pi}\int_S \Big\{\varepsilon\,[\boldsymbol{n}\times\dot{\boldsymbol{E}}] + [\boldsymbol{n}\times\boldsymbol{H}]\times\frac{\boldsymbol{R}}{R^2} + [\boldsymbol{n}\times\dot{\boldsymbol{H}}]\times\frac{\boldsymbol{R}}{cR}$$
$$+ [\boldsymbol{n}\cdot\boldsymbol{H}]\frac{\boldsymbol{R}}{R^2} + [\boldsymbol{n}\cdot\dot{\boldsymbol{H}}]\frac{\boldsymbol{R}}{cR}\Big\}\frac{\mathrm{d}s_q}{R}. \tag{10.55}$$

The formulas above for $\boldsymbol{E}_{\text{sc}}$ and $\boldsymbol{H}_{\text{sc}}$ simplify when the scattering object B is a perfect conductor. In that case, the relevant boundary condition is

$$\boldsymbol{n}\times\boldsymbol{E} = \boldsymbol{0} \quad \text{on } S. \tag{10.56}$$

This eliminates some of the terms in (10.53) and (10.55). Define a surface current $\boldsymbol{J}$ by

$$\boldsymbol{J} = \boldsymbol{n}\times\boldsymbol{H}.$$

Evidently, $\boldsymbol{J}$ is a tangential vector field ($\boldsymbol{n}\cdot\boldsymbol{J}=0$); its surface divergence is

$$\operatorname{Div}\boldsymbol{J} = \operatorname{Div}(\boldsymbol{n}\times\boldsymbol{H}) = -\boldsymbol{n}\cdot\operatorname{curl}\boldsymbol{H} = -\varepsilon\,\boldsymbol{n}\cdot\dot{\boldsymbol{E}},$$

where we have used [189, eqn (2.75)] and $(1.84)_2$. Thus, if we define $\sigma(\boldsymbol{r},t) = \varepsilon\,\boldsymbol{n}\cdot\boldsymbol{E}$, we obtain

$$\operatorname{Div}\boldsymbol{J} + \dot{\sigma} = 0. \tag{10.57}$$

Similarly, if we compute the surface divergence of (10.56), we find that $\boldsymbol{n}\cdot\boldsymbol{H}=0$, assuming that $\boldsymbol{H}=\boldsymbol{0}$ initially. Using all these results, we find that (10.53) and (10.55) reduce to

$$\boldsymbol{E}_{\text{sc}}(\boldsymbol{r},t) = -\frac{1}{4\pi}\int_S \Big\{\mu\,[\dot{\boldsymbol{J}}] - [\sigma]\frac{\boldsymbol{R}}{\varepsilon R^2} - [\dot{\sigma}]\frac{\boldsymbol{R}}{\varepsilon cR}\Big\}\frac{\mathrm{d}s_q}{R}, \tag{10.58}$$

$$\boldsymbol{H}_{\text{sc}}(\boldsymbol{r},t) = \frac{1}{4\pi}\int_S \Big\{\frac{1}{R}[\boldsymbol{J}] + \frac{1}{c}[\dot{\boldsymbol{J}}]\Big\}\times\frac{\boldsymbol{R}}{R^2}\,\mathrm{d}s_q. \tag{10.59}$$

These representations can be found in [628, eqn (2.11)] and [454, §7.4], for example.

Let us derive time-domain boundary integral equations. From (10.59), taking account of the jump behaviour, we obtain

$$\boldsymbol{J}(\boldsymbol{r},t) = 2\,\boldsymbol{n}\times\boldsymbol{H}_{\text{inc}} + \frac{1}{2\pi}\boldsymbol{n}(\boldsymbol{r})\times\int_S \Big\{\frac{1}{R}[\boldsymbol{J}] + \frac{1}{c}[\dot{\boldsymbol{J}}]\Big\}\times\frac{\boldsymbol{R}}{R^2}\,\mathrm{d}s_q. \tag{10.60}$$

This is known as the (time dependent) *magnetic field integral equation* (MFIE); it is an integral equation for the surface current $\boldsymbol{J}$. See, for example, [700, eqn $(4.38)_2$], [628, eqn (2.15)], [454, eqn (7.21)] and [783, eqn (1)]. The time derivative of (10.60) has also been used [845].

Similarly, if we start from (10.58), we obtain

$$\boldsymbol{n}\times\boldsymbol{E}_{\text{inc}}(\boldsymbol{r},t) = \frac{1}{4\pi}\boldsymbol{n}(\boldsymbol{r})\times\int_S \Big\{\mu\,[\dot{\boldsymbol{J}}] - [\sigma]\frac{\boldsymbol{R}}{\varepsilon R^2} - [\dot{\sigma}]\frac{\boldsymbol{R}}{\varepsilon cR}\Big\}\frac{\mathrm{d}s_q}{R}, \tag{10.61}$$

which is to be solved for $\boldsymbol{J}$ and σ subject to (10.57). This pair defines what is known as the *electric field integral equation* (EFIE). See, for example, [700, eqn $(4.38)_1$] and [729].

The solvability of the EFIE has been proved by Rynne [729]; see also [33, 182] and [742, Appendix A].

Time-stepping methods (Section 10.3.1) for electromagnetic integral equations were first used by Bennett & Weeks [93] for a two-dimensional version of the MFIE; Poggio & Miller [700, §4.3.2] and Mittra [628, §2.3.1] describe the method for the MFIE (10.60) itself. Mittra [628, §2.3.3] also describes a time-stepping method for the EFIE (10.61). In their 1978 review, Bennett & Ross [92, p. 316] wrote 'The space-time integral equation technique provides a powerful tool for calculation of the transient scattered response of targets'; although true, that was before the instability problem was clearly identified in the context of acoustic scattering [90, 399]. Subsequent work, including [824, 730, 783, 708, 852, 731, 745, 219], leads to the following summary:

> *Early work in acoustics and electrodynamics concentrated on explicit formulations. These are grossly unstable, especially when problems of other than trivial size are addressed, and much of the literature concerns ways to stabilise the computation, using various smoothing and averaging procedures. Of late, implicit methods have been applied to electromagnetic analyses* [110, 757]. *The method has been successfully applied to large problems with no need for such* [procedures]. (Walker et al. [862])

Let us now parallel the discussion in Section 10.3.2. Time-domain versions of frequency-domain integral equations without irregular frequencies have been investigated [758, 20, 87, 169]. It has been shown numerically 'that the inaccuracy of naive spatial integral computations may affect the stability' [552, p. 4929]. Various temporal approximations have been used, including continuous piecewise quadratics [110], B-splines [896], Knab's [489] approximate prolate spheroidal wavefunctions [876, 889], Laguerre polynomials [183, 803, 863, 908, 909] and a localised exponential function [429]. For comparative reviews and comments, see [336, 850, 851]. For spatial discontinuous Galerkin schemes, see [689, 907]. Space-time Galerkin schemes have been devised [703]. For fast methods, see [757, 759, 34], for example.

Convolution quadrature methods (Section 10.4) have been developed for electromagnetic problems [866, 170, 41, 158]. For a review, see [550]. For a combination of Fourier transform in t and frequency-domain boundary integral equations (as discussed in Section 5.6.1), see [488]. For time-domain boundary integral equations in terms of electromagnetic potentials (1.87), see [721, 722].

10.6 Elastodynamics

The analogue of Kirchhoff's formula for elastodynamic problems (Section 1.6.2) was obtained by Love [573, §§14–17]. The actual formulas are complicated; see [880, Theorem 3.3], [5, Theorem 3.4], [269, §5.11], [682, §VI] and [492, §2.2]. However, they can be used to derive time-domain boundary integral equations (direct method, Section 10.2) [269, §5.14], For two-dimensional plane-strain problems, these equations were first solved numerically, using time-stepping methods (Section 10.3.1), by Niwa et al. [662]; for subsequent computations, see [589, 590, 23]. For three-dimensional implementations, see [400, 464, 8, 413, 716, 811, 679]. Numerical instability and what to do about it have been discussed [688, 101, 862, 678, 37].

One way to derive Love's time-domain integral representation is to use Laplace transforms. Assuming zero initial conditions, the governing equation for the displacement $\boldsymbol{u}(\boldsymbol{r},t)$ (1.91) becomes

$$(\lambda + 2\mu)\,\mathrm{grad}\,\mathrm{div}\,\boldsymbol{U} - \mu\,\mathrm{curl}\,\mathrm{curl}\,\boldsymbol{U} - \rho s^2 \boldsymbol{U} = \boldsymbol{0}, \tag{10.62}$$

where $\boldsymbol{U}(\boldsymbol{r},s) = \mathscr{L}\{\boldsymbol{u}(\boldsymbol{r},t)\} = \int_0^\infty \boldsymbol{u}(\boldsymbol{r},t)\,\mathrm{e}^{-st}\,\mathrm{d}t$ is the Laplace transform of $\boldsymbol{u}(\boldsymbol{r},t)$.

The basic fundamental solution for (10.62) is a 3×3 symmetric matrix with entries $\widetilde{G}_{ij}$. It can be obtained from the well-known frequency-domain version [599, §6.5.1] by replacing the frequency ω by is. The result is

$$\widetilde{G}_{ij}(\boldsymbol{r}-\boldsymbol{r}';s) = \Psi\delta_{ij} + \frac{c_s^2}{s^2}\frac{\partial^2}{\partial x_i\,\partial x_j}(\Phi - \Psi)$$

where $\Phi = \mathrm{e}^{-sR/c_p}/R$, $\Psi = \mathrm{e}^{-sR/c_s}/R$, $R = |\boldsymbol{r}-\boldsymbol{r}'|$, c_p is the compressional wave speed, c_s is the shear wave speed and δ_{ij} is the Kronecker delta.

We have

$$\frac{\partial^2\Phi}{\partial x_i\,\partial x_j} = \left[\left(\frac{3}{R^2} + \frac{3s}{c_pR} + \frac{s^2}{c_p^2}\right)\frac{R_iR_j}{R^2} - \left(\frac{1}{R} + \frac{s}{c_p}\right)\frac{\delta_{ij}}{R}\right]\Phi,$$

where $R_i = x_i - x_i'$. There is a similar formula for $\partial^2\Psi/\partial x_i\,\partial x_j$. Hence

$$\widetilde{G}_{ij} = \left(A_{ij}^\Phi + s^{-1}B_{ij}^\Phi + s^{-2}C_{ij}^\Phi\right)\mathrm{e}^{-sR/c_p} + \left(A_{ij}^\Psi + s^{-1}B_{ij}^\Psi + s^{-2}C_{ij}^\Psi\right)\mathrm{e}^{-sR/c_s} \tag{10.63}$$

where

$$A_{ij}^\Phi = \frac{c_s^2}{c_p^2}\frac{R_iR_j}{R^3}, \quad A_{ij}^\Psi = \frac{\delta_{ij}}{R} - \frac{R_iR_j}{R^3} = \frac{\partial^2 R}{\partial x_i\,\partial x_j},$$

$$B_{ij}^\Phi = -\frac{c_s}{c_p}B_{ij}^\Psi = \frac{c_s^2}{c_pR^2}\left(\frac{3R_iR_j}{R^2} - \delta_{ij}\right) = \frac{c_s^2 R}{c_p}\frac{\partial^2 R^{-1}}{\partial x_i\,\partial x_j},$$

$$C_{ij}^\Phi = -C_{ij}^\Psi = c_s^2\frac{\partial^2 R^{-1}}{\partial x_i\,\partial x_j}.$$

These six quantities depend on $\boldsymbol{r}-\boldsymbol{r}'$ but not on s. The dependence of $\widetilde{G}_{ij}$ on s is seen explicitly in (10.63). The stresses corresponding to $\widetilde{G}_{ij}$ can be calculated using Hooke's law (1.90).

Love's integral representation consists of elastodynamic single layers and double layers. The simpler constituent is the single layer, defined in the Laplace-transform domain by

$$U_i(\boldsymbol{r},s) = \int_S \widetilde{G}_{ij}(\boldsymbol{r}-\boldsymbol{r}',s)\,F_j(\boldsymbol{r}',s)\,\mathrm{d}s(\boldsymbol{r}'), \tag{10.64}$$

where $F_j(\boldsymbol{r},s) = \mathscr{L}\{f_j(\boldsymbol{r},t)\}$ and $\boldsymbol{f}(\boldsymbol{r},t)$ is a causal vector density function, which means $\boldsymbol{f}(\boldsymbol{r},t) = \boldsymbol{0}$ for $t \leq 0$. We substitute (10.63) in (10.64) and then invert. The A-terms are straightforward; use of (1.74) gives

$$\mathrm{e}^{-sR/c} F_j(\boldsymbol{r}',s) = \mathscr{L}\{f_j(\boldsymbol{r}',t-R/c)\}, \quad c = c_p, c_s.$$

A typical B-term is

$$\frac{\mathrm{e}^{-sR/c}}{s} F_j(\boldsymbol{r}',s) = \mathrm{e}^{-sR/c} \mathscr{L}\left\{ \int_0^t f_j(\boldsymbol{r}',\tau)\,\mathrm{d}\tau \right\} = \mathscr{L}\left\{ \int_0^{t-R/c} f_j(\boldsymbol{r}',\tau)\,\mathrm{d}\tau \right\},$$

whereas a typical C-term is

$$\frac{\mathrm{e}^{-sR/c}}{s^2} F_j(\boldsymbol{r}',s) = \mathrm{e}^{-sR/c} \mathscr{L}\left\{ \int_0^t \tau f_j(\boldsymbol{r}',\tau)\,\mathrm{d}\tau \right\} = \mathscr{L}\left\{ \int_0^{t-R/c} \tau f_j(\boldsymbol{r}',\tau)\,\mathrm{d}\tau \right\}.$$

Here, we have used (1.77). Hence, inverting (10.64), we obtain

$$\begin{aligned} u_i(\boldsymbol{r},t) = {} & \int_S \left\{ A_{ij}^{\Phi} f_j(\boldsymbol{r}',t-R/c_p) + A_{ij}^{\Psi} f_j(\boldsymbol{r}',t-R/c_s) \right\} \mathrm{d}s(\boldsymbol{r}') \\ & + \int_S \left\{ B_{ij}^{\Phi} \int_0^{t-R/c_p} f_j(\boldsymbol{r}',\tau)\,\mathrm{d}\tau + B_{ij}^{\Psi} \int_0^{t-R/c_s} f_j(\boldsymbol{r}',\tau)\,\mathrm{d}\tau \right. \\ & \qquad \left. + C_{ij}^{\Phi} \int_{t-R/c_s}^{t-R/c_p} \tau f_j(\boldsymbol{r}',\tau)\,\mathrm{d}\tau \right\} \mathrm{d}s(\boldsymbol{r}'), \end{aligned} \tag{10.65}$$

where we have used $C_{ij}^{\Phi} = -C_{ij}^{\Psi}$. Noting that $B_{ij}^{\Phi} = (R/c_p)C_{ij}^{\Phi}$ and $B_{ij}^{\Psi} = -(R/c_s)C_{ij}^{\Phi}$, we can rewrite the last two lines of (10.65) as

$$\int_S C_{ij}^{\Phi} \left\{ \int_{R/c_p}^t \tau f_j(\boldsymbol{r}',\tau - R/c_p)\,\mathrm{d}\tau - \int_{R/c_s}^t \tau f_j(\boldsymbol{r}',\tau - R/c_s)\,\mathrm{d}\tau \right\} \mathrm{d}s(\boldsymbol{r}').$$

Once elastodynamic single layers and double layers have been defined, we can derive integral equations using the indirect method (Section 10.1). For reviews of this approach, see [492, §3.3.1] and [181].

The earliest computations were done by Rizzo and his student Cruse; they solved a Laplace-transformed problem and then inverted the transform numerically [210, 209]. Analogous treatments using Fourier transforms instead of Laplace transforms have also been reported [493, 589, 663].

Another option is to reformulate problems in terms of scalar and vector potentials (see Section 1.6.2 and discussions in [490, 682]); for two early attempts in this direction, see [491, 762]. This approach was abandoned (mainly in favour of direct methods), but it has seen a revival [281].

Finally, we note that CQMs (Section 10.4) have been used for elastodynamic problems by Schanz and his co-authors [743, §4.2], [478, 618, 460] and by Hirose and his co-authors [319, 605].

10.7 Hydrodynamics

Consider a body B immersed in deep water. For simplicity, assume that B is submerged so that its boundary S does not intersect the mean free surface at $z=0$. The velocity potential u satisfies Laplace's equation, $\nabla^2 u=0$ in B_e, the domain occupied by the water. We want to solve the IBVP for u with a Neumann boundary condition on S, $\partial u/\partial n=v$, a given function, together with the free-surface boundary condition (1.93). We can assume that u satisfies zero initial conditions on $z=0$ ($u_0=u_1=0$ in (4.29)); if not, subtract an appropriate solution of the Cauchy–Poisson problem (Section 4.6.5).

Introduce the Laplace transform of u, $U(\boldsymbol{r},s)=\mathscr{L}\{u(\boldsymbol{r},t)\}$. It satisfies $\nabla^2 U=0$ in B_e and $\partial U/\partial n=V$ on S, with $V=\mathscr{L}\{v\}$. The free-surface condition (1.93) becomes

$$s^2 U+g\,\partial U/\partial z=0 \qquad \text{at } z=0.$$

The boundary-value problem for U can be solved using a boundary integral equation and an appropriate fundamental solution $\mathscr{G}(\boldsymbol{r},\boldsymbol{r}';s)$, which we take in the form

$$\mathscr{G}(\boldsymbol{r},\boldsymbol{r}';s)=\mathscr{G}(P,Q;s)=R^{-1}+\mathscr{H}(\boldsymbol{r},\boldsymbol{r}';s).$$

Here $R=|\boldsymbol{r}-\boldsymbol{r}'|$ and $\nabla_Q^2\mathscr{H}=0$ when $z'<0$. We choose $\mathscr{H}$ so that the free-surface condition is satisfied. Write

$$\mathscr{G}(\boldsymbol{r},\boldsymbol{r}';s)=\int_0^\infty J_0(k\rho)\left(\mathrm{e}^{-k|z-z'|}+h(s)\,\mathrm{e}^{k(z+z')}\right)\mathrm{d}k, \quad z<0,\ z'<0,$$

where J_0 is a Bessel function and $\rho=\{(x-x')^2+(y-y')^2\}^{1/2}$. Applying the free-surface condition at $z'=0$ gives

$$s^2(1+h)+gk(h-1)=0$$

whence

$$h(s)=\frac{gk-s^2}{gk+s^2}=-1+\frac{2gk}{s^2+gk}=-1+2\sqrt{gk}\,\mathscr{L}\left\{\sin(t\sqrt{gk})\right\}$$

and

$$\mathscr{G}(\boldsymbol{r},\boldsymbol{r}';s)=\frac{1}{R}-\frac{1}{R_1}+\int_0^\infty J_0(k\rho)\,\mathrm{e}^{k(z+z')}\frac{2gk\,\mathrm{d}k}{s^2+gk}, \quad z<0,\ z'<0, \tag{10.66}$$

with $R_1=\{\rho^2+(z+z')^2\}^{1/2}$.

Formal inversion of (10.66) introduces a Dirac delta function in t (from the first two terms) and gives a known formula for $\mathscr{L}^{-1}\{\mathscr{G}\}$ [84, eqn (4)], [83, eqn (26)]; this has been used for time-domain computations of u. For water of constant finite depth, see [299, 897].

Alternatively, we can solve for U and then invert. For example, we can seek U in the form of a single-layer potential, writing

$$U(\boldsymbol{r},s)=\int_S M(\boldsymbol{r}',s)\,\mathscr{G}(\boldsymbol{r},\boldsymbol{r}';s)\,\mathrm{d}s(\boldsymbol{r}'). \tag{10.67}$$

Application of the boundary condition yields a boundary integral equation for M. Then inversion of (10.67) gives [84, eqn (7)]

$$u(\boldsymbol{r},t) = \int_S \left(\frac{1}{R} - \frac{1}{R_1} \right) \mu(\boldsymbol{r}',t)\, \mathrm{d}s(\boldsymbol{r}') + 2\int_0^t \int_S \mu(\boldsymbol{r}',t')\, f(\rho, z+z', t-t')\, \mathrm{d}s(\boldsymbol{r}')\, \mathrm{d}t', \qquad (10.68)$$

where $\mu = \mathscr{L}^{-1}\{M\}$ and f is defined by

$$f(\rho,z,t) = \int_0^\infty \omega\, \mathrm{e}^{kz} J_0(k\rho) \sin \omega t \, \mathrm{d}k \quad \text{with} \quad \omega = \omega(k) = \sqrt{gk}. \qquad (10.69)$$

Wehausen [872, §4], [873, §3] has given alternative integral representations for u motivated by Volterra's analysis [859]. He has also investigated primitive causality (C1 in Section 1.5) for floating bodies [874]. Subsequently, Falnes [282] showed that the dispersive nature of water waves can lead to non-causal impulse response functions; see (1.79) and [211].

10.7.1 Clément's Equation

The function f, defined by (10.69), satisfies the axisymmetric form of Laplace's equation in cylindrical polar coordinates,

$$\rho \frac{\partial}{\partial \rho}\left(\rho \frac{\partial f}{\partial \rho} \right) + \rho^2 \frac{\partial^2 f}{\partial z^2} = 0, \qquad (10.70)$$

precisely because $\mathrm{e}^{kz} J_0(k\rho)$ satisfies the same equation. Let us express the spatial derivatives in (10.70) in terms of time derivatives. As $k^2 = \omega^4/g^2$, we see immediately from (10.69) that

$$\frac{\partial^2 f}{\partial z^2} = \frac{1}{g^2} \frac{\partial^4 f}{\partial t^4}. \qquad (10.71)$$

Next, as

$$\frac{\partial}{\partial \rho} J_0(k\rho) = k J_0'(k\rho) = \frac{k}{\rho} \frac{\partial}{\partial k} J_0(k\rho),$$

we have

$$\begin{aligned} \rho \frac{\partial f}{\partial \rho} &= \int_0^\infty \omega k\, \mathrm{e}^{kz} \sin \omega t \, \frac{\partial}{\partial k} J_0(k\rho)\, \mathrm{d}k \\ &= -\int_0^\infty \frac{\partial}{\partial k}\left(\omega k\, \mathrm{e}^{kz} \sin \omega t \right) J_0(k\rho)\, \mathrm{d}k \\ &= -\int_0^\infty \omega\, \mathrm{e}^{kz} \left\{ \left(\frac{3}{2} + kz \right) \sin \omega t + \frac{1}{2} \omega t \cos \omega t \right\} J_0(k\rho)\, \mathrm{d}k, \end{aligned}$$

after an integration by parts and use of $2\omega(\mathrm{d}\omega/\mathrm{d}k) = g$. Similarly,

$$\rho\frac{\partial}{\partial\rho}\left(\rho\frac{\partial f}{\partial\rho}\right) = \int_0^\infty \frac{\partial}{\partial k}\left(\omega k\,\mathrm{e}^{kz}\left\{\left(\frac{3}{2}+kz\right)\sin\omega t + \frac{1}{2}\omega t\cos\omega t\right\}\right)J_0(k\rho)\,\mathrm{d}k$$

$$= \int_0^\infty \omega\,\mathrm{e}^{kz}\left\{\frac{9}{4}+\omega^2\left(\frac{4z}{g}-\frac{t^2}{4}\right)+\omega^4\frac{z^2}{g^2}\right\}J_0(k\rho)\sin\omega t\,\mathrm{d}k$$

$$+\int_0^\infty \omega^2\,\mathrm{e}^{kz}\left(\frac{7t}{4}+\omega^2\frac{zt}{g}\right)J_0(k\rho)\cos\omega t\,\mathrm{d}k$$

$$= \frac{9}{4}f - \left(\frac{4z}{g}-\frac{t^2}{4}\right)\frac{\partial^2 f}{\partial t^2} + \frac{z^2}{g^2}\frac{\partial^4 f}{\partial t^4} + \frac{7t}{4}\frac{\partial f}{\partial t} - \frac{zt}{g}\frac{\partial^3 f}{\partial t^3}.$$

Combining this formula with (10.70) and (10.71), we obtain

$$\frac{R_0^2}{g^2}\frac{\partial^4 f}{\partial t^4} - \frac{zt}{g}\frac{\partial^3 f}{\partial t^3} + \left(\frac{t^2}{4}-\frac{4z}{g}\right)\frac{\partial^2 f}{\partial t^2} + \frac{7t}{4}\frac{\partial f}{\partial t} + \frac{9}{4}f = 0, \qquad (10.72)$$

where $R_0 = (\rho^2+z^2)^{1/2}$. This is *Clément's equation* [185, eqn (5.5)]. It is a fourth-order ordinary differential equation for $f(\rho,z,t)$ with t as the independent variable. Initial conditions can be inferred from (10.69):

$$f = 0, \quad \frac{\partial f}{\partial t} = \int_0^\infty \omega^2\,\mathrm{e}^{kz}J_0(k\rho)\,\mathrm{d}k = -\frac{gz}{R_0^3}, \quad \frac{\partial^2 f}{\partial t^2} = 0,$$

$$\frac{\partial^3 f}{\partial t^3} = -\int_0^\infty \omega^4\,\mathrm{e}^{kz}J_0(k\rho)\,\mathrm{d}k = -\frac{g^2}{R_0^5}(2z^2-\rho^2).$$

Using these together with (10.72) gives an initial-value problem that can be solved for f, using established numerical methods [381]. This provides an alternative to direct computation of the integral (10.69). For some comparisons, see [100].

Our derivation of (10.72) is based on a recent paper by Newman [656].

10.8 Method of Fundamental Solutions (MFS, ESM)

The standard *method of fundamental solutions* (MFS) is a numerical method used for solving elliptic boundary-value problems. In particular, it has been used extensively for scattering problems in the frequency domain [277]. There are versions of the MFS for time-domain problems.

To develop these methods, let us consider Problem DI_0 again. In Section 10.3, we sought u in the form of a single-layer potential, leading to the integral equation (10.14). Thus, we distributed simple sources over the surface S. Instead, let us distribute simple sources over an auxiliary surface S' inside S; we may think of S' as being a closed surface similar to but smaller than S (see Fig. 10.1). This gives the representation

$$u(\boldsymbol{r},t) = -\frac{1}{2\pi}\int_{S'}\mu\left(\boldsymbol{r}',t-\frac{|\boldsymbol{r}-\boldsymbol{r}'|}{c}\right)\frac{\mathrm{d}s(\boldsymbol{r}')}{|\boldsymbol{r}-\boldsymbol{r}'|}, \qquad (10.73)$$

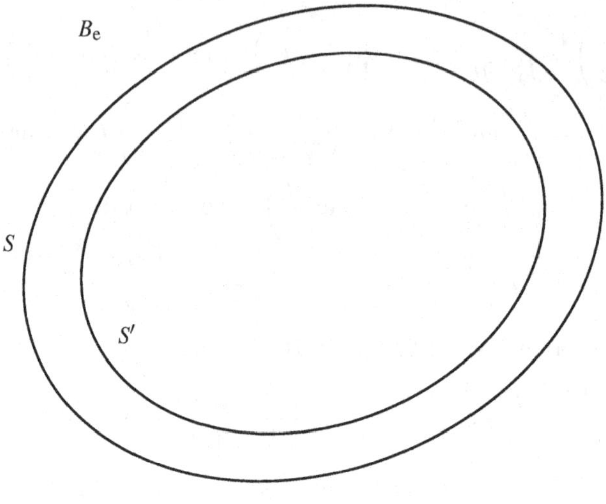

Figure 10.1 Simple sources are distributed over the auxiliary surface S' inside S.

where $\boldsymbol{r}$ locates a point outside S and $\boldsymbol{r}'$ locates the integration point on S'. Applying the boundary condition, $u = d$ on S, gives

$$\frac{1}{2\pi}\int_{S'} \mu\left(\boldsymbol{r}', t - \frac{|\boldsymbol{r}-\boldsymbol{r}'|}{c}\right) \frac{\mathrm{d}s(\boldsymbol{r}')}{|\boldsymbol{r}-\boldsymbol{r}'|} = -d(\boldsymbol{r},t), \quad \boldsymbol{r} \in S. \tag{10.74}$$

This is the same as (10.14) except that the integration is over S', not S, which means we always have $\boldsymbol{r} \neq \boldsymbol{r}'$ and so a simpler numerical scheme can be devised. Proceeding as in Section 10.3.1, break the auxiliary surface S' into N disjoint elements, $S' = \cup_{j=1}^{N} S'_j$, and then use a piecewise-constant spatial approximation, $\mu(\boldsymbol{r}', t) = \mu_j(t)$ for all $\boldsymbol{r}' \in S'_j$. In addition, choose a central point $\boldsymbol{r}'_j \in S'_j$ so that we can use the midpoint rule, $\int_{S'_j} g(\boldsymbol{r}')\,\mathrm{d}s(\boldsymbol{r}') \simeq g(\boldsymbol{r}'_j)\,|S'_j|$, where $|S'_j|$ is the surface area of S'_j. This reduces the representation (10.73) to

$$u(\boldsymbol{r},t) = \sum_{j=1}^{N} \frac{1}{|\boldsymbol{r}-\boldsymbol{r}'_j|} f_j\left(t - \frac{|\boldsymbol{r}-\boldsymbol{r}'_j|}{c}\right), \tag{10.75}$$

where $f_j(t) = -(2\pi)^{-1}|S'_j|\,\mu_j(t)$. Equation (10.75) gives a representation for u as a finite number of simple sources (2.11) located at points inside S; it is the usual starting point for the MFS.

To find the strengths of the N sources $f_j(t)$, $j = 1, 2, \ldots, N$, we apply the boundary condition at N points $\boldsymbol{r}_i \in S$, $i = 1, 2, \ldots, N$, marching forward in time, in steps of size Δ. This gives

$$\sum_{j=1}^{N} \frac{1}{|\boldsymbol{r}_i - \boldsymbol{r}'_j|} f_j\left(t_n - \frac{|\boldsymbol{r}_i - \boldsymbol{r}'_j|}{c}\right) = d(\boldsymbol{r}_i, t_n), \quad i = 1, 2, \ldots, N, \tag{10.76}$$

where $t_n = n\Delta$. Then, as in Section 10.3.1, use a piecewise-linear temporal approximation for $f_j(t)$,

$$f_j(t) = \Delta^{-1}\left\{f_j^{\ell}\,(t - t_{\ell-1}) + f_j^{\ell-1}\,(t_\ell - t)\right\}, \quad t_{\ell-1} \le t \le t_\ell, \tag{10.77}$$

where $f_j^{\ell} = f_j(t_\ell)$. Define an integer m_{ij} by

$$t_{m_{ij}} \le |\boldsymbol{r}_i - \boldsymbol{r}'_j|/c \ < t_{m_{ij}+1} \tag{10.78}$$

whence $t_{n-1-m_{ij}} < t_n - |\boldsymbol{r}_i - \boldsymbol{r}'_j|/c \le t_{n-m_{ij}}$ and then $f_j(t_n - |\boldsymbol{r}_i - \boldsymbol{r}'_j|/c)$ can be evaluated using (10.77) with $\ell = n - m_{ij}$. Equation (10.76) becomes

$$\sum_{j=1}^{N} \frac{1}{|\boldsymbol{r}_i - \boldsymbol{r}'_j|}\left\{(1 - \varepsilon_{ij})\, f_j^{n-m_{ij}} + \varepsilon_{ij}\, f_j^{n-1-m_{ij}}\right\} = d(\boldsymbol{r}_i, t_n), \quad i = 1, 2, \ldots, N, \tag{10.79}$$

where $\varepsilon_{ij} = |\boldsymbol{r}_i - \boldsymbol{r}'_j|/(c\Delta) - m_{ij}$; from (10.78), we have $0 \le \varepsilon_{ij} < 1$.

Suppose that S' is so close to S that $m_{ii} = 0$; thus $|\boldsymbol{r}_i - \boldsymbol{r}'_i| < c\Delta$ for $i = 1, 2, \ldots, N$. Then (10.79) can be written as an explicit formula for f_i^n in terms of quantities computed from previous time steps.

The method described above is a time-domain version of the MFS. It was developed by Kropp & Svensson in two papers from 1995 [508, 507]. Further developments followed [537, 538]; there is also a useful review by Lee [536]. All these papers use the terminology *equivalent source method* (ESM), but this is perhaps misleading because we do not know that we have equivalence: we do not know that u can be represented exactly as (10.73) or approximately as (10.75). (The same difficulty arises when the MFS is used in the frequency domain.)

There are variants of the time-domain MFS. For example, we could use a different temporal approximation instead of (10.77), or we could collocate at more points on S than there are source points. Also, we should expect instabilities with explicit schemes [538, 536]. Finally, some applications to moving scatterers have been made [537, 867].

10.9 Use of Fourier Transforms

We have already noted that a Fourier transform with respect to t converts a time-domain problem into a frequency-domain problem, and we outlined a basic strategy for exploiting this conversion in Section 5.6.1. This strategy can be effective [487] but developing implementations that are both accurate and efficient requires more effort. We describe such an implementation [19] for plane-wave scattering by a sound-soft obstacle, Problem DI_0. Thus $u = d$ on S with $d = -u_{\mathrm{inc}}$,

$$u_{\mathrm{inc}}(\boldsymbol{r},t) = f(t - \widehat{\boldsymbol{e}}\cdot\boldsymbol{r}/c) \quad \text{and} \quad \widehat{u}_{\mathrm{inc}}(\boldsymbol{r},\omega) = \widehat{f}(\omega)\exp\{\mathrm{i}(\omega/c)\widehat{\boldsymbol{e}}\cdot\boldsymbol{r}\}; \tag{10.80}$$

see (5.22) and (5.23). Suppose that the given function f is infinitely differentiable and compactly supported, with $f(t) = 0$ for $t < 0$ and for $t > T$; these conditions are very restrictive because they mean, in particular, that all derivatives of f are zero at $t = 0$ and at $t = T$. Then, see (5.13),

$$\widehat{f}(\omega) = \int_0^T f(t)\,\mathrm{e}^{\mathrm{i}\omega t}\,\mathrm{d}t. \tag{10.81}$$

Break the range of integration into N subintervals, each of length $3h$, using the points $t = t_n = 3nh$, $n = 0,1,2,\ldots,N$, where $h = \frac{1}{3}T$; the factor of 3 is introduced for algebraic convenience. Next, introduce a smooth 'flat-topped' partition of unity (PoU), subordinate to the points at t_n, $n = 0,1,2,\ldots,N$; this means we have a set of functions $w_n(t)$, one for each n, with the following *PoU properties*:

- w_n is real, non-negative and infinitely differentiable;
- $w_n(t) = 1$ for $|t - t_n| \le h$ (so 'flat-topped');
- $w_n(t) = 0$ for $|t - t_n| \ge 2h$ (which defines the support of w_n); and
- $\sum_n w_n(t) = 1$ for all t in the range of integration, $0 \le t \le T$.

It remains to define $w_n(t)$ for $h < |t - t_n| < 2h$; see below for prescriptions.

Let $f_n(t) = w_n(t)f(t)$. Then, as $f = \sum_n f_n$,

$$\widehat{f}(\omega) = \sum_{n=0}^{N} \widehat{f_n}(\omega), \quad \text{where} \quad \widehat{f_n}(\omega) = \int_{t_n-2h}^{t_n+2h} f_n(t)\,\mathrm{e}^{\mathrm{i}\omega t}\,\mathrm{d}t, \tag{10.82}$$

where we have integrated over the support of f_n. A simple substitution gives

$$\widehat{f_n}(\omega) = \mathrm{e}^{\mathrm{i}\omega t_n}\widehat{f_n^{\mathrm{slow}}}(\omega), \quad \text{where} \quad \widehat{f_n^{\mathrm{slow}}}(\omega) = \int_{-2h}^{2h} f_n(t+t_n)\,\mathrm{e}^{\mathrm{i}\omega t}\,\mathrm{d}t. \tag{10.83}$$

This function oscillates slowly (because the range of integration is short). Thus the formula $(10.83)_1$ factors out an explicit term that typically oscillates rapidly ($\mathrm{e}^{\mathrm{i}\omega t_n}$), leaving a computable term ($\widehat{f_n^{\mathrm{slow}}}$) that oscillates slowly.

As $f(t)$ and $f_n(t)$ are both smooth and compactly supported, repeated integration by parts in (10.81), $(10.82)_2$ and $(10.83)_2$ shows that $|\widehat{f}(\omega)|$, $|\widehat{f_n}(\omega)|$ and $|\widehat{f_n^{\mathrm{slow}}}(\omega)|$ all decay rapidly as $|\omega| \to \infty$, faster than any inverse power of $|\omega|$. Consequently, $f(t)$, $f_n(t)$ and $f_n^{\mathrm{slow}}(t)$ can be well approximated by band-limited functions, giving, for example,

$$f_n^{\mathrm{slow}}(t) = \frac{1}{2\pi}\int_{-\infty}^{\infty} \widehat{f_n^{\mathrm{slow}}}(\omega)\,\mathrm{e}^{-\mathrm{i}\omega t}\,\mathrm{d}\omega \simeq \frac{1}{2\pi}\int_{-\Omega}^{\Omega} \widehat{f_n^{\mathrm{slow}}}(\omega)\,\mathrm{e}^{-\mathrm{i}\omega t}\,\mathrm{d}\omega, \tag{10.84}$$

where the error decreases rapidly as the bandwidth Ω increases. There are good quadrature rules for approximating the integral over ω using M samples of the integrand at $\omega = \omega_m$, $m = 1,2,\ldots,M$, where $-\Omega \le \omega_1 < \omega_2 < \cdots < \omega_M \le \Omega$. It is at these M frequencies that we shall solve frequency-domain scattering problems.

Let $\widehat{u}_{\mathrm{pw}}(\boldsymbol{r},\omega)$ be the outgoing frequency-domain solution for scattering of a plane wave, with $\widehat{u}_{\mathrm{pw}}(\boldsymbol{r},\omega) = -\exp\{\mathrm{i}(\omega/c)\widehat{\boldsymbol{e}}\cdot\boldsymbol{r}\}$ on the boundary S; see (10.80) and (8.4). Linearity gives

$$\widehat{u}(\boldsymbol{r},\omega) = \widehat{f}(\omega)\,\widehat{u}_{\mathrm{pw}}(\boldsymbol{r},\omega).$$

Using the PoU to decompose $\widehat{f}$ (see (10.82)) gives

$$\widehat{u}(\boldsymbol{r},\omega) = \sum_{n=0}^{N} \widehat{u}_n(\boldsymbol{r},\omega) \quad \text{with} \quad \widehat{u}_n(\boldsymbol{r},\omega) = \widehat{f}_n(\omega)\,\widehat{u}_{\mathrm{pw}}(\boldsymbol{r},\omega)$$

whence $u = \sum_{n=0}^{N} u_n$ where

$$u_n(\boldsymbol{r},t) \simeq \frac{1}{2\pi}\int_{-\Omega}^{\Omega} \widehat{f}_n^{\mathrm{slow}}(\omega)\,\widehat{u}_{\mathrm{pw}}(\boldsymbol{r},\omega)\,\mathrm{e}^{-\mathrm{i}\omega(t-t_n)}\,\mathrm{d}\omega,$$

an integral that can be well approximated using M frequencies ω_m. It is notable that the expensive frequency-domain computations are separated from the PoU calculations, and this contributes significantly to the algorithm's performance; see [19] for details and numerical results.

Smooth Flat-Topped Partitions of Unity

Partitions of unity are found in books on analysis or differential topology (where they are smooth but seldom constructed) and in books on splines or wavelets (where they are constructed but they are not infinitely differentiable). For a smooth flat-topped PoU, we require the properties listed above on page 184. If we focus on one subinterval, $t_n \le t \le t_{n+1} = t_n + 3h$, the PoU properties imply the following:

- for $t_n \le t \le t_n + h$, $w_n(t) = 1$ and $w_{n+1}(t) = 0$;
- for $t_n + h \le t \le t_n + 2h$, we require $w_n(t) + w_{n+1}(t) = 1$; and
- for $t_n + 2h \le t \le t_{n+1}$, $w_n(t) = 0$ and $w_{n+1}(t) = 1$.

Notice that, in the interval $t_n \le t \le t_{n+1}$, all the functions w_m are identically zero, except for two, namely, w_n and w_{n+1}. We shall build each w_n using a translated (and scaled, if necessary) version of a single function w.

Let us give some examples. Bruno and his co-authors have used several different but similar functions. One is

$$S_1(x;x_0,x_1) = \begin{cases} 1, & |x| \le x_0, \\ \mathscr{E}(u), & x_0 < |x| < x_1, \quad u = \dfrac{|x|-x_0}{x_1-x_0}, \\ 0, & |x| \ge x_1. \end{cases} \tag{10.85}$$

where $0 < x_0 < x_1$ and

$$\mathscr{E}(\xi) = \exp\left(\frac{2\,\mathrm{e}^{-1/\xi}}{\xi - 1}\right), \quad 0 < \xi < 1, \text{ with limiting values } \mathscr{E}(0) = 1 \text{ and } \mathscr{E}(1) = 0.$$

We see that $S_1(x;x_0,x_1)$ is an even function of x; it is a symmetric flat-topped 'bump' function. It was used by Bruno & Delourme [134, eqn (17)]. If we put $x_0 = h$ and $x_1 = 2h$, we obtain the function

$$W(t) = \begin{cases} 1, & |t| \le h, \\ \mathscr{E}(u), & h < |t| < 2h, \quad u = (|t|/h) - 1, \\ 0, & |t| \ge 2h. \end{cases}$$

Then, to obtain a partition of unity, define

$$w_n(t) = \frac{W(t-t_n)}{\sum_m W(t-t_m)}; \tag{10.86}$$

the sum in the denominator has at most two non-trivial terms.

If we replace $|x|$ in (10.85) by x, we obtain

$$S_2(x;x_0,x_1) = \begin{cases} 1, & x \le x_0, \\ \mathscr{E}(u), & x_0 < x < x_1, \quad u = \dfrac{x - x_0}{x_1 - x_0}, \\ 0, & x \ge x_1. \end{cases} \tag{10.87}$$

This is a smoothed step function. It was used by Bruno & Kunyansky (see $E(t)$ on p. 2925 of [136], but replace $|t|$ by t in the definition of x) and by Bruno et al. (see $S(x,x_0,x_1)$ on p. 635 of [135] but replace $|x|$ by x in the definition of u).

Next, consider

$$S_3(x;x_0,x_1) = \begin{cases} 0, & x \le -x_1, \\ 1 - \mathscr{E}(u_1), & -x_1 < x < -x_0, \quad u_1 = \dfrac{x + x_1}{x_1 - x_0}, \\ 1, & -x_0 \le x \le x_0, \\ \mathscr{E}(u_2), & x_0 < x < x_1, \quad u_2 = \dfrac{x - x_0}{x_1 - x_0}, \\ 0, & x \ge x_1. \end{cases} \tag{10.88}$$

Unlike S_1, $S_3(x;x_0,x_1)$ is not an even function of x. If we put $x_0 = h$ and $x_1 = 2h$, we obtain the function

$$w(t) = \begin{cases} 0, & t \le -2h, \\ 1 - \mathscr{E}(u_1), & -2h < t < -h, \quad u_1 = (t/h) + 2, \\ 1, & -h \le t \le h, \\ \mathscr{E}(u_2), & h < t < 2h, \quad u_2 = (t/h) - 1, \\ 0, & t \ge 2h. \end{cases} \tag{10.89}$$

This is the function used by Anderson et al. [19, eqn (12)]. The benefit of introducing asymmetry into w is that the set of functions $w_n(t) = w(t - t_n)$, integer n, is a partition of unity; the normalisation (10.86) is not required.

Section 13 of the book by Tu [836] contains a step-by-step construction of an infinitely differentiable flat-topped bump function, starting with [836, eqn (13.1)],

$$g(x) = \frac{f(x)}{f(x) + f(1-x)} \quad \text{with} \quad f(x) = \begin{cases} \mathrm{e}^{-1/x}, & x > 0, \\ 0, & x \le 0. \end{cases}$$

These define a smooth step function, with $g(x) = 0$ for $x \leq 0$ and $g(x) = 1$ for $x \geq 1$. From (10.87), $1 - S_2(x;0,1)$ is another smooth step function, but they differ in the interval $0 < x < 1$.

Then, for a flat-topped bump function, Tu obtains [836, p. 143]

$$W(t) = 1 - g\left(\frac{t^2 - a^2}{b^2 - a^2}\right).$$

This even function satisfies $W(t) = 1$ for $|t| \leq a$ and $W(t) = 0$ for $|t| \geq b > a > 0$. Thus, it is comparable with $S_1(t;a,b)$; see (10.85). Tu goes on [836, p. 146] to discuss partitions of unity using the normalisation (10.86).

We conclude that smooth, flat-topped partitions of unity can be constructed explicitly but, of course, they are not uniquely defined.

10.10 Cracks, Screens and Other Thin Scatterers

The interactions between waves and thin structures have been studied for more than a century. Early work involves solving the two-dimensional wave equation in the presence of a thin semi-infinite screen:

> *More papers have probably been written about the diffraction of a harmonic wave or a pulse by wedges and half-planes than about any other boundary-value problem for the wave equation.... The diffraction of pulses was first considered by Sommerfeld* [788] *and by Lamb* [515].
>
> (Friedlander [310, p. 108])

See [310, Chapter 5] for details and references. For scattering by a thin straight screen of finite width, see [304, 820, 80, 359].

Three-dimensional problems involving a thin semi-infinite screen (or wedge) with an infinite straight edge have also been studied, perhaps starting with Cagniard's 1935 paper [141] on the scattering of a spherical pulse by a sound-hard screen. See also [471, 861, 822, 902]. In many cases, simple explicit solutions can be obtained, sometimes from applications of the *Cagniard–de Hoop method*. This method was developed by Cagniard in the 1930s, simplified by de Hoop [225], and popularised by Cagniard's book [142]. It is useful when a problem can be reduced to a pair of integral transforms in such a way that they 'are played against each other to the point of their mutual annihilation' [329, p. 533]. For expositions of the method, see [5, §7.9] and [868, Chapter 6].

For three-dimensional problems involving a bounded thin screen, the prototypical problem is scattering by a disc. We discuss such problems in more detail. Thus consider a flat screen, Ω, in the plane $z = 0$. We assume that the screen is sound-hard, but other boundary conditions could be used; we shall give the relevant formulas for a sound-soft screen.

There is an incident field, $u_{\text{inc}}(\boldsymbol{r},t)$, and the problem is to compute the scattered field, $u_{sc}(\boldsymbol{r},t)$. Thus we seek a bounded wavefunction satisfying the boundary condition

$$\frac{\partial u_{sc}}{\partial z} = g_{inc} \quad \text{on both sides of } \Omega, \tag{10.90}$$

where $g_{inc}(x,y) = -\partial u_{inc}/\partial z$ evaluated at $z=0$. We also impose zero initial conditions.

It can be shown that the solution must be an odd function of z (see (10.108) below), so the problem can be reduced to one in the half-space $z>0$. (If the screen had been soft, with $u_{sc}+u_{inc}=0$ on Ω, then u_{sc} would be an even function of z.)

Let $\mathscr{L}\{u_{sc}\} = U(\boldsymbol{r},s)$ denote the Laplace transform of $u_{sc}(\boldsymbol{r},t)$ with respect to t. Thus U solves the modified Helmholtz equation (5.2) and the Laplace transform of (10.90).

10.10.1 Derivation of Integral Equations Using Fourier Transforms

Take the Fourier transform of $\nabla^2 U = (s/c)^2 U$ with respect to x and y, with, for example,

$$\mathscr{U}(\xi,\eta,z,s) = \mathscr{F}\{U\} = \int U(x,y,z,s)\,\mathrm{e}^{\mathrm{i}(\xi x+\eta y)}\,\mathrm{d}x\,\mathrm{d}y;$$

the integration is over the whole xy-plane. Then, with $\kappa^2 = \xi^2+\eta^2$, and writing, for example, $\mathscr{U}' = \partial\mathscr{U}/\partial z$, we obtain $\mathscr{U}'' = \left(\kappa^2+(s/c)^2\right)\mathscr{U}$. Hence,

$$\mathscr{U}(\xi,\eta,z,s) = B(\xi,\eta,s)\,\mathrm{e}^{-\gamma z}, \quad z>0, \tag{10.91}$$

for some function B, where

$$\gamma = \left(\kappa^2+(s/c)^2\right)^{1/2}, \quad \operatorname{Re}\gamma \geq 0, \tag{10.92}$$

and we have discarded a solution proportional to $\mathrm{e}^{+\gamma z}$.

We have a screen, Ω, in the xy-plane. The rest of the xy-plane is denoted by Ω'. As we have split the problem into two half-space problems, we must also impose continuity of U across Ω'. Thus, let

$$V(x,y,s) = U(x,y,0+,s) - U(x,y,0-,s);$$

this gives the discontinuity in U across the plane $z=0$. Hence $V(x,y,s)=0$ for $(x,y)\in\Omega'$. The Fourier transform of V is

$$\mathscr{V}(\xi,\eta,s) = \int_\Omega V(x,y,s)\,\mathrm{e}^{\mathrm{i}(\xi x+\eta y)}\,\mathrm{d}x\,\mathrm{d}y = 2B(\xi,\eta,s) = 2\mathscr{U}(\xi,\eta,0,s). \tag{10.93}$$

We obtain an integral equation by inverting, $U = \mathscr{F}^{-1}\mathscr{U}$, and imposing the Laplace transform of (10.90),

$$\frac{1}{4\pi^2}\int \mathscr{U}'(\xi,\eta,0,s)\,\mathrm{e}^{-\mathrm{i}(\xi x+\eta y)}\,\mathrm{d}\xi\,\mathrm{d}\eta = G_{inc}(x,y,s), \quad (x,y)\in\Omega, \tag{10.94}$$

where the integration is over the whole $\xi\eta$-plane, $\mathscr{U}$ is given by (10.91) and $G_{\rm inc} = \mathscr{L}\{g_{\rm inc}\}$. Symbolically, we have

$$\mathscr{F}^{-1}\{\gamma\mathscr{F}\{V\}\} = -2G_{\rm inc}. \tag{10.95}$$

This integral equation holds for flat screens Ω of any shape.

If Ω had been sound-soft, with $u_{\rm sc} = -u_{\rm inc}$ on Ω instead of (10.90), we would have obtained

$$\mathscr{F}^{-1}\{\gamma^{-1}\mathscr{F}\{T\}\} = 2\mathscr{L}\{u_{\rm inc}(x,y,0,t)\}, \tag{10.96}$$

where T is the discontinuity in $\partial U/\partial z$ across Ω,

$$T(x,y,s) = \left.\frac{\partial U}{\partial z}\right|_{z=0+} - \left.\frac{\partial U}{\partial z}\right|_{z=0-}.$$

Returning to (10.95), this equation can be written as a hypersingular integral equation, as follows. Let $\mathbb{L} = \nabla_2^2 - (s/c)^2$, where ∇_2^2 is the two-dimensional Laplacian with respect to x and y. We have $\mathbb{L}\,{\rm e}^{{\rm i}(\xi x+\eta y)} = -\gamma^2{\rm e}^{{\rm i}(\xi x+\eta y)}$, using (10.92). Thus

$$\mathscr{F}^{-1}\{\gamma\mathscr{F}\{V\}\} = -\mathbb{L}\,\mathscr{F}^{-1}\{\gamma^{-1}\mathscr{F}\{V\}\}.$$

Then, changing the order of integration (which is now permissible), we obtain

$$(\mathscr{F}^{-1}\{\gamma^{-1}\mathscr{F}\{V\}\})(x,y,s) = \int_\Omega M(x-x',y-y',s)\,V(x',y',s)\,{\rm d}A' \tag{10.97}$$

where ${\rm d}A' = {\rm d}x'\,{\rm d}y'$ and

$$M(x,y,s) = \int \frac{1}{4\pi^2\gamma}{\rm e}^{-{\rm i}(\xi x+\eta y)}\,{\rm d}\xi\,{\rm d}\eta = \int_0^\infty \frac{\kappa}{2\pi\gamma}J_0(\kappa r)\,{\rm d}\kappa = \frac{{\rm e}^{-sr/c}}{2\pi r}. \tag{10.98}$$

To evaluate M, we used polar coordinates,

$$x = r\cos\theta, \quad y = r\sin\theta, \quad \xi = \kappa\cos\beta, \quad \eta = \kappa\sin\beta, \tag{10.99}$$

${\rm e}^{{\rm i}\kappa x} = \sum_{n=-\infty}^{\infty} {\rm i}^n J_n(\kappa r)\,{\rm e}^{{\rm i}n\theta}$ and [353, 6.554 (1)]. Thus, (10.95) becomes

$$\mathbb{L}\int_\Omega \frac{{\rm e}^{-sR/c}}{4\pi R}V(x',y',s)\,{\rm d}A' = G_{\rm inc}(x,y,s), \quad (x,y)\in\Omega, \tag{10.100}$$

with $R = \{(x-x')^2+(y-y')^2\}^{1/2}$. We have

$$\begin{aligned}\mathbb{L}\left(\frac{{\rm e}^{-sR/c}}{R}\right) &= \nabla_2^2\left(\frac{1}{R}\right) + \nabla_2^2\left(\frac{{\rm e}^{-sR/c}-1}{R}\right) - \frac{s^2}{c^2}\frac{{\rm e}^{-sR/c}}{R} \\ &= \nabla_2^2\left(\frac{1}{R}\right) + \frac{(1-sR/c)\,{\rm e}^{-sR/c}-1}{R^3},\end{aligned} \tag{10.101}$$

where the last term is $O(R^{-1})$ as $R\to 0$. When (10.101) is combined with (10.100), we see that V solves a hypersingular integral equation over Ω; here, we are recalling the definition

$$⨎_\Omega g(x',y')\,\frac{{\rm d}A'}{R^3} = \nabla_2^2\int_\Omega g(x',y')\,\frac{{\rm d}A'}{R}, \tag{10.102}$$

which defines the hypersingular integral on the left.

For an alternative strategy, write $\gamma = \kappa + (\gamma - \kappa)$ in (10.95) giving

$$-\mathscr{F}^{-1}\{\kappa\mathscr{F}\{V\}\} - \mathscr{F}^{-1}\{(\gamma-\kappa)\mathscr{F}\{V\}\} = 2G_{\text{inc}}.$$

A calculation similar to that leading to (10.97) with (10.98) gives

$$-\mathscr{F}^{-1}\{\kappa\mathscr{F}\{V\}\} = \nabla_2^2\mathscr{F}^{-1}\{\kappa^{-1}\mathscr{F}\{V\}\} = \nabla_2^2\int_\Omega \frac{V\,\mathrm{d}A'}{2\pi R}.$$

Thus we obtain

$$\frac{1}{4\pi}\,⨎_\Omega \frac{V}{R^3}\,\mathrm{d}A' + (\mathscr{K}_\Omega V)(x,y,s) = G_{\text{inc}}(x,y,s), \quad (x,y)\in\Omega$$

where

$$(\mathscr{K}_\Omega v)(x,y,s) = -\frac{1}{2}\mathscr{F}^{-1}\{(\gamma-\kappa)\mathscr{F}\{V\}\} = \int_\Omega \mathscr{K}(x-x',y-y',s)V(x',y',s)\,\mathrm{d}A'$$

and

$$\mathscr{K}(X,Y,s) = \frac{1}{8\pi^2}\int(\kappa-\gamma)\mathrm{e}^{-\mathrm{i}(\xi X+\eta Y)}\,\mathrm{d}\xi\,\mathrm{d}\eta = \frac{1}{4\pi}\int_0^\infty \kappa(\kappa-\gamma)\,J_0(\kappa R)\,\mathrm{d}\kappa.$$

Adopting similar procedures for the sound-soft problem, we find that (10.96) can be written as a Fredholm integral equation of the first kind over Ω with a weakly-singular kernel:

$$\int_\Omega T(x',y',s)\,\frac{\mathrm{e}^{-sR/c}}{4\pi R}\,\mathrm{d}A' = \mathscr{L}\{u_{\text{inc}}\}, \quad (x,y)\in\Omega. \tag{10.103}$$

Here, we have used (10.97) and (10.98).

The integral equations for V and T can be solved using any of the methods available for frequency-domain problems; for methods and references, see [287] and [599, §6.7]. If such a method is adopted, there remains the problem of inverting the Laplace transform.

Alternatively, we can invert the Laplace transform directly, so as to obtain a time-domain integral equation over Ω. If we do that with (10.103) for sound-soft screens, we obtain

$$\int_\Omega \tau(x',y',t-R/c)\,\frac{\mathrm{d}A'}{4\pi R} = u_{\text{inc}}(x,y,t), \quad (x,y)\in\Omega, \tag{10.104}$$

where $T(x,y,s) = \mathscr{L}\{\tau(x,y,t)\}$ and, as usual, it is assumed that $\tau(x,y,t) = 0$ for $t \leq 0$. A similar calculation can be made for sound-hard screens, starting with (10.100) and (10.101).

10.10.2 Derivation of Integral Equations Using Layer Potentials

When solving $\nabla^2 U = (s/c)^2 U$, we can use the integral representation (9.17). When this is specialised to a thin flat screen, we obtain

$$U(x,y,z,s) = \int_\Omega V(x',y',s)\left\{\lim_{z'\to 0}\frac{\partial}{\partial z'}\left(\frac{\mathrm{e}^{-s\mathscr{R}/c}}{4\pi\mathscr{R}}\right)\right\}\mathrm{d}A' - \int_\Omega T(x',y',s)\,\frac{\mathrm{e}^{-s\mathscr{R}_0/c}}{4\pi\mathscr{R}_0}\,\mathrm{d}A', \qquad (10.105)$$

where $\mathscr{R}^2 = R^2 + (z-z')^2$ and $\mathscr{R}_0^2 = R^2 + z^2$.

For a sound-soft screen, $V = 0$ and then (10.105) shows that U can be written as a single-layer potential; application of the boundary condition $U = -\mathscr{L}\{u_{\mathrm{inc}}\}$ on Ω reduces (10.105) to (10.103).

For a sound-hard screen, $T = 0$ and then (10.105) shows that U can be written as a double-layer potential. Application of the boundary condition will lead to a hypersingular integral equation for V; here, we recall an equivalent definition to (10.102),

$$\oint_\Omega g(x',y')\,\frac{\mathrm{d}A'}{R^3} = \lim_{z\to 0}\frac{\partial}{\partial z}\int_\Omega g(x',y')\left\{\lim_{z'\to 0}\frac{\partial}{\partial z'}\left(\frac{1}{\mathscr{R}}\right)\right\}\mathrm{d}A'. \qquad (10.106)$$

However, as the basic representation for U is the Laplace transform of the Kirchhoff formula (9.12), we can write down equivalent formulas in the time domain. For the sound-soft case, the relevant integral representation is

$$u_{\mathrm{sc}}(x,y,z,t) = -\int_\Omega \tau(x',y',t-\mathscr{R}_0/c)\,\frac{\mathrm{d}A'}{4\pi\mathscr{R}_0} \qquad (10.107)$$

leading back to the integral equation (10.104).

For the sound-hard case, we have the double-layer representation

$$u_{\mathrm{sc}}(x,y,z,t) = \frac{z}{4\pi}\int_\Omega \left\{\frac{1}{\mathscr{R}_0^3}\,v(x',y',t-\mathscr{R}_0/c) + \frac{1}{c\mathscr{R}_0^2}\,\dot{v}(x',y',t-\mathscr{R}_0/c)\right\}\mathrm{d}A' \qquad (10.108)$$

where $V(x,y,s) = \mathscr{L}\{v(x,y,t)\}$, $\dot{v}(x,y,t) = \partial v/\partial t$ and $v(x,y,t) = 0$ for $t \le 0$. Application of the boundary condition (10.90) then leads to a hypersingular equation for v.

10.10.3 Comments and Literature

The main virtue of the Fourier-transform derivation (Section 10.10.1) is that it does not require access to the underlying fundamental solution (free-space Green's function), which for acoustic problems is simply $\mathrm{e}^{-s\mathscr{R}/c}/\mathscr{R}$. Consequently, similar methods could be developed for more complicated problems. On the other hand, Green's function techniques (Section 10.10.2) can be used for non-flat screens.

Another option is to solve for B instead of V. From (10.93), we have $B(\xi,\eta,s) =$

$\mathscr{U}(\xi,\eta,0,s)$. Imposing (10.90) on Ω and $V=0$ on Ω' gives a pair of *dual integral equations* for B:

$$\frac{1}{4\pi^2}\int \gamma B(\xi,\eta,s)\,\mathrm{e}^{-\mathrm{i}(\xi x+\eta y)}\,\mathrm{d}\xi\,\mathrm{d}\eta = -G_{\mathrm{inc}}(x,y,s), \qquad (x,y)\in\Omega, \qquad (10.109)$$

$$\frac{1}{4\pi^2}\int B(\xi,\eta,s)\,\mathrm{e}^{-\mathrm{i}(\xi x+\eta y)}\,\mathrm{d}\xi\,\mathrm{d}\eta = 0, \qquad (x,y)\in\Omega'. \qquad (10.110)$$

Methods for handling dual integral equations are available; see [287] for references.

The integral representations in terms of time-domain layer potentials, (10.107) and (10.108), have been known for many years. Fox [304, p. 74] derived them in 1948 and wrote, 'The relations we have given in this section are, of course, the pulse analogues of well-known relations [516, Chapter X, §299] for sinusoidal waves'; see also [305, §2], where multiple screens are considered. The same time-domain representations were derived and analysed by Ha-Duong [376]. Numerical solutions for sound-hard screens, using a time-stepping method (Section 10.3.1), were obtained by Hirose & Achenbach [414] and by Kawai & Terai [466]. For more complicated applications involving thin structures, see [684, 388, 339, 340].

Although related methods could be developed for electromagnetic scattering by thin objects, little has been done. Bennett & Mieras [91] used a time-domain integral equation and gave results for scattering by several objects including a circular disc; results for a disc were also given by Dominek [242] using a Fourier synthesis of frequency-domain solutions.

On the other hand, the literature on elastodynamic problems is extensive, mainly because such time-domain crack problems are encountered in studies of dynamic fracture mechanics [307, 685]. Early work on two-dimensional plane-strain problems includes [821, 777, 439]. These papers consider flat cracks and use dual integral equations. Similar methods have been developed for flat cracks in three dimensions [776, 166, 438, 771]. Time-domain integral equations have been used for plane-strain problems [81, 321, 808, 323] and for three-dimensional problems [415, 904, 807, 806].

References

[1] T. Abboud, P. Joly, J. Rodríguez & I. Terrasse, Coupling discontinuous Galerkin methods and retarded potentials for transient wave propagation on unbounded domains. *J. Comp. Phys.* **230** (2011) 5877–5907. Cited: pp. 28 & 166.

[2] I. Åberg, G. Kristensson & D.J.N. Wall, Transient waves in nonstationary media. *J. Math. Phys.* **37** (1996) 2229–2252. Cited: p. 6.

[3] M. Abramowitz & I.A. Stegun (ed.), *Handbook of Mathematical Functions*. New York: Dover, 1965. Cited: pp. 52, 54, 55, 114 & 117.

[4] A.I. Abreu, J.A.M. Carrer & W.J. Mansur, Scalar wave propagation in 2D: a BEM formulation based on the operational quadrature method. *Eng. Anal. Bound. Elem.* **27** (2003) 101–105. Cited: p. 171.

[5] J.D. Achenbach, *Wave Propagation in Elastic Solids*. Amsterdam: North-Holland, 1973. Cited: pp. 26, 176 & 187.

[6] R. Adelman, N.A. Gumerov & R. Duraiswami, Semi-analytical computation of acoustic scattering by spheroids and disks. *J. Acoust. Soc. Amer.* **136** (2014) EL405–EL410. Cited: p. 128.

[7] R. Adelman, N.A. Gumerov & R. Duraiswami, Software for computing the spheroidal wave functions using arbitrary precision arithmetic. arXiv:1408.0074, August 2014. Cited: p. 55.

[8] S. Ahmad & P.K. Banerjee, Time-domain transient elastodynamic analysis of 3-D solids by BEM. *Int. J. Numer. Meth. Eng.* **26** (1988) 1709–1728. Cited: p. 177.

[9] A. Aimi, M. Diligenti, A. Frangi & C. Guardasoni, Neumann exterior wave propagation problems: computational aspects of 3D energetic Galerkin BEM. *Comput. Mech.* **51** (2013) 475–493. Cited: p. 167.

[10] A. Aimi, M. Diligenti, C. Guardasoni, I. Mazzieri & S. Panizzi, An energy approach to space-time Galerkin BEM for wave propagation problems. *Int. J. Numer. Meth. Eng.* **80** (2009) 1196–1240. Cited: p. 167.

[11] N. Akkas, Residual potential method in spherical coordinates and related approximations. *Mechanics Research Comm.* **6** (1979) 257–262. Cited: pp. 120 & 121.

[12] N. Akkaş & A.E. Engin, Transient response of a spherical shell in an acoustic medium—Comparison of exact and approximate solutions. *J. Sound Vib.* **73** (1980) 447–460. Cited: p. 121.

[13] N. Akkas & U. Zakout, Transient response of an infinite elastic medium containing a spherical cavity with and without a shell embedment. *Int. J. Eng. Sci.* **35** (1997) 89–112. Cited: p. 121.

[14] N. Akkas, U. Zakout & G.E. Tupholme, Propagation of waves from a spherical cavity with and without a shell embedment. *Acta Mechanica* **142** (2000) 1–11. Cited: p. 121.

[15] B. Alpert, L. Greengard & T. Hagstrom, Rapid evaluation of nonreflecting boundary kernels for time-domain wave propagation. *SIAM J. Numer. Anal.* **37** (2000) 1138–1164. Cited: p. 121.

[16] B. Alpert, L. Greengard & T. Hagstrom, Nonreflecting boundary conditions for the time-dependent wave equation. *J. Comp. Phys.* **180** (2002) 270–296. Cited: p. 121.

[17] M.S. Aly & T.T.Y. Wong, Scattering of a transient electromagnetic wave by a dielectric sphere. *IEE Proc. H: Microwaves, Antennas & Propagation* **138** (1991) 192–198. Cited: p. 120.

[18] B.E. Anderson, M. Griffa, C. Larmat, T.J. Ulrich & P.A. Johnson, Time reversal. *Acoustics Today* **4**, 1 (2008) 5–16. Cited: p. 12.

[19] T.G. Anderson, O.P. Bruno & M. Lyon, High-order, dispersionless 'fast-hybrid' wave equation solver. Part I: $\mathscr{O}(1)$ sampling cost via incident-field windowing and recentering. *SIAM J. Sci. Comput.* **42** (2020) A1348–A1379. Cited: pp. 98, 99, 170, 184, 185 & 186.

[20] F.P. Andriulli, K. Cools, F. Olyslager & E. Michielssen, Time domain Calderón identities and their application to the integral equation analysis of scattering by PEC objects Part II: stability. *IEEE Trans. Antennas & Propag.* **57** (2009) 2365–2375. Cited: p. 176.

[21] J.H. Ansell & G.E. Tupholme, Use of Clemmow functions in the study of an acoustic pulse generated by a deformable sphere. *J. Sound Vib.* **25** (1972) 185–195. Cited: p. 119.

[22] F. Anselmet & P.-O. Mattei, *Acoustics, Aeroacoustics and Vibrations*. Hoboken, NJ: Wiley, 2016. Cited: p. 87.

[23] H. Antes, A boundary element procedure for transient wave propagations in two-dimensional isotropic elastic media. *Finite Elements in Analysis & Design* **1** (1985) 313–322. Cited: p. 177.

[24] H. Antes & J. Baaran, Noise radiation from moving surfaces. *Eng. Anal. Bound. Elem.* **25** (2001) 725–740. Cited: pp. 150 & 156.

[25] S.S. Antman, The equations for large vibrations of strings. *American Mathematical Monthly* **87** (1980) 359–370. Cited: p. 12.

[26] I. Arnaoudov & G. Venkov, Scattering of electromagnetic plane waves by a spheroid uniformly moving in free space. *Math. Meth. Appl. Sci.* **29** (2006) 1423–1433. Cited: p. 125.

[27] F.M. Arscott, *Periodic Differential Equations*. New York: Macmillan, 1964. Cited: pp. 54 & 55.

[28] J. Asakura, T. Sakurai, H. Tadano, T. Ikegami & K. Kimura, A numerical method for nonlinear eigenvalue problems using contour integrals. *JSIAM Letters* **1** (2009) 52–55. Cited: p. 135.

[29] R.J. Astley & J.G. Bain, A three-dimensional boundary element scheme for acoustic radiation in low Mach number flows. *J. Sound Vib.* **109** (1986) 445–465. Cited: p. 125.

[30] M. Atiyah, M. Dunajski & L.J. Mason, Twistor theory at fifty: from contour integrals to twistor strings. *Proc. Roy. Soc.* A **473** (2017) 20170530. Cited: p. 46.

[31] M. Auphan & J. Matthys, Reflection of a plane impulsive acoustic pressure wave by a rigid sphere. *J. Sound Vib.* **66** (1979) 227–237. Cited: p. 40.

[32] S.A. Azizoglu, S.S. Koc & O.M. Buyukdura, Time domain scattering of scalar waves by two spheres in free-space. *SIAM J. Appl. Math.* **70** (2009) 694–709. Cited: p. 36.

[33] A. Bachelot & A. Pujols, Équations intégrales espace-temps pour le système de Maxwell. *C. R. Acad. Sci. Paris Sér. I Math.* **314** (1992) 639–644. Cited: p. 176.

[34] H. Bağcı, A.E. Yılmaz, J.-M. Jin & E. Michielssen, Time domain adaptive integral method for surface integral equations. In: *Modeling and Computations in Electromagnetics: A Volume Dedicated to Jean-Claude Nédélec* (ed. H. Ammari) pp. 65–104.

Lecture Notes in Computational Science and Engineering **59**. Berlin: Springer, 2008. Cited: pp. 167 & 176.

[35] L.Y. Bahar, The Laplace transform of the derivative of a function with finite jumps. *J. Franklin Institute* **288** (1969) 275–289. Cited: p. 97.

[36] W. Bai & G.J. Diebold, Moving photoacoustic sources: acoustic waveforms in one, two, and three dimensions and application to trace gas detection. *J. Appl. Phys.* **125** (2019) 060902. Cited: p. 85.

[37] X. Bai & R.Y.S. Pak, On the stability of direct time-domain boundary element methods for elastodynamics. *Eng. Anal. Bound. Elem.* **96** (2018) 138–149. Cited: p. 177.

[38] B.B. Baker & E.T. Copson, *The Mathematical Theory of Huygens' Principle*, 3rd edition. New York: Chelsea, 1987. Cited: pp. 143, 146, 147, 150 & 156.

[39] G. Bal, M. Fink & O. Pinaud, Time-reversal by time-dependent perturbations. *SIAM J. Appl. Math.* **79** (2019) 754–780. Cited: p. 7.

[40] G.R. Baldock & T. Bridgeman, *Mathematical Theory of Wave Motion*. Chichester: Ellis Horwood Ltd., 1981. Cited: pp. 13 & 15.

[41] J. Ballani, L. Banjai, S. Sauter & A. Veit, Numerical solution of exterior Maxwell problems by Galerkin BEM and Runge–Kutta convolution quadrature. *Numer. Math.* **123** (2013) 643–670. Cited: p. 176.

[42] A. Bamberger & T. Ha Duong, Formulation variationnelle espace-temps pour le calcul par potentiel retardé de la diffraction d'une onde acoustique (I). *Math. Meth. Appl. Sci.* **8** (1986) 405–435. Cited: pp. 89, 91, 92 & 167.

[43] A. Bamberger & T. Ha Duong, Formulation variationnelle pour le calcul de la diffraction d'une onde acoustique par une surface rigide. *Math. Meth. Appl. Sci.* **8** (1986) 598–608. Cited: pp. 89, 92 & 167.

[44] L. Banjai, Multistep and multistage convolution quadrature for the wave equation: algorithms and experiments. *SIAM J. Sci. Comput.* **32** (2010) 2964–2994. Cited: p. 171.

[45] L. Banjai & M. Kachanovska, Fast convolution quadrature for the wave equation in three dimensions. *J. Comp. Phys.* **279** (2014) 103–126. Cited: p. 171.

[46] L. Banjai & M. Schanz, Wave propagation problems treated with convolution quadrature and BEM. In: *Fast Boundary Element Methods in Engineering and Industrial Applications* (ed. U. Langer, M. Schanz, O. Steinbach & W.L. Wendland) pp. 145–184. *Lecture Notes in Applied and Computational Mechanics* **63**. Berlin: Springer, 2012. Cited: p. 170.

[47] L. Banz, H. Gimperlein, Z. Nezhi & E.P. Stephan, Time domain BEM for sound radiation of tires. *Comput. Mech.* **58** (2016) 45–57. Cited: p. 167.

[48] G. Bao, Y. Gao & P. Li, Time-domain analysis of an acoustic–elastic interaction problem. *Arch. Rational Mech. Anal.* **229** (2018) 835–884. Cited: p. 82.

[49] R.G. Barakat, Transient diffraction of scalar waves by a fixed sphere. *J. Acoust. Soc. Amer.* **32** (1960) 61–66. Cited: p. 119.

[50] P.E. Barbone & D.G. Crighton, Vibrational modes of submerged elastic bodies. *Applied Acoustics* **43** (1994) 295–317. Cited: pp. 108 & 137.

[51] C. Bardos, M. Concordel & G. Lebeau, Extension de la théorie de la diffusion pour un corps élastique immergé dans un fluide. Comportement asymptotique des résonances. *Journal d'Acoustique* **2** (1989) 31–38. Cited: p. 73.

[52] C. Bardos & M. Fink, Mathematical foundations of the time reversal mirror. *Asymptotic Analysis* **29** (2002) 157–182. Cited: p. 12.

[53] C. Barnes & D.V. Anderson, The sound field from a pulsating sphere and the development of a tail in pulse propagation. *J. Acoust. Soc. Amer.* **24** (1952) 229. Cited: p. 108.

[54] A. Barnett, L. Greengard & T. Hagstrom, High-order discretization of a stable time-domain integral equation for 3D acoustic scattering. *J. Comp. Phys.* **402** (2020) 109047. Cited: p. 166.

[55] B.E. Barrowes, K. O'Neill, T.M. Grzegorczyk & J.A. Kong, On the asymptotic expansion of the spheroidal wave function and its eigenvalues for complex size parameter. *Stud. Appl. Math.* **113** (2004) 271–301. Cited: pp. 53, 54 & 55.

[56] G. Barton, *Elements of Green's Functions and Propagation: Potentials, Diffusion and Waves*. Oxford: Clarendon Press, 1989. Cited: pp. 11, 13, 56, 57, 73 & 86.

[57] G.K. Batchelor, *An Introduction to Fluid Dynamics*. Cambridge: Cambridge University Press, 1967. Cited: pp. 1, 2 & 9.

[58] H. Bateman, The solution of partial differential equations by means of definite integrals. *Proc. London Math. Soc.*, Ser. 2, **1** (1904) 451–458. Cited: pp. 45 & 46.

[59] H. Bateman, A generalisation of the Legendre polynomial. *Proc. London Math. Soc.*, Ser. 2, **3** (1905) 111–123. Cited: p. 51.

[60] H. Bateman, The conformal transformations of a space of four dimensions and their applications to geometrical optics. *Proc. London Math. Soc.*, Ser. 2, **7** (1909) 70–89. Cited: p. 47.

[61] H. Bateman, The determination of solutions of the equation of wave motion involving an arbitrary function of three variables which satisfies a partial differential equation. *Trans. Camb. Phil. Soc.* **21** (1910) 257–280. Cited: pp. 47 & 51.

[62] H. Bateman, *The Mathematical Analysis of Electrical and Optical Wave-Motion*. Cambridge: Cambridge University Press, 1915. Reprint: New York: Dover, 1955. Cited: pp. 11, 44, 45, 47, 49, 50, 55, 56, 57, 63 & 150.

[63] H. Bateman, The mean value of a function of spherical polar coordinates round a circle on a sphere. *Terrestrial Magnetism & Atmospheric Electricity* **20** (1915) 127–129. Cited: p. 79.

[64] H. Bateman, *Differential Equations*. London: Longmans, Green & Co., 1918. New impression, 1926. Reprint: New York: Chelsea, 1966. Cited: pp. 44 & 47.

[65] H. Bateman, A solution of the wave-equation. *Annals of Math.*, Ser. 2, **31** (1930) 158–162. Cited: p. 79.

[66] H. Bateman, Physical problems with discontinuous initial conditions. *Proc. Nat. Acad. Sci.* **16** (1930) 205–211. Cited: p. 79.

[67] H. Bateman, *Partial Differential Equations of Mathematical Physics*. Cambridge: Cambridge University Press, 1932. Cited: pp. 19, 44, 45, 46, 47, 50, 51, 55, 79, 146, 149 & 150.

[68] H. Bateman, A partial differential equation associated with Poisson's work on the theory of sound. *Amer. J. Math.* **60** (1938) 293–296. Cited: p. 36.

[69] C.E. Baum, The singularity expansion method. In: *Transient Electromagnetic Fields* (ed. L.B. Felsen) pp. 129–179. *Topics in Applied Physics* **10**. Berlin: Springer, 1976. Cited: pp. 137 & 138.

[70] C.E. Baum, Emerging technology for transient and broad-band analysis and synthesis of antennas and scatterers. *Proc. IEEE* **64** (1976) 1598–1616. Cited: pp. 137, 141 & 142.

[71] C.E. Baum, Toward an engineering theory of electromagnetic scattering: the singularity and eigenmode expansion methods. In: [847], pp. 571–651. Cited: pp. 138 & 141.

[72] C.E. Baum, Discrimination of buried targets via the singularity expansion. *Inverse Prob.* **13** (1997) 557–570. Cited: p. 138.

[73] C.E. Baum & L. Carin, Singularity expansion method, symmetry and target identification. In: *Scattering* (ed. R. Pike & P. Sabatier) pp. 431–447. San Diego, CA: Academic Press, 2002. Cited: pp. 138 & 142.

[74] C.E. Baum, E.J. Rothwell, Y.F. Chen & D.P. Nyquist, The singularity expansion method and its application to target identification. *Proc. IEEE* **79** (1991) 1481–1492. Cited: p. 138.

[75] A. Bayliss & E. Turkel, Radiation boundary conditions for wave-like equations. *Comm. Pure Appl. Math.* **33** (1980) 707–725. Cited: p. 32.

[76] J.T. Beale, Acoustic scattering from locally reacting surfaces. *Indiana Univ. Math. J.* **26** (1977) 199–222. Cited: p. 81.

[77] J.T. Beale & S.I. Rosencrans, Acoustic boundary conditions. *Bull. Amer. Math. Soc.* **80** (1974) 1276–1278. Cited: p. 81.

[78] R. Beals, Laplace transform methods for evolution equations. In: [328], pp. 1–26. Cited: p. 97.

[79] R. Beals & P.C. Greiner, Strings, waves, drums: spectra and inverse problems. *Analysis & Applications* **7** (2009) 131–183. Cited: p. 1.

[80] E. Becache, A variational boundary integral equation method for an elastodynamic antiplane crack. *Int. J. Numer. Meth. Eng.* **36** (1993) 969–984. Cited: p. 187.

[81] E. Bécache & T. Ha Duong, A space-time variational formulation for the boundary integral equation in a 2D elastic crack problem. *Mathematical Modelling & Numerical Analysis* **28** (1994) 141–176. Cited: p. 192.

[82] H. Bech & A. Leder, Particle sizing by ultrashort laser pulses – numerical simulation. *Optik* **115** (2004) 205–217. Cited: p. 120.

[83] R.F. Beck, Time-domain computations for floating bodies. *Appl. Ocean Res.* **16** (1994) 267–282. Cited: p. 179.

[84] R.F. Beck & S. Liapis, Transient motions of floating bodies at zero forward speed. *J. Ship Res.* **31** (1987) 164–176. Cited: pp. 179 & 180.

[85] A. Bedford & D.S. Drumheller, *Introduction to Elastic Wave Propagation*. Chichester: Wiley, 1994. Cited: p. 13.

[86] B. Bedrosian & F.L. DiMaggio, Transient response of submerged spheroidal shells. *Int. J. Solids Struct.* **8** (1972) 111–129. Cited: p. 130.

[87] Y. Beghein, K. Cools, H. Bagcı & D. De Zutter, A space-time mixed Galerkin marching-on-in-time scheme for the time-domain combined field integral equation. *IEEE Trans. Antennas & Propag.* **61** (2013) 1228–1238. Cited: p. 176.

[88] E. Beltrami, Sul teorema di Kirchhoff. *Rendiconti della Reale Accademia dei Lincei*, Ser. 5, **4**, 2° semestre (1895) 51–52. Cited: p. 150.

[89] C. Ben Amar & C. Hazard, Time reversal and scattering theory for time-dependent acoustic waves in a homogeneous medium. *IMA J. Appl. Math.* **76** (2011) 938–955. Cited: p. 11.

[90] C.L. Bennett & H. Mieras, Time domain integral equation solution for acoustic scattering from fluid targets. *J. Acoust. Soc. Amer.* **69** (1981) 1261–1265. Cited: pp. 166 & 176.

[91] C.L. Bennett & H. Mieras, Time domain scattering from open thin conducting surfaces. *Radio Sci.* **16** (1981) 1231–1239. Cited: p. 192.

[92] C.L. Bennett & G.F. Ross, Time-domain electromagnetics and its applications. *Proc. IEEE* **66** (1978) 299–318. Cited: p. 176.

[93] C.L. Bennett Jr & W.L. Weeks, Transient scattering from conducting cylinders. *IEEE Trans. Antennas & Propag.* **AP-18** (1970) 627–633. Cited: p. 176.

[94] S. Benzoni-Gavage & D. Serre, *Multidimensional Hyperbolic Partial Differential Equations*. Oxford: Oxford University Press, 2007. Cited: pp. 59, 80 & 88.

[95] B.S. Berger, The dynamic response of a prolate spheroidal shell submerged in an acoustical medium. *J. Appl. Mech.* **41** (1974) 925–929. Cited: p. 130.

[96] B.S. Berger & D. Klein, Application of the Cesaro mean to the transient interaction of a spherical acoustic wave and a spherical elastic shell. *J. Appl. Mech.* **39** (1972) 623–625. Cited: p. 119.

[97] P.G. Bergmann, The wave equation in a medium with a variable index of refraction. *J. Acoust. Soc. Amer.* **17** (1946) 329–333. Cited: p. 4.

[98] T. Betcke, J. Phillips & E.A. Spence, Spectral decompositions and nonnormality of boundary integral operators in acoustic scattering. *IMA J. Numer. Anal.* **34** (2014) 700–731. Cited: p. 140.

[99] T. Betcke, N. Salles & W. Śmigaj, Overresolving in the Laplace domain for convolution quadrature methods. *SIAM J. Sci. Comput.* **39** (2017) A188–A213. Cited: pp. 172, 173 & 174.

[100] H.B. Bingham, A note on the relative efficiency of methods for computing the transient free-surface Green function. *Ocean Engng.* **120** (2016) 15–20. Cited: p. 181.

[101] B. Birgisson, E. Siebrits & A.P. Peirce, Elastodynamic direct boundary element methods with enhanced numerical stability properties. *Int. J. Numer. Meth. Eng.* **46** (1999) 871–888. Cited: p. 177.

[102] G. Birkhoff, Sound waves in fluids. *Appl. Numer. Math.* **3** (1987) 3–24. Cited: p. 61.

[103] D.T. Blackstock, Transient solution for sound radiated into a viscous fluid. *J. Acoust. Soc. Amer.* **41** (1967) 1312–1319. Cited: p. 19.

[104] D.R. Bland, *Vibrating Strings*. London: Routledge & Kegan Paul, 1960. Cited: pp. 12 & 15.

[105] N. Bleistein, *Mathematical Methods for Wave Phenomena*. San Diego, CA: Academic Press, 1984. Cited: pp. 13, 20, 61 & 73.

[106] N. Bleistein & J.K. Cohen, Nonuniqueness in the inverse source problem in acoustics and electromagnetics. *J. Math. Phys.* **18** (1977) 194–201. Cited: p. 88.

[107] N. Bleistein & R.A. Handelsman, *Asymptotic Expansions of Integrals*, revised edition. New York: Dover, 1986. Cited: p. 15.

[108] D.I. Blokhintsev, Acoustics of a nonhomogeneous moving medium. National Advisory Committee for Aeronautics, Technical Memorandum 1399, 1956. Translation from Russian, 1946. Cited: pp. 10 & 57.

[109] M.J. Bluck & S.P. Walker, Analysis of three-dimensional transient acoustic wave propagation using the boundary integral equation method. *Int. J. Numer. Meth. Eng.* **39** (1996) 1419–1431. Cited: pp. 166 & 167.

[110] M.J. Bluck & S.P. Walker, Time-domain BIE analysis of large three-dimensional electromagnetic scattering problems. *IEEE Trans. Antennas & Propag.* **45** (1997) 894–901. Cited: p. 176.

[111] N.N. Bojarski, The k-space formulation of the scattering problem in the time domain. *J. Acoust. Soc. Amer.* **72** (1982) 570–584. Cited: p. 86.

[112] B.A. Boley, Discontinuities in integral-transform solutions. *Quart. Appl. Math.* **19** (1962) 273–284. Cited: p. 97.

[113] G. Bollig & K.J. Langenberg, The singularity expansion method as applied to the elastodynamic scattering problem. *Wave Motion* **5** (1983) 331–354. Cited: p. 130.

[114] A. Bóna & M.A. Slawinski, *Wavefronts and Rays as Characteristics and Asymptotics*, 2nd edition. Singapore: World Scientific, 2015. Cited: p. 59.

[115] A.D. Boozer, A toy model of electrodynamics in $(1+1)$ dimensions. *European J. Phys.* **28** (2007) 447–464. Cited: p. 13.

[116] L. Borcea, Imaging with waves in random media. *Notices Amer. Math. Soc.* **66** (2019) 1800–1812. Cited: p. 5.

[117] L. Borcea, J. Garnier & K. Solna, Wave propagation and imaging in moving random media. *Multiscale Modeling & Simulation* **17** (2019) 31–67. Cited: p. 6.

[118] B. Borden, *Radar Imaging of Airborne Targets*. Bristol: Institute of Physics Publishing, 1999. Cited: p. 138.

[119] V.V. Borisov, A.V. Manankova & A.B. Utkin, Spherical harmonic representation of the electromagnetic field produced by a moving pulse of current density. *J. Phys. A: Math. & General* **29** (1996) 4493–4514. Cited: p. 39.

[120] A. Boström, Time-dependent scattering by a bounded obstacle in three dimensions. *J. Math. Phys.* **23** (1982) 1444–1450. Cited: pp. 34 & 130.

[121] J.J. Bowman, T.B.A. Senior & P.L.E. Uslenghi (ed.), *Electromagnetic and Acoustic Scattering by Simple Shapes*. Amsterdam: North-Holland, 1969. Revised printing: Levittown, PA: Hemisphere, 1987. Cited: p. 228.

[122] J.P. Boyd, The optimization of convergence for Chebyshev polynomial methods in an unbounded domain. *J. Comp. Phys.* **45** (1982) 43–79. Cited: p. 42.

[123] J.P. Boyd, *Chebyshev and Fourier Spectral Methods*, 2nd edition. New York: Dover, 2001. Cited: p. 42.

[124] J.P. Boyd & N. Flyer, Compatibility conditions for time-dependent partial differential equations and the rate of convergence of Chebyshev and Fourier spectral methods. *Comput. Meth. Appl. Mech. Eng.* **175** (1999) 281–309. Cited: pp. 78 & 79.

[125] L.M. Brekhovskikh, *Waves in Layered Media*. New York: Academic Press, 1960. Cited: pp. 30 & 81.

[126] L.M. Brekhovskikh & O.A. Godin, *Acoustics of Layered Media I*. Berlin: Springer, 1990. Cited: pp. 30 & 81.

[127] H. Bremmer, The jumps of discontinuous solutions of the wave equation. *Comm. Pure Appl. Math.* **4** (1951) 419–426. Cited: p. 67.

[128] K.S. Brentner & F. Farassat, Modeling aerodynamically generated sound of helicopter rotors. *Prog. Aerospace Sci.* **39** (2003) 83–120. Cited: p. 156.

[129] J. Brillouin, Rayonnement transitoire des sources sonores et problèmes connexes. *Annales des Télécommunications* **5** (1950) 160–172 & 179–194. Cited: p. 119.

[130] T.J.I'A. Bromwich, Normal coordinates in dynamical systems. *Proc. London Math. Soc.*, Ser. 2, **15** (1916) 401–448. Cited: p. 81.

[131] T.J.I'A. Bromwich, Electromagnetic waves. *Phil. Mag.*, Ser. 6, **38** (1919) 143–164. Cited: pp. 39 & 40.

[132] T.J.I'A. Bromwich, The scattering of plane electric waves by spheres. *Phil. Trans. Roy. Soc.* A **220** (1920) 175–206. Cited: p. 39.

[133] H. Brunner, *Volterra Integral Equations: An Introduction to Theory and Applications*. Cambridge: Cambridge University Press, 2017. Cited: p. 122.

[134] O.P. Bruno & B. Delourme, Rapidly convergent two-dimensional quasi-periodic Green function throughout the spectrum—including Wood anomalies. *J. Comp. Phys.* **262** (2014) 262–290. Cited: p. 186.

[135] O.P. Bruno, C.A. Geuzaine, J.A. Monro Jr & F. Reitich, Prescribed error tolerances within fixed computational times for scattering problems of arbitrarily high frequency: the convex case. *Phil. Trans. Roy. Soc.* A **362** (2004) 629–645. Cited: p. 186.

[136] O.P. Bruno & L.A. Kunyansky, Surface scattering in three dimensions: an accelerated high-order solver. *Proc. Roy. Soc.* A **457** (2001) 2921–2934. Cited: p. 186.

[137] R.N. Buchal, The approach to steady state of solutions of exterior boundary value problems for the wave equation. *J. Math. & Mech.* **12** (1963) 225–234. Cited: p. 87.

[138] J.Z. Buchwald & C.-P. Yeang, Kirchhoff's theory for optical diffraction, its predecessor and subsequent development: the resilience of an inconsistent theory. *Archive for History of Exact Sciences* **70** (2016) 463–511. Cited: p. 150.

[139] R. Burridge & Z. Alterman, The elastic radiation from an expanding spherical cavity. *Geophysical J. Royal Astronomical Society* **30** (1972) 451–477. Cited: p. 40.

[140] O.M. Buyukdura & S.S. Koc, Two alternative expressions for the spherical wave expansion of the time domain scalar free-space Green's function and an application: scattering by a soft sphere. *J. Acoust. Soc. Amer.* **101** (1997) 87–91. Cited: pp. 36 & 39.

[141] L. Cagniard, Diffraction d'une onde progressive par un écran en forme de demi-plan. *Journal de Physique et Le Radium* **6** (1935) 310–318. Cited: p. 187.

[142] L. Cagniard, *Reflection and Refraction of Progressive Seismic Waves*. New York: McGraw-Hill, 1962. Translated and revised edition of book originally published in French in 1939. Cited: p. 187.

[143] F. Cakoni, H. Haddar & A. Lechleiter, On the factorization method for a far field inverse scattering problem in the time domain. *SIAM J. Math. Anal.* **51** (2019) 854–872. Cited: p. 33.

[144] F. Cakoni & J.D. Rezac, Direct imaging of small scatterers using reduced time dependent data. *J. Comp. Phys.* **338** (2017) 371–387. Cited: pp. 33 & 91.

[145] W.B. Campbell, J. Macek & T.A. Morgan, Relativistic time-dependent multipole analysis for scalar, electromagnetic, and gravitational fields. *Phys. Rev.* D **15** (1977) 2156–2164. Cited: pp. 39 & 40.

[146] L.M.B.C. Campos, On 36 forms of the acoustic wave equation in potential flows and inhomogeneous media. *Appl. Mech. Rev.* **60** (2007) 149–171. Cited: pp. 6 & 7.

[147] M. Carley, Fast evaluation of transient acoustic fields. *J. Acoust. Soc. Amer.* **139** (2016) 630–635. Cited: p. 40.

[148] H.S. Carslaw & J.C. Jaeger, *Operational Methods in Applied Mathematics*, 2nd edition. London: Oxford University Press, 1948. Cited: pp. 18 & 116.

[149] G. Caviglia & A. Morro, A closed-form solution for reflection and transmission of transient waves in multilayers. *J. Acoust. Soc. Amer.* **116** (2004) 643–654. Cited: p. 18.

[150] D. Censor, Scattering by time varying obstacles. *J. Sound Vib.* **25** (1972) 101–110. See [719]. Cited: pp. 127 & 227.

[151] D. Censor, Harmonic and transient scattering from time varying obstacles. *J. Acoust. Soc. Amer.* **76** (1984) 1527–1534. Cited: p. 128.

[152] P. Cermelli, E. Fried & M.E. Gurtin, Transport relations for surface integrals arising in the formulation of balance laws for evolving fluid interfaces. *J. Fluid Mech.* **544** (2005) 339–351. Cited: p. 156.

[153] M. Cessenat, *Mathematical Methods in Electromagnetism*. Singapore: World Scientific, 1996. Cited: p. 73.

[154] P. Chadwick, *Continuum Mechanics*. London: George Allen & Unwin, 1976. Cited: p. 69.

[155] P. Chadwick & B. Powdrill, Application of the Laplace transform methods to wave motions involving strong discontinuities. *Proc. Camb. Phil. Soc.* **60** (1964) 313–324. Cited: p. 95.

[156] P. Chadwick & B. Powdrill, Singular surfaces in linear thermoelasticity. *Int. J. Eng. Sci.* **3** (1965) 561–595. Cited: p. 64.

[157] A. Chambolle & F. Santosa, Control of the wave equation by time-dependent coefficient. *ESAIM: Control, Optimisation & Calculus of Variations* **8** (2002) 375–392. Cited: p. 6.

[158] J.F.-C. Chan & P. Monk, Time dependent electromagnetic scattering by a penetrable obstacle. *BIT Numer. Math.* **55** (2015) 5–31. Cited: p. 176.

[159] R. Chapko & R. Kress, On the numerical solution of initial boundary value problems by the Laguerre transformation and boundary integral equations. In: *Integral and Integro-differential Equations* (ed. R.P. Agarwal & D. O'Regan) pp. 55–69. Amsterdam: Gordon & Breach, 2000. Cited: p. 101.

[160] C.J. Chapman, The spiral Green function in acoustics and electromagnetism. *Proc. Roy. Soc.* A **431** (1990) 157–167. Cited: p. 57.

[161] C.J. Chapman, *High Speed Flow*. Cambridge: Cambridge University Press, 2000. Cited: pp. 10 & 59.

[162] D.J. Chappell, A convolution quadrature Galerkin boundary element method for the exterior Neumann problem of the wave equation. *Math. Meth. Appl. Sci.* **32** (2009) 1585–1608. Cited: pp. 92 & 171.

[163] D.J. Chappell & P.J. Harris, On the choice of coupling parameter in the time domain Burton–Miller formulation. *Quart. J. Mech. Appl. Math.* **62** (2009) 431–450. Cited: p. 166.

[164] D.J. Chappell, P.J. Harris, D. Henwood & R. Chakrabarti, A stable boundary element method for modeling transient acoustic radiation. *J. Acoust. Soc. Amer.* **120** (2006) 74–80. Cited: pp. 151 & 166.

[165] J. Chazarain & A. Piriou, *Introduction to the Theory of Linear Partial Differential Equations*. Amsterdam: North-Holland, 1982. Cited: pp. 71, 73, 88 & 89.

[166] E.P. Chen & G.C. Sih, Transient response of cracks to impact loads. In: *Elastodynamic Crack Problems* (ed. G.C. Sih) pp. 1–58. Leyden: Noordhoff, 1977. Cited: p. 192.

[167] Q. Chen, H. Haddar, A. Lechleiter & P. Monk, A sampling method for inverse scattering in the time domain. *Inverse Prob.* **26** (2010) 085001. Cited: pp. 33, 91 & 92.

[168] Q. Chen & P. Monk, Discretization of the time domain CFIE for acoustic scattering problems using convolution quadrature. *SIAM J. Math. Anal.* **46** (2014) 3107–3130. Cited: p. 171.

[169] Q. Chen & P. Monk, Time domain CFIEs for electromagnetic scattering problems. *Appl. Numer. Math.* **79** (2014) 62–78. Cited: p. 176.

[170] Q. Chen, P. Monk, X. Wang & D. Weile, Analysis of convolution quadrature applied to the time-domain electric field integral equation. *Communications in Computational Physics* **11** (2012) 383–399. Cited: p. 176.

[171] R. Chen, S.B. Sayed, N. Alharthi, D. Keyes & H. Bagci, An explicit marching-on-in-time scheme for solving the time domain Kirchhoff integral equation. *J. Acoust. Soc. Amer.* **146** (2019) 2068–2079. Cited: p. 171.

[172] V.C. Chen, *The Micro-Doppler Effect in Radar*. Norwood, MA: Artech House, 2011. Cited: p. 125.

[173] M. Cheney & B. Borden, *Fundamentals of Radar Imaging*. Philadelphia: SIAM, 2009. Cited: p. 13.

[174] G. Chertock, Sound radiation from prolate spheroids. *J. Acoust. Soc. Amer.* **33** (1961) 871–876. See [778]. Cited: pp. 55 & 230.

[175] W.C. Chew, Vector potential electromagnetics with generalized gauge for inhomogeneous media: formulation. *Progress in Electromagnetics Research* **149** (2014) 69–84. Cited: p. 25.

[176] W.C. Chew & W.H. Weedon, A 3D perfectly matched medium from modified Maxwell's equations with stretched coordinates. *Microwave & Optical Technology Lett.* **7** (1994) 599–604. Cited: p. 42.

[177] H.A. Cho, M.A. Golberg, A.S. Muleshkov & X. Li, Trefftz methods for time dependent partial differential equations. *CMC: Computers, Materials & Continua* **1** (2004) 1–37. Cited: p. 101.

[178] S.K. Cho, *Electromagnetic Scattering*. New York: Springer, 1990. Cited: p. 133.

[179] P.-L. Chow, *Stochastic Partial Differential Equations*, 2nd edition. Boca Raton, FL: CRC Press, 2015. Cited: p. 85.

[180] E.B. Christoffel, Untersuchungen über die mit dem Fortbestehen linearer partieller Differentialgleichungen verträglichen Unstetigkeiten. *Annali di Matematica Pura ed Applicata*, Ser. 2, **8** (1877) 81–112. Cited: p. 69.

[181] I. Chudinovich, Boundary equations in dynamic problems of the theory of elasticity. *Acta Applicandae Mathematica* **65** (2001) 169–183. Cited: p. 178.

[182] I.Yu. Chudinovich, The solvability of boundary equations in mixed problems for non-stationary Maxwell's system. *Math. Meth. Appl. Sci.* **20** (1997) 425–448. Cited: p. 176.

[183] Y.-S. Chung, T.K. Sarkar, B.H. Jung, M. Salazar-Palma, Z. Ji, S. Jang & K. Kim, Solution of time domain electric field integral equation using the Laguerre polynomials. *IEEE Trans. Antennas & Propag.* **52** (2004) 2319–2328. Cited: p. 176.

[184] M. Cianferra, S. Ianniello & V. Armenio, Assessment of methodologies for the solution of the Ffowcs Williams and Hawkings equation using LES of incompressible single-phase flow around a finite-size square cylinder. *J. Sound Vib.* **453** (2019) 1–24. Cited: p. 87.

[185] A.H. Clément, An ordinary differential equation for the Green function of time-domain free-surface hydrodynamics. *J. Engng. Math.* **33** (1998) 201–217. Cited: p. 181.

[186] D.S. Cohen & G.H. Handelman, Scattering of a plane acoustical wave by a spherical obstacle. *J. Acoust. Soc. Amer.* **38** (1965) 827–834. Cited: p. 119.

[187] D.M. Cole, D.D. Kosloff & J.B. Minster, A numerical boundary integral equation method for elastodynamics. I. *Bull. Seismological Soc. Amer.* **68** (1978) 1331–1357. Cited: p. 166.

[188] F. Collino & P. Monk, The perfectly matched layer in curvilinear coordinates. *SIAM J. Sci. Comput.* **19** (1998) 2061–2090. Cited: p. 42.

[189] D. Colton & R. Kress, *Integral Equation Methods in Scattering Theory*. New York: Wiley, 1983. Cited: pp. 98, 133, 173 & 175.

[190] D. Colton & R. Kress, *Inverse Acoustic and Electromagnetic Scattering Theory*, 3rd edition. New York: Springer, 2013. Cited: pp. 45, 85 & 100.

[191] A.W. Conway, The field of force due to a moving electron. *Proc. London Math. Soc.*, Ser. 2, **1** (1903) 154–165. Cited: p. 55.

[192] A.W. Conway, On an expansion of the point-potential. *Proc. Roy. Soc.* A **94** (1918) 436–452. Cited: p. 55.

[193] J.B. Conway, *A Course in Functional Analysis*, 2nd edition. New York: Springer, 1990. Cited: pp. 139 & 140.

[194] J. Cooper, Scattering of plane waves by a moving obstacle. *Arch. Rational Mech. Anal.* **71** (1979) 113–141. Cited: p. 125.

[195] J. Cooper, Scattering by moving bodies: the quasi stationary approximation. *Math. Meth. Appl. Sci.* **2** (1980) 131–148. Cited: p. 125.

[196] E.T. Copson, On the Riemann–Green function. *Arch. Rational Mech. Anal.* **1** (1958) 324–348. Cited: p. 39.

[197] E.T. Copson, *Partial Differential Equations*. Cambridge: Cambridge University Press, 1975. Cited: pp. 19, 34 & 73.

[198] M. Costabel, Time-dependent problems with the boundary integral equation method. In: *Encyclopedia of Computational Mechanics* (ed. E. Stein, R. de Borst & T.J.R. Hughes), vol. 1, pp. 703–721. New York: Wiley, 2004. See [199]. Cited: pp. 89, 151, 156, 159 & 202.

[199] M. Costabel & F.-J. Sayas, Time-dependent problems with the boundary integral equation method. In: *Encyclopedia of Computational Mechanics*, 2nd edition (ed. E. Stein, R. de Borst & T.J.R. Hughes), vol. 2, 24 pp. New York: Wiley, 2017. Updated version of [198]. Cited: pp. 156, 159 & 202.

[200] C.A. Coulson, *Waves*, 5th edition. Edinburgh: Oliver & Boyd, 1949. Cited: p. 146.

[201] R. Courant, Hyperbolic partial differential equations and applications. In: *Modern Mathematics for the Engineer* (ed. E.F. Beckenbach), pp. 92–109. New York: McGraw-Hill, 1956. Cited: p. 32.

[202] R. Courant & D. Hilbert, *Methoden der Mathematischen Physik*, vol. 2. Berlin: Springer, 1937. Cited: p. 65.

[203] R. Courant & D. Hilbert, *Methods of Mathematical Physics*, vol. 2. New York: Interscience, 1962. Cited: pp. 19, 20, 49, 58, 61, 68, 71 & 73.

[204] W. Craig, *A Course on Partial Differential Equations*. Providence, RI: American Mathematical Society, 2018. Cited: pp. 16, 17 & 20.

[205] D.G. Crighton, Basic principles of aerodynamic noise generation. *Prog. Aerospace Sci.* **16** (1975) 31–96. Cited: pp. 85 & 87.

[206] D.G. Crighton, Scattering and diffraction of sound by moving bodies. *J. Fluid Mech.* **72** (1975) 209–227. Cited: p. 127.

[207] D.G. Crighton, A.P. Dowling, J.E. Ffowcs Williams, M. Heckl & F.G. Leppington, *Modern Methods in Analytical Acoustics*. London: Springer, 1992. Cited: pp. 1 & 205.

[208] S.C. Crow, Aerodynamic sound emission as a singular perturbation problem. *Stud. Appl. Math.* **49** (1970) 21–44. Cited: p. 87.

[209] T.A. Cruse, A direct formulation and numerical solution of the general transient elastodynamic problem. II. *J. Math. Anal. Appl.* **22** (1968) 341–355. Cited: p. 178.

[210] T.A. Cruse & F.J. Rizzo, A direct formulation and numerical solution of the general transient elastodynamic problem. I. *J. Math. Anal. Appl.* **22** (1968) 244–259. Cited: p. 178.

[211] W.E. Cummins, The impulse response function and ship motions. *Schiffstechnik* **9** (1962) 101–109. Cited: p. 180.

[212] N. Curle, The influence of solid boundaries upon aerodynamic sound. *Proc. Roy. Soc.* A **231** (1955) 505–514. Cited: pp. 87 & 150.

[213] R.C. Dalang & L. Quer-Sardanyons, Stochastic integrals for spde's: a comparison. *Expositiones Mathematicae* **29** (2011) 67–109. Cited: p. 85.

[214] J.M. D'Archangelo, P. Savage, H. Überall, K.B. Yoo, S.H. Brown & J.W. Dickey, Complex eigenfrequencies of rigid and soft spheroids. *J. Acoust. Soc. Amer.* **77** (1985) 6–10. Cited: p. 130.

[215] W.C. Davidon, Time-dependent multipole analysis. *J. Phys. A: Math., Nuclear & General* **6** (1973) 1635–1646. Cited: p. 40.

[216] B. Davies, *Integral Transforms and Their Applications*, 2nd edition. New York: Springer, 1985. Cited: pp. 17, 21 & 116.

[217] E.B. Davies, Non-self-adjoint differential operators. *Bull. London Math. Soc.* **34** (2002) 513–532. Cited: p. 139.

[218] P.J. Davies, Numerical stability and convergence of approximations of retarded potential integral equations. *SIAM J. Numer. Anal.* **31** (1994) 856–875. Cited: p. 166.

[219] P.J. Davies, On the stability of time-marching schemes for the general surface electric-field integral equation. *IEEE Trans. Antennas & Propag.* **44** (1996) 1467–1473. Cited: p. 176.

[220] P.J. Davies & D.B. Duncan, Averaging techniques for time-marching schemes for retarded potential integral equations. *Appl. Numer. Math.* **23** (1997) 291–310. Cited: pp. 164 & 166.

[221] P.J. Davies & D.B. Duncan, Stability and convergence of collocation schemes for retarded potential integral equations. *SIAM J. Numer. Anal.* **42** (2004) 1167–1188. Cited: p. 166.

[222] P.J. Davies & D.B. Duncan, Convolution-in-time approximations of time domain boundary integral equations. *SIAM J. Sci. Comput.* **35** (2013) B43–B61. Cited: p. 167.

[223] P.J. Davies & D.B. Duncan, Convolution spline approximations for time domain boundary integral equations. *J. Integ. Eqns Appl.* **26** (2014) 369–410. Cited: pp. 167 & 168.

[224] L. Debnath & D. Bhatta, *Integral Transforms and Their Applications*, 3rd edition. Boca Raton, FL: CRC Press, 2015. Cited: p. 101.

[225] A.T. de Hoop, A modification of Cagniard's method for solving seismic pulse problems. *Applied Scientific Research, Section B* **8** (1960) 349–356. Cited: p. 187.

[226] A.T. de Hoop, A time-domain energy theorem for scattering of plane acoustic waves in fluids. *J. Acoust. Soc. Amer.* **77** (1985) 11–14. Cited: p. 148.

[227] A.T. de Hoop, *Handbook of Radiation and Scattering of Waves*. London: Academic Press, 1995. Cited: pp. 1 & 148.

[228] A.T. de Hoop, Fields and waves excited by impulsive point sources in motion—the general 3D time-domain Doppler effect. *Wave Motion* **43** (2005) 116–122. Cited: p. 57.

[229] A.J. Devaney, *Mathematical Foundations of Imaging, Tomography and Wavefield Inversion*. Cambridge: Cambridge University Press, 2012. Cited: pp. 85, 86 & 88.

[230] A.J. Devaney & G.C. Sherman, Plane-wave representations for scalar wave fields. *SIAM Rev.* **15** (1973) 765–786. Cited: pp. 45, 85 & 86.

[231] E. Di Nezza, G. Palatucci & E. Valdinoci, Hitchhiker's guide to the fractional Sobolev spaces. *Bulletin des Sciences Mathématiques* **136** (2012) 521–573. Cited: p. 89.

[232] Y. Ding, A. Forestier & T. Ha Duong, A Galerkin scheme for the time domain integral equation of acoustic scattering from a hard surface. *J. Acoust. Soc. Amer.* **86** (1989) 1566–1572. Cited: p. 167.

[233] P.A.M. Dirac, Classical theory of radiating electrons. *Proc. Roy. Soc.* A **167** (1938) 148–169. Cited: p. 57.

[234] P.E. Doak, Fundamentals of aerodynamic sound theory and flow duct acoustics. *J. Sound Vib.* **28** (1973) 527–561. Cited: pp. 85 & 87.

[235] S.J. Dodson, S.P. Walker & M.J. Bluck, Costs and cost scaling in time-domain integral-equation analysis of electromagnetic scattering. *IEEE Antennas & Propagation Magazine* **40**, No. 4 (1998) 12–21. Cited: p. 167.

[236] S.J. Dodson, S.P. Walker & M.J. Bluck, Implicitness and stability of time domain integral equation scattering analyses. *Appl. Computational Electromagnetics Soc. J.* **13** (1998) 291–301. Cited: p. 166.

[237] J.L. Dohner, R. Shoureshi & R.J. Bernhard, Transient analysis of three-dimensional wave propagation using the boundary element method. *Int. J. Numer. Meth. Eng.* **24** (1987) 621–634. Cited: p. 166.

[238] C.L. Dolph, On some mathematical aspects of SEM, EEM and scattering. *Electromagnetics* **1** (1981) 375–383. The bibliography for this paper is contained in [621]. Cited: p. 140.

[239] C.L. Dolph & S.K. Cho, On the relationship between the singularity expansion method and the mathematical theory of scattering. *IEEE Trans. Antennas & Propag.* **AP-28** (1980) 888–897. Cited: pp. 135, 138 & 142.

[240] C.L. Dolph, V. Komkov & R.A. Scott, A critique of the singularity expansion and eigenmode expansion methods. In: *Acoustic, Electromagnetic and Elastic Wave Scattering—Focus on the T-matrix Approach* (ed. V.K. Varadan & V.V. Varadan) pp. 453–461. New York: Pergamon, 1980. Cited: p. 138.

[241] C.L. Dolph & R.A. Scott, Recent developments in the use of complex singularities in electromagnetic theory and elastic wave propagation. In: [847], pp. 503–570. Cited: pp. 133 & 135.

[242] A.K. Dominek, Transient scattering analysis for a circular disk. *IEEE Trans. Antennas & Propag.* **39** (1991) 815–819. Cited: p. 192.

[243] A. Dowling, Convective amplification of real simple sources. *J. Fluid Mech.* **74** (1976) 529–546. Cited: p. 127.

[244] A.P. Dowling, Effects of motion on acoustic sources. In: [207], pp. 406–427. Cited: p. 127.

[245] A.P. Dowling & J.E. Ffowcs Williams, *Sound and Sources of Sound*. Chichester: Ellis Horwood, 1983. Cited: pp. 57, 85, 87 & 127.

[246] D.G. Dudley & J.P. Quintenz, Transient electromagnetic penetration of a spherical shell. *J. Appl. Phys.* **46** (1975) 173–177. Cited: p. 120.

[247] G.F.D. Duff, Hyperbolic differential equations and waves. In: [328], pp. 27–155. Cited: pp. 13, 71 & 73.

[248] G.F.D. Duff & D. Naylor, *Differential Equations of Applied Mathematics*. New York: Wiley, 1966. Cited: pp. 13, 61 & 86.

[249] P. Duhem, *Hydrodynamique, Élasticité, Acoustique*. Paris: Librairie Scientifique Hermann, 1891. Cited: p. 108.

[250] S. Dyatlov & M. Zworski, *Mathematical Theory of Scattering Resonances*. Providence, RI: American Mathematical Society, 2019. Cited: p. 133.

[251] C.T. Dyka, R.P. Ingel & G.C. Kirby, Stabilizing the retarded potential method for transient fluid–structure interaction problems. *Int. J. Numer. Meth. Eng.* **40** (1997) 3767–3783. Cited: p. 166.

[252] M. Eastwood, Introduction to Penrose transform. In: *The Penrose Transform and Analytic Cohomology in Representation Theory* (ed. M. Eastwood, J. Wolf & R. Zierau) pp. 71–75. Providence, RI: American Mathematical Society, 1993. Cited: p. 46.

[253] D.E. Edmunds, L.E. Fraenkel & M. Pemberton, Frederick Gerard Friedlander. 25 December 1917–20 May 2001. *Biographical Memoirs of Fellows of the Royal Society* **63** (2017) 273–307. Cited: p. 77.

[254] H.A. Eide, J.J. Stamnes, K. Stamnes & F.M. Schulz, New method for computing expansion coefficients for spheroidal functions. *J. Quant. Spectrosc. Radiat. Transfer* **63** (1999) 191–203. Cited: p. 53.

[255] D.M. Eidus, The principle of limit amplitude. *Russian Math. Surveys* **24** (1969) 97–167. Cited: p. 87.

[256] M. Eller, Loss of derivatives for hyperbolic boundary problems with constant coefficients. *Discrete & Continuous Dynamical Systems*, Ser. B **23** (2018) 1347–1361. Cited: p. 88.

[257] B.O. Enflo & C.M. Hedberg, *Theory of Nonlinear Acoustics in Fluids*. Dordrecht: Kluwer, 2002. Cited: p. 10.

[258] A.E. Engin & Y.K. Liu, Axisymmetric response of a fluid-filled spherical shell in free vibrations. *J. Biomechanics* **3** (1970) 11–22. Cited: p. 83.

[259] C.L. Epstein, L. Greengard & T. Hagstrom, On the stability of time-domain integral equations for acoustic wave propagation. *Discrete & Continuous Dynamical Systems*, Ser. A **36** (2016) 4367–4382. Cited: p. 166.

[260] M. Epstein, Theories of growth. In: *Constitutive Modelling of Solid Continua* (ed. J. Merodio & R. Ogden) pp. 257–284. Cham: Springer, 2020. Cited: p. 7.

[261] A. Erdélyi, On certain discontinuous wave functions. *Proc. Edinburgh Math. Soc.*, Ser. 3, **8** (1947) 39–42. Cited: p. 80.

[262] A. Erdélyi, W. Magnus, F. Oberhettinger & F.G. Tricomi, *Higher Transcendental Functions*, vol. 3. New York: McGraw-Hill, 1955. Cited: pp. 53, 54, 55 & 169.

[263] A.A. Ergin, B. Shanker & E. Michielssen, Fast evaluation of three-dimensional transient wave fields using diagonal translation operators. *J. Comp. Phys.* **146** (1998) 157–180. Cited: p. 167.

[264] A.A. Ergin, B. Shanker & E. Michielssen, Analysis of transient wave scattering from rigid bodies using a Burton–Miller approach. *J. Acoust. Soc. Amer.* **106** (1999) 2396–2404. Cited: pp. 163 & 166.

[265] A.A. Ergin, B. Shanker & E. Michielssen, Fast transient analysis of acoustic wave scattering from rigid bodies using a two-level plane wave time domain algorithm. *J. Acoust. Soc. Amer.* **106** (1999) 2405–2416. Cited: p. 167.

[266] A.A. Ergin, B. Shanker & E. Michielssen, The plane-wave time-domain algorithm for the fast analysis of transient wave phenomena. *IEEE Antennas & Propagation Magazine* **41**, No. 4 (1999) 39–52. Cited: p. 167.

[267] A.A. Ergin, B. Shanker & E. Michielssen, Fast analysis of transient acoustic wave scattering from rigid bodies using the multilevel plane wave time domain algorithm. *J. Acoust. Soc. Amer.* **107** (2000) 1168–1178. Cited: p. 167.

[268] A.C. Eringen, Elasto-dynamic problem concerning the spherical cavity. *Quart. J. Mech. Appl. Math.* **10** (1957) 257–270. Cited: p. 120.

[269] A.C. Eringen & E.S. Şuhubi, *Elastodynamics*, vol. 2: *Linear Theory*. New York: Academic Press, 1975. Cited: pp. 120 & 176.

[270] J.D. Eshelby, The elastic field of a crack extending non-uniformly under general anti-plane loading. *J. Mechanics & Physics of Solids* **17** (1969) 177–199. Cited: p. 50.

[271] G. Eskin, *Lectures on Linear Partial Differential Equations*. Providence, RI: American Mathematical Society, 2011. Cited: pp. 85, 86, 87, 89 & 90.

[272] R. Estrada & R.P. Kanwal, Applications of distributional derivatives to wave propagation. *J. IMA* **26** (1980) 39–63. Cited: p. 58.

[273] R. Estrada & R.P. Kanwal, Non-classical derivation of the transport theorems for wave fronts. *J. Math. Anal. Appl.* **159** (1991) 290–297. Cited: p. 58.

[274] L. Euler, Eclaircissemens sur le mouvement des cordes vibrantes, 1766. This is paper E317 in the Euler Archive, eulerarchive.maa.org. Cited: p. 13.

[275] L.C. Evans, *Partial Differential Equations*, 2nd edition. Providence, RI: American Mathematical Society, 2010. Cited: pp. 13, 17, 19, 20, 34, 59, 61, 65, 73 & 89.

[276] J.B. Fahnline, Solving transient acoustic boundary value problems with equivalent sources using a lumped parameter approach. *J. Acoust. Soc. Amer.* **140** (2016) 4115–4129. Cited: p. 166.

[277] G. Fairweather, A. Karageorghis & P.A. Martin, The method of fundamental solutions for scattering and radiation problems. *Eng. Anal. Bound. Elem.* **27** (2003) 759–769. Cited: p. 181.

[278] S. Falletta, G. Monegato & L. Scuderi, A space-time BIE method for nonhomogeneous exterior wave equation problems. The Dirichlet case. *IMA J. Numer. Anal.* **32** (2012) 202–226. Cited: pp. 92 & 171.

[279] S. Falletta, G. Monegato & L. Scuderi, A space-time BIE method for wave equation problems: the (two-dimensional) Neumann case. *IMA J. Numer. Anal.* **34** (2014) 390–434. Cited: p. 93.

[280] S. Falletta, G. Monegato & L. Scuderi, On the discretization and application of two space-time boundary integral equations for 3D wave propagation problems in unbounded domains. *Appl. Numer. Math.* **124** (2018) 22–43. Cited: p. 171.

[281] S. Falletta, G. Monegato & L. Scuderi, Two boundary integral equation methods for linear elastodynamics problems on unbounded domains. *Computers & Math. with Applications* **78** (2019) 3841–3861. Cited: p. 178.

[282] J. Falnes, On non-causal impulse response functions related to propagating water waves. *Appl. Ocean Res.* **17** (1995) 379–389. Cited: p. 180.

[283] F. Farassat, Discontinuities in aerodynamics and aeroacoustics: the concept and applications of generalized derivatives. *J. Sound Vib.* **55** (1977) 165–193. Cited: p. 58.

[284] F. Farassat, Acoustic radiation from rotating blades—the Kirchhoff method in aeroacoustics. *J. Sound Vib.* **239** (2001) 785–800. Cited: p. 156.

[285] F. Farassat & M.H. Dunn, A simple derivation of the acoustic boundary condition in the presence of flow. *J. Sound Vib.* **224** (1999) 384–386. Cited: p. 126.

[286] F. Farassat & M.K. Myers, Extension of Kirchhoff's formula to radiation from moving surfaces. *J. Sound Vib.* **123** (1988) 451–460. Comments by D.L. Hawkings: **132** (1989) 160. Authors' reply: **132** (1989) 511. Cited: pp. 154 & 156.

[287] L. Farina, P.A. Martin & V. Péron, Hypersingular integral equations over a disc: convergence of a spectral method and connection with Tranter's method. *J. Comp. Appl. Math.* **269** (2014) 118–131. Cited: pp. 190 & 192.

[288] C.L.S. Farn & H. Huang, Transient acoustic fields generated by a body of arbitrary shape. *J. Acoust. Soc. Amer.* **43** (1968) 252–257. Cited: p. 166.

[289] L. Fatone, G. Pacelli, M.C. Recchioni & F. Zirilli, Optimal-control methods for two new classes of smart obstacles in time-dependent acoustic scattering. *J. Engng. Math.* **56** (2006) 385–413. Cited: p. 80.

[290] A.T. Fedorchenko, On some fundamental flaws in present aeroacoustic theory. *J. Sound Vib.* **232** (2000) 719–782. Cited: p. 87.

[291] L.B. Felsen, Progressing and oscillatory waves for hybrid synthesis of source excited propagation and diffraction. *IEEE Trans. Antennas & Propag.* **AP-32** (1984) 775–796. Cited: p. 138.

[292] L.B. Felsen & G.M. Whitman, Wave propagation in time-varying media. *IEEE Trans. Antennas & Propag.* **AP-18** (1970) 242–253. Cited: pp. 6 & 12.

[293] R.P. Feynman, R.B. Leighton & M. Sands, *The Feynman Lectures on Physics*, vol. 2, Reading, MA: Addison-Wesley, 1964. Cited: p. 57.

[294] J.E. Ffowcs Williams, Aeroacoustics. *J. Sound Vib.* **190** (1996) 387–398. Cited: p. 87.

[295] J.E. Ffowcs Williams & D.L. Hawkings, Sound generation by turbulence and surfaces in arbitrary motion. *Phil. Trans. Roy. Soc.* A **264** (1969) 321–342. Cited: pp. 153, 154, 155 & 156.

[296] S.E. Field & S.R. Lau, Fast evaluation of far-field signals for time-domain wave propagation. *J. Scientific Computing* **64** (2015) 647–669. Cited: p. 117.

[297] M. Fink, Time reversed acoustics. *Physics Today* **50**, 3 (1997) 34–40. Cited: p. 12.

[298] M. Fink, D. Cassereau, A. Derode, C. Prada, P. Roux, M. Tanter, J.-L. Thomas & F. Wu, Time-reversed acoustics. *Reports on Progress in Physics* **63** (2000) 1933–1995. Cited: p. 12.

[299] A.B. Finkelstein, The initial value problem for transient water waves. *Comm. Pure Appl. Math.* **10** (1957) 511–522. Cited: pp. 84 & 179.

[300] F.A. Fischer, Über die Totalreflexion von ebenen Impulswellen. *Annalen der Physik*, 6 Folge, **2** (1948) 211–224. Cited: p. 30.

[301] C. Flammer, *Spheroidal Wave Functions*. Stanford, CA: Stanford University Press, 1957. Cited: pp. 52 & 55.

[302] S.M. Flatté (ed.), *Sound Transmission through a Fluctuating Ocean*. Cambridge: Cambridge University Press, 1979. Cited: pp. 6 & 7.

[303] J.-P. Fouque, J. Garnier, G. Papanicolaou & K. Sølna, *Wave Propagation and Time Reversal in Randomly Layered Media*. New York: Springer, 2007. Cited: pp. 18, 59 & 85.

[304] E.N. Fox, The diffraction of sound pulses by an infinitely long strip. *Phil. Trans. Roy. Soc.* A **241** (1948) 71–103. Cited: pp. 187 & 192.

[305] E.N. Fox, The diffraction of two-dimensional sound pulses incident on an infinite uniform slit in a perfectly reflecting screen. *Phil. Trans. Roy. Soc.* A **242** (1949) 1–32. Cited: p. 192.

[306] G. Franceschetti, A canonical problem in transient radiation—the spherical antenna. *IEEE Trans. Antennas & Propag.* **AP-26** (1978) 551–555. Cited: p. 120.

[307] L.B. Freund, *Dynamic Fracture Mechanics*. Cambridge: Cambridge University Press, 1989. Cited: p. 192.

[308] F.G. Friedlander, On the solutions of the wave equation with discontinuous derivatives. *Proc. Camb. Phil. Soc.* **38** (1942) 378–382. Cited: p. 67.

[309] F.G. Friedlander, Simple progressive solutions of the wave equation. *Proc. Camb. Phil. Soc.* **43** (1947) 360–373. Cited: p. 30.

[310] F.G. Friedlander, *Sound Pulses*. Cambridge: Cambridge University Press, 1958. Cited: pp. 60, 61, 64, 65, 67, 69, 70, 73, 74, 119 & 187.

[311] F.G. Friedlander, On the radiation field of pulse solutions of the wave equation. *Proc. Roy. Soc.* A **269** (1962) 53–65. Cited: pp. 32, 33, 36, 85, 88 & 123.

[312] F.G. Friedlander, On the radiation field of pulse solutions of the wave equation. II. *Proc. Roy. Soc.* A **279** (1964) 386–394. Cited: pp. 32 & 88.

[313] F.G. Friedlander, On the radiation field of pulse solutions of the wave equation. III. *Proc. Roy. Soc.* A **299** (1967) 264–278. Cited: p. 32.

[314] F.G. Friedlander, An inverse problem for radiation fields. *Proc. London Math. Soc.*, Ser. 3, **27** (1973) 551–576. Cited: pp. 32 & 88.

[315] A. Friedman & M. Shinbrot, The initial value problem for the linearized equations of water waves. *J. Math. & Mech.* **17** (1967) 107–180. Cited: p. 84.

[316] M.B. Friedman & R. Shaw, Diffraction of pulses by cylindrical obstacles of arbitrary cross section. *J. Appl. Mech.* **29** (1962) 40–46. Cited: p. 165.

[317] P.A. Frost & E.Y. Harper, Acoustic radiation from surfaces oscillating at large amplitude and small Mach number. *J. Acoust. Soc. Amer.* **58** (1975) 318–325. Cited: pp. 10 & 127.

[318] M. Fukuhara, R. Misawa, K. Niino & N. Nishimura, Stability of boundary element methods for the two dimensional wave equation in time domain revisited. *Eng. Anal. Bound. Elem.* **108** (2019) 321–338. Cited: p. 166.

[319] A. Furukawa, T. Saitoh & S. Hirose, Convolution quadrature time-domain boundary element method for 2-D and 3-D elastodynamic analyses in general anisotropic elastic solids. *Eng. Anal. Bound. Elem.* **39** (2014) 64–74. Cited: p. 178.

[320] C.G. Gal, G.R. Goldstein & J.A. Goldstein, Oscillatory boundary conditions for acoustic wave equations. *J. Evolution Equations* **3** (2003) 623–635. Cited: pp. 80 & 81.

[321] R. Gallego & J. Domínguez, Hypersingular BEM for transient elastodynamics. *Int. J. Numer. Meth. Eng.* **39** (1996) 1681–1705. Cited: p. 192.

[322] P.R. Garabedian, *Partial Differential Equations*. New York: Wiley, 1964. Cited: pp. 13, 41 & 59.

[323] F. García-Sánchez & C. Zhang, A comparative study of three BEM for transient dynamic crack analysis of 2-D anisotropic solids. *Comput. Mech.* **40** (2007) 753–769. Cited: p. 192.

[324] L. Garding, Review of Wilcox [887]. *Bull. London Math. Soc.* **9** (1977) 122–123. Cited: p. 73.

[325] T.J. Garner, A. Lakhtakia, J.K. Breakall & C.F. Bohren, Time-domain electromagnetic scattering by a sphere in uniform translational motion. *J. Opt. Soc. Amer.* A **34** (2017) 270–279. Cited: p. 125.

[326] J. Garnier & G. Papanicolaou, *Passive Imaging with Ambient Noise*. Cambridge: Cambridge University Press, 2016. Cited: p. 5.

[327] H. Garnir, Sur la transformation de Laplace des distributions. *Comptes Rendus de l'Académie des Sciences, Paris* **234** (1952) 583–585. Cited: p. 97.

[328] H.G. Garnir (ed.), *Boundary Value Problems for Linear Evolution Partial Differential Equations*. Dordrecht: Reidel, 1977. Cited: pp. 197, 205 & 235.

[329] W.W. Garvin, Exact transient solution of the buried line source problem. *Proc. Roy. Soc.* A **234** (1956) 528–541. Cited: p. 187.

[330] G.C. Gaunaurd & H.C. Strifors, Frequency- and time-domain analysis of the transient resonance scattering resulting from the interaction of a sound pulse with submerged elastic shells. *IEEE Trans. Ultrasonics, Ferroelectrics, & Frequency Control* **40** (1993) 313–324. Cited: p. 119.

[331] G.C. Gaunaurd & H.C. Strifors, Transient resonance scattering and target identification. *Appl. Mech. Rev.* **50** (1997) 131–148. Cited: p. 119.

[332] T.L. Geers, Excitation of an elastic cylindrical shell by a transient acoustic wave. *J. Appl. Mech.* **36** (1969) 459–469. Cited: pp. 120 & 121.

[333] T.L. Geers, Residual potential and approximate methods for three-dimensional fluid–structure interaction problems. *J. Acoust. Soc. Amer.* **49** (1971) 1505–1510. Cited: p. 121.

[334] T.L. Geers & M.A. Sprague, A residual-potential boundary for time-dependent, infinite-domain problems in computational acoustics. *J. Acoust. Soc. Amer.* **127** (2010) 675–682. Cited: pp. 120 & 121.

[335] A. Gelb, The resolution of the Gibbs phenomenon for spherical harmonics. *Mathematics of Computation* **66** (1997) 699–717. Cited: p. 119.

[336] A. Geranmayeh, W. Ackermann & T. Weiland, Temporal discretization choices for stable boundary element methods in electromagnetic scattering problems. *Appl. Numer. Math.* **59** (2009) 2751–2773. Cited: p. 176.

[337] P.C. Gibson, The combinatorics of scattering in layered media. *SIAM J. Appl. Math.* **74** (2014) 919–938. Cited: p. 18.

[338] H. Gimperlein, M. Maischak & E.P. Stephan, Adaptive time domain boundary element methods with engineering applications. *J. Integ. Eqns Appl.* **29** (2017) 75–105. Cited: p. 167.

[339] H. Gimperlein, F. Meyer, C. Özdemir, D. Stark & E.P. Stephan, Boundary elements with mesh refinements for the wave equation. *Numer. Math.* **139** (2018) 867–912. Cited: p. 192.

[340] H. Gimperlein, C. Özdemir, D. Stark & E.P. Stephan, hp-version time domain boundary elements for the wave equation on quasi-uniform meshes. *Comput. Meth. Appl. Mech. Eng.* **356** (2019) 145–174. Cited: p. 192.

[341] H. Gimperlein, C. Özdemir & E.P. Stephan, Time domain boundary element methods for the Neumann problem: error estimates and acoustic problems. *J. Comp. Math.* **36** (2018) 70–89. Cited: p. 167.

[342] H. Gimperlein, C. Özdemir & E.P. Stephan, A time-dependent FEM-BEM coupling method for fluid-structure interaction in 3*d*. *Appl. Numer. Math.* **152** (2020) 49–65. Cited: p. 82.

[343] D. Givoli, *Numerical Methods for Problems in Infinite Domains*. Amsterdam: Elsevier, 1992. Cited: pp. 32 & 42.

[344] D. Givoli & D. Cohen, Nonreflecting boundary conditions based on Kirchhoff-type formulae. *J. Comp. Phys.* **117** (1995) 102–113. Cited: p. 150.

[345] S. Glegg & W. Devenport, *Aeroacoustics of Low Mach Number Flows*. Oxford: Academic Press, 2017. Cited: pp. 87 & 154.

[346] I.C. Gohberg & M.G. Kreĭn, *Introduction to the Theory of Linear Nonselfadjoint Operators*. Providence, RI: American Mathematical Society, 1969. Cited: p. 140.

[347] G.R. Goldstein, Derivation and physical interpretation of general boundary conditions. *Advances in Differential Equations* **11** (2006) 457–480. Cited: p. 80.

[348] J.D. Gonzalez, E.F. Lavia & S. Blanc, A computational method to calculate the exact solution for acoustic scattering by fluid spheroids. *Acta Acustica united with Acustica* **102** (2016) 1061–1071. Cited: p. 128.

[349] A. Goriely, *The Mathematics and Mechanics of Biological Growth*. New York: Springer, 2017. Cited: pp. 7 & 9.

[350] A.G. Gorshkov & D.V. Tarlakovsky, *Transient Aerohydroelasticity of Spherical Bodies*. Berlin: Springer, 2001. Cited: pp. 37 & 108.

[351] D. Gottlieb & C.-W. Shu, On the Gibbs phenomenon and its resolution. *SIAM Rev.* **39** (1997) 644–668. Cited: p. 119.

[352] G. Gouesbet & G. Gréhan, Generic formulation of a generalized Lorenz–Mie theory for a particle illuminated by laser pulses. *Particle & Particle Systems Characterization* **17** (2000) 213–224. Cited: p. 120.

[353] I.S. Gradshteyn & I.M. Ryzhik, *Table of Integrals, Series, and Products*, 4th edition. New York: Academic Press, 1980. Cited: pp. 122 & 189.

[354] K.D. Granzow, Multipole theory in the time domain. *J. Math. Phys.* **7** (1966) 634–640. Erratum: **7** (1966) 2280. Cited: pp. 40 & 114.

[355] K.D. Granzow, Time-domain treatment of a spherical boundary-value problem. *J. Appl. Phys.* **39** (1968) 3435–3441. Cited: pp. 40 & 114.

[356] L. Greengard, T. Hagstrom & S. Jiang, The solution of the scalar wave equation in the exterior of a sphere. *J. Comp. Phys.* **274** (2014) 191–207. Cited: pp. 78, 114, 115, 117, 119 & 120.

[357] D.H. Griffel, *Applied Functional Analysis*. Mineola, NY: Dover, 2002. Cited: p. 90.

[358] P. Grinfeld, Small oscillations of a soap bubble. *Stud. Appl. Math.* **128** (2011) 30–39. Cited: p. 84.

[359] P. Grob & P. Joly, Conservative coupling between finite elements and retarded potentials. Application to vibroacoustics. *SIAM J. Sci. Comput.* **29** (2007) 1127–1159. Cited: p. 187.

[360] P.H.L. Groenenboom, The application of boundary elements to steady and unsteady potential fluid flow problems in two and three dimensions. *Applied Mathematical Modelling* **6** (1982) 35–40. Cited: pp. 151 & 166.

[361] P.H.L. Groenenboom, Wave propagation phenomena. In: *Progress in Boundary Element Methods*, vol. 2 (ed. C.A. Brebbia) pp. 24–52. London: Pentech Press, 1983. Cited: p. 166.

[362] P.H.L. Groenenboom, C.A. Brebbia & J.J. De Jong, New developments and engineering applications of boundary elements in the field of transient wave propagation. *Engineering Analysis* **3** (1986) 201–207. Cited: p. 166.

[363] C.E. Grosch & S.A. Orszag, Numerical solution of problems in unbounded regions: coordinate transforms. *J. Comp. Phys.* **25** (1977) 273–295. Cited: p. 42.

[364] E. Grosswald, Bessel polynomials. *Lect. Notes Math.* **698**. Berlin: Springer, 1978. Cited: p. 114.

[365] M.J. Grote & J.B. Keller, Exact nonreflecting boundary conditions for the time dependent wave equation. *SIAM J. Appl. Math.* **55** (1995) 280–297. Cited: p. 40.

[366] M.J. Grote & C. Kirsch, Nonreflecting boundary condition for time-dependent multiple scattering. *J. Comp. Phys.* **221** (2007) 41–62. Cited: p. 32.

[367] Y. Guo, P. Monk & D. Colton, Toward a time domain approach to the linear sampling method. *Inverse Prob.* **29** (2013) 095016. Cited: pp. 31, 33 & 89.

[368] M.E. Gurtin, E. Fried & L. Anand, *The Mechanics and Thermodynamics of Continua*. Cambridge: Cambridge University Press, 2010. Cited: p. 69.

[369] B. Gustafsson, H.-O. Kreiss & J. Oliger, *Time-Dependent Problems and Difference Methods*, 2nd edition. Hoboken, NJ: Wiley, 2013. Cited: pp. 88 & 89.

[370] S. Güttel & F. Tisseur, The nonlinear eigenvalue problem. *Acta Numerica* **26** (2017) 1–94. Cited: p. 135.

[371] A. Gutzmer, Ueber den analytischen Ausdruck des Huygensschen Princips. *Journal für die reine und angewandte Mathematik* **114** (1895) 333–337. Cited: pp. 143 & 150.

[372] W. Hackbusch, W. Kress & S.A. Sauter, Sparse convolution quadrature for time domain boundary integral formulations of the wave equation. *IMA J. Numer. Anal.* **29** (2009) 158–179. Cited: pp. 158 & 171.

[373] J. Hadamard, *Leçons sur la propagation des ondes et les équations de l'hydrodynamique*. Paris: Hermann, 1903. Reprint: New York, Chelsea, 1949. Cited: pp. 64 & 69.

[374] J. Hadamard, *Lectures on Cauchy's Problem in Linear Partial Differential Equations*. New Haven, CT: Yale University Press, 1923. Cited: pp. 16, 71 & 146.

[375] H. Haddar, A. Lechleiter & S. Marmorat, An improved time domain linear sampling method for Robin and Neumann obstacles. *Applicable Analysis* **93** (2014) 369–390. Cited: pp. 31, 80, 91 & 93.

[376] T. Ha-Duong, On the transient acoustic scattering by a flat object. *Japan Journal of Applied Mathematics* **7** (1990) 489–513. Cited: p. 192.

[377] T. Ha-Duong, On retarded potential boundary integral equations and their discretisation. In: *Topics in Computational Wave Propagation: Direct and Inverse Problems* (ed. M. Ainsworth, P. Davies, D. Duncan, B. Rynne & P. Martin) pp. 301–336. *Lecture Notes in Computational Science and Engineering* **31**. Berlin: Springer, 2003. Cited: pp. 24, 91, 147, 156, 159, 162 & 167.

[378] T. Ha-Duong, B. Ludwig & I. Terrasse, A Galerkin BEM for transient acoustic scattering by an absorbing obstacle. *Int. J. Numer. Meth. Eng.* **57** (2003) 1845–1882. Cited: p. 167.

[379] T. Hagstrom, Radiation boundary conditions for the numerical simulation of waves. *Acta Numerica* **8** (1999) 47–106. Cited: pp. 32, 40, 80 & 121.

[380] T. Hagstrom, New results on absorbing layers and radiation boundary conditions. In: *Topics in Computational Wave Propagation: Direct and Inverse Problems* (ed. M. Ainsworth, P. Davies, D. Duncan, B. Rynne & P. Martin) pp. 1–42. *Lecture Notes in Computational Science and Engineering* **31**. Berlin: Springer, 2003. Cited: pp. 32, 40, 42 & 121.

[381] E. Hairer, S.P. Nørsett & G. Wanner, *Solving Ordinary Differential Equations I*, 2nd revised edition. Berlin: Springer, 1993. Cited: pp. 169, 170, 171 & 181.

[382] J.A. Hamilton & R.J. Astley, Exact solutions for transient spherical radiation. *J. Acoust. Soc. Amer.* **109** (2001) 1848–1858. Cited: pp. 119 & 120.

[383] M.F. Hamilton & D.T. Blackstock (ed.), *Nonlinear Acoustics*. San Diego, CA: Academic Press, 1998. Cited: pp. 10 & 211.

[384] M.F. Hamilton & C.L. Morfey, Model equations. In: [383], pp. 41–63. Cited: p. 10.

[385] S. Hanish, *A Treatise on Acoustic Radiation*, 3rd edition. Washington, DC: Naval Research Laboratory, 1989. This is volume 1 of a 5-volume set. Cited: p. 119.

[386] T.B. Hansen, Spherical expansions of time-domain acoustic fields: application to nearfield scanning. *J. Acoust. Soc. Amer.* **98** (1995) 1204–1215. Cited: p. 115.

[387] T.B. Hansen & A.D. Yaghjian, *Plane-Wave Theory of Time-Domain Fields*. New York: IEEE Press, 1999. Cited: pp. 33, 88 & 148.

[388] J.A. Hargreaves & T. Cox, A transient boundary element method for acoustic scattering from mixed regular and thin rigid bodies. *Acta Acustica united with Acustica* **95** (2009) 678–689. Cited: p. 192.

[389] S.M. Hasheminejad, A. Bahari & S. Abbasion, Modelling and simulation of acoustic pulse interaction with a fluid-filled hollow elastic sphere through numerical Laplace inversion. *Applied Mathematical Modelling* **35** (2011) 22–49. Cited: p. 119.

[390] M. Hassell & F.-J. Sayas, Convolution quadrature for wave simulations. In: *Numerical Simulation in Physics and Engineering* (ed. I. Higueras, T. Roldán & J.J. Torrens) pp. 71–159. Cham: Springer, 2016. Cited: p. 170.

[391] M.E. Hassell & F.-J. Sayas, A fully discrete BEM–FEM scheme for transient acoustic waves. *Comput. Meth. Appl. Mech. Eng.* **309** (2016) 106–130. Cited: p. 171.

[392] S. Hayek, Vibration of a spherical shell in an acoustic medium. *J. Acoust. Soc. Amer.* **40** (1966) 342–348. Cited: p. 83.

[393] S. Hayek & F.L. DiMaggio, Complex natural frequencies of vibrating submerged spheroidal shells. *Int. J. Solids Struct.* **6** (1970) 333–351. Cited: p. 130.

[394] S.I. Hayek & J.E. Boisvert, Vibration of prolate spheroidal shells with shear deformation and rotatory inertia: axisymmetric case. *J. Acoust. Soc. Amer.* **114** (2003) 2799–2811. Cited: p. 84.

[395] C. Hazard & M. Lenoir, Determination of scattering frequencies for an elastic floating body. *SIAM J. Math. Anal.* **24** (1993) 1458–1514. Cited: p. 133.

[396] C. Hazard & M. Lenoir, Surface water waves. In: *Scattering*, vol. 1 (ed. R. Pike & P. Sabatier) pp. 618–636. London: Academic Press, 2002. Cited: p. 131.

[397] O. Heaviside, On the extra current. *Phil. Mag.*, Ser. 5, **2** (1876) 135–145. Also: *Electrical Papers*, vol. 1, pp. 53–61. New York: Chelsea, 1970. Cited: p. 19.

[398] G.S. Heller, Propagation of acoustic discontinuities in an inhomogeneous moving liquid medium. *J. Acoust. Soc. Amer.* **25** (1953) 950–951. Cited: p. 69.

[399] G.C. Herman, Scattering of transient acoustic waves by an inhomogeneous obstacle. *J. Acoust. Soc. Amer.* **69** (1981) 909–915. Cited: pp. 166 & 176.

[400] G.C. Herman, Scattering of transient elastic waves by an inhomogeneous obstacle: contrast in volume density of mass. *J. Acoust. Soc. Amer.* **71** (1982) 264–272. Cited: p. 177.

[401] G.C. Herman & P.M. van den Berg, A least-square iterative technique for solving time-domain scattering problems. *J. Acoust. Soc. Amer.* **72** (1982) 1947–1953. Cited: p. 167.

[402] R. Hersh, Mixed problems in several variables. *J. Mathematics & Mechanics* **12** (1963) 317–334. Cited: p. 88.

[403] H.G. Heuser, *Functional Analysis*. Chichester: Wiley, 1982. Cited: pp. 139 & 140.

[404] E. Heyman, Focus wave modes: a dilemma with causality. *IEEE Trans. Antennas & Propag.* **37** (1989) 1604–1608. Cited: p. 49.

[405] E. Heyman & A.J. Devaney, Time-dependent multipoles and their application for radiation from volume source distributions, *J. Math. Phys.* **37** (1996) 682–692. Cited: pp. 40 & 85.

[406] E. Heyman & L.B. Felsen, A wavefront interpretation of the singularity expansion method. *IEEE Trans. Antennas & Propag.* **AP-33** (1985) 706–718. Cited: p. 138.

[407] E. Heyman & L.B. Felsen, Comments on Hillion [412] with author's reply. *IEEE Trans. Antennas & Propag.* **42** (1994) 1668–1670. Cited: pp. 49 & 212.

[408] R.L. Higdon, Initial-boundary value problems for linear hyperbolic systems. *SIAM Rev.* **28** (1986) 177–217. Cited: p. 71.

[409] D.A. Hill & J.R. Wait, The transient electromagnetic response of a spherical shell of arbitrary thickness. *Radio Sci.* **7** (1972) 931–935. Cited: p. 120.

[410] P. Hillion, The Courant–Hilbert solutions of the wave equation. *J. Math. Phys.* **33** (1992) 2749–2753. Cited: pp. 49 & 50.

[411] P. Hillion, Diffraction and Weber functions. *SIAM J. Appl. Math.* **57** (1997) 1702–1715. Cited: p. 46.

[412] P.T.M. Hillion, Nondispersive waves: interpretation and causality. *IEEE Trans. Antennas & Propag.* **40** (1992) 1031–1035. See [407]. Cited: pp. 49 & 212.

[413] S. Hirose, Boundary integral equation method for transient analysis of 3-D cavities and inclusions. *Eng. Anal. Bound. Elem.* **8** (1991) 146–154. Cited: p. 177.

[414] S. Hirose & J.D. Achenbach, BEM method to analyze the interaction of an acoustic pulse with a rigid circular disk. *Wave Motion* **10** (1988) 267–275. Cited: p. 192.

[415] S. Hirose & J.D. Achenbach, Time-domain boundary element analysis of elastic wave interaction with a crack. *Int. J. Numer. Meth. Eng.* **28** (1989) 629–644. Cited: p. 192.

[416] D.B. Hodge, Eigenvalues and eigenfunctions of the spheroidal wave equation. *J. Math. Phys.* **11** (1970) 2308–2312. Cited: p. 53.

[417] S. Holm, *Waves with Power-Law Attenuation*. Cham: Springer, 2019. Cited: p. 19.

[418] S. Holm, S.P. Näsholm, F. Prieur & R. Sinkus, Deriving fractional acoustic wave equations from mechanical and thermal constitutive equations. *Computers & Math. with Applications* **66** (2013) 621–629. Cited: p. 19.

[419] H.G. Hopkins, Dynamic expansion of spherical cavities in metals. In: *Progress in Solid Mechanics*, vol. 1 (ed. I.N. Sneddon & R. Hill) pp. 83–164. Amsterdam: North-Holland, 1960. Cited: p. 108.

[420] M. Howe, *Acoustics and Aerodynamic Sound*. Cambridge: Cambridge University Press, 2015. Cited: pp. 87, 153 & 154.

[421] M.S. Howe, *Acoustics of Fluid–Structure Interactions*. Cambridge: Cambridge University Press, 1998. Cited: pp. 6, 86, 127, 154, 155 & 156.

[422] M.S. Howe, *Theory of Vortex Sound*. Cambridge: Cambridge University Press, 2003. Cited: pp. 85, 87, 153 & 154.

[423] G.C. Hsiao, T. Sánchez-Vizuet & F.-J. Sayas, Boundary and coupled boundary–finite element methods for transient wave–structure interaction. *IMA J. Numer. Anal.* **37** (2017) 237–265. Cited: p. 82.

[424] G.C. Hsiao, F.-J. Sayas & R.J. Weinacht, Time-dependent fluid–structure interaction. *Math. Meth. Appl. Sci.* **40** (2017) 486–500. Cited: p. 82.

[425] G.C. Hsiao & R.J. Weinacht, A representation formula for the wave equation revisited. *Applicable Analysis* **91** (2012) 371–380. Cited: pp. 146 & 150.

[426] G.C. Hsiao & W.L. Wendland, *Boundary Integral Equations*. Berlin: Springer, 2008. Cited: p. 92.

[427] F.Q. Hu, M.E. Pizzo & D.M. Nark, On a time domain boundary integral equation formulation for acoustic scattering by rigid bodies in uniform mean flow. *J. Acoust. Soc. Amer.* **142** (2017) 3624–3636. Cited: pp. 125 & 150.

[428] F.Q. Hu, M.E. Pizzo & D.M. Nark, On the use of a Prandtl–Glauert–Lorentz transformation for acoustic scattering by rigid bodies with a uniform flow. *J. Sound Vib.* **443** (2019) 198–211. Cited: p. 125.

[429] J.-L. Hu, C.H. Chan & Y. Xu, A new temporal basis function for the time-domain integral equation method. *IEEE Microwave & Wireless Components Lett.* **11** (2001) 465–466. Cited: p. 176.

[430] R. Huan & L.L. Thompson, Accurate radiation boundary conditions for the time-dependent wave equation on unbounded domains. *Int. J. Numer. Meth. Eng.* **47** (2000) 1569–1603. Cited: p. 39.

[431] H. Huang, Transient interaction of plane acoustic waves with a spherical elastic shell. *J. Acoust. Soc. Amer.* **45** (1969) 661–670. Cited: p. 119.

[432] H. Huang & G.C. Gaunaurd, Transient diffraction of a plane step pressure pulse by a hard sphere: neoclassical solution. *J. Acoust. Soc. Amer.* **104** (1998) 3236–3244. Cited: p. 119.

[433] H. Huang, Y.P. Lu & Y.F. Wang, Transient interaction of spherical acoustic waves and a spherical elastic shell. *J. Appl. Mech.* **38** (1971) 71–74. Cited: p. 119.

[434] H. Huang & H.U. Mair, Neoclassical solution of transient interaction of plane acoustic waves with a spherical elastic shell. *Shock & Vibration* **3** (1996) 85–98. Cited: p. 119.

[435] C. Hunter & B. Guerrieri, The eigenvalues of the angular spheroidal wave equation. *Stud. Appl. Math.* **66** (1982) 217–240. Cited: p. 53.

[436] A. Inselberg, Cochlear dynamics: the evolution of a mathematical model. *SIAM Rev.* **20** (1978) 301–351. Cited: p. 81.

[437] M.E.H. Ismail, *Classical and Quantum Orthogonal Polynomials in One Variable*. Cambridge: Cambridge University Press, 2005. Cited: p. 114.

[438] S. Itou, Transient analysis of stress waves around a rectangular crack under impact load. *J. Appl. Mech.* **47** (1980) 958–959. Cited: p. 192.

[439] S. Itou, Transient analysis of stress waves around two coplanar Griffith cracks under impact load. *Engineering Fracture Mech.* **13** (1980) 349–356. Cited: p. 192.

[440] J.D. Jackson, *Classical Electrodynamics*, 2nd edition. New York: Wiley, 1975. Cited: pp. 24, 25 & 57.

[441] H.-W. Jang & J.-G. Ih, Stabilization of time domain acoustic boundary element method for the exterior problem avoiding the nonuniqueness. *J. Acoust. Soc. Amer.* **133** (2013) 1237–1244. Cited: p. 166.

[442] H. Jeffreys, *Operational Methods in Mathematical Physics*. Cambridge: Cambridge University Press, 1927. Cited: p. 94.

[443] F.B. Jensen, W.A. Kuperman, M.B. Porter & H. Schmidt, *Computational Ocean Acoustics*, 2nd edition. New York: Springer, 2011. Cited: pp. 78 & 98.

[444] F. John, On the motion of floating bodies. I. *Comm. Pure Appl. Math.* **2** (1949) 13–57. Cited: pp. 27 & 84.

[445] F. John, Hyperbolic and parabolic equations. Part I of *Partial Differential Equations* (L. Bers, F. John & M. Schechter), pp. 1–129. Providence, RI: American Mathematical Society, 1964. Cited: pp. 16, 61, 62, 78, 85, 106 & 108.

[446] F. John, *Partial Differential Equations*, 4th edition. New York: Springer, 1982. Cited: pp. 11, 20 & 41.

[447] L.R. Johnson, Scattering of elastic waves by a spheroidal inclusion. *Geophys. J. Int.* **212** (2018) 1829–1858. Cited: p. 128.

[448] P. Joly, An elementary introduction to the construction and the analysis of perfectly matched layers for time domain wave propagation. *S⃗eMA J.* **57** (2012) 5–48. Cited: pp. 19 & 42.

[449] P. Joly & J. Rodríguez, Mathematical aspects of variational boundary integral equations for time dependent wave propagation. *J. Integ. Eqns Appl.* **29** (2017) 137–187. Cited: pp. 92 & 167.

[450] A.R. Jones, Some calculations on the scattering efficiencies of a sphere illuminated by an optical pulse. *J. Phys. D: Appl. Phys.* **40** (2007) 7306–7312. Cited: p. 120.

[451] D.S. Jones, *The Theory of Electromagnetism*. Oxford: Pergamon Press, 1964. Cited: pp. 24, 25, 57, 151 & 174.

[452] D.S. Jones, *Generalised Functions*. New York: McGraw-Hill, 1966. Cited: pp. 90, 97, 153 & 155.

[453] D.S. Jones, The mathematical theory of noise shielding. *Prog. Aerospace Sci.* **17** (1977) 149–229. Cited: p. 156.

[454] D.S. Jones, *Methods in Electromagnetic Wave Propagation*. Oxford: Clarendon Press, 1987. Cited: pp. 22, 112, 132, 133 & 175.

[455] J.B. Jones-Oliveira, Transient analytic and numerical results for the fluid–solid interaction of prolate spheroidal shells. *J. Acoust. Soc. Amer.* **99** (1996) 392–407. Cited: p. 130.

[456] J.B. Jones-Oliveira & L.P. Harten, Transient fluid–solid interaction of submerged spherical shells revisited: proliferation of frequencies and acoustic radiation effects. *J. Acoust. Soc. Amer.* **96** (1994) 918–925. Cited: p. 119.

[457] M.C. Junger & D. Feit, *Sound, Structures, and Their Interaction*, 2nd edition. New York: Acoustical Society of America, 1993. Cited: p. 83.

[458] M.C. Junger & W. Thompson, Jr, Oscillatory acoustic transients radiated by impulsively accelerated bodies. *J. Acoust. Soc. Amer.* **38** (1965) 978–986. Cited: p. 119.

[459] E.I. Jury, *Theory and Application of the z-Transform Method*. New York: Wiley, 1964. Cited: p. 169.

[460] B. Kager & M. Schanz, Fast and data sparse time domain BEM for elastodynamics. *Eng. Anal. Bound. Elem.* **50** (2015) 212–223. Cited: p. 178.

[461] T. Kailath, *Linear Systems*. Englewood Cliffs, NJ: Prentice-Hall, 1980. Cited: p. 24.

[462] E.G. Kalnins & W. Miller, Jr, Lie theory and the wave equation in space-time. 5. *R*-separable solutions of the wave equation $\psi_{tt} - \Delta_3 \psi = 0$. *J. Math. Phys.* **19** (1978) 1247–1257. Cited: p. 47.

[463] K. Kaouri, D.J. Allwright, C.J. Chapman & J.R. Ockendon, Singularities of wavefields and sonic boom. *Wave Motion* **45** (2008) 217–237. Cited: p. 57.

[464] D.L. Karabalis & D.E. Beskos, Dynamic response of 3-D rigid surface foundations by time domain boundary element method. *Earthquake Engineering & Structural Dynamics* **12** (1984) 73–93. Cited: p. 177.

[465] A. Karlsson & G. Kristensson, Wave splitting in the time domain for a radially symmetric geometry. *Wave Motion* **12** (1990) 197–211. Cited: p. 123.

[466] Y. Kawai & T. Terai, A numerical method for the calculation of transient acoustic scattering from thin rigid plates. *J. Sound Vib.* **141** (1990) 83–96. Cited: p. 192.

[467] M. Kawashita, W. Kawashita & H. Soga, Relation between scattering theories of the Wilcox and Lax–Phillips types and a concrete construction of the translation representation. *Comm. Partial Diff. Eqns.* **28** (2003) 1437–1470. Cited: p. 73.

[468] J. Keener & J. Sneyd, *Mathematical Physiology*, 2nd edition. New York: Springer, 2009. Cited: p. 81.

[469] J. Keilson & W.R. Nunn, Laguerre transformation as a tool for the numerical solution of integral equations of convolution type. *Appl. Math. & Computation* **5** (1979) 313–359. Cited: p. 101.

[470] J.B. Keller, Geometrical acoustics. I. The theory of weak shock waves. *J. Appl. Phys.* **25** (1954) 938–947. Cited: p. 69.

[471] J.B. Keller & A. Blank, Diffraction and reflection of pulses by wedges and corners. *Comm. Pure Appl. Math.* **4** (1951) 75–94. Cited: p. 187.

[472] J.B. Keller & I.I. Kolodner, Damping of underwater explosion bubble oscillations. *J. Appl. Phys.* **27** (1956) 1152–1161. Cited: p. 108.

[473] J.B. Keller & M. Miksis, Bubble oscillations of large amplitude. *J. Acoust. Soc. Amer.* **68** (1980) 628–633. Cited: p. 108.

[474] O.D. Kellogg, *Foundations of Potential Theory*. Berlin: Springer, 1929. Cited: pp. 47 & 157.

[475] E.M. Kennaugh, The scattering of short electromagnetic pulses by a conducting sphere. *Proc. IRE* **49** (1961) 380. Cited: p. 120.

[476] E.M. Kennaugh & D.L. Moffatt, Transient and impulse response approximations. *Proc. IEEE* **53** (1965) 893–901. Cited: p. 120.

[477] V.A. Khromov, Generalization of Kirchhoff's theorem for the case of a surface moving in an arbitrary way. *Soviet Physics-Acoustics* **9** (1963) 68–71. Cited: p. 156.

[478] L. Kielhorn & M. Schanz, Convolution quadrature method-based symmetric Galerkin boundary element method for 3-d elastodynamics. *Int. J. Numer. Meth. Eng.* **76** (2008) 1724–1746. Cited: p. 178.

[479] J.G. Kingston, One-way waves. *SIAM Rev.* **30** (1988) 645–649. Cited: p. 20.

[480] P. Kirby, Calculation of spheroidal wave functions. *Computer Phys. Comm.* **175** (2006) 465–472. Cited: p. 55.

[481] P. Kirby, Calculation of radial prolate spheroidal wave functions of the second kind. *Computer Phys. Comm.* **181** (2010) 514–519. Cited: p. 55.

[482] G. Kirchhoff, Zur Theorie der Lichtstrahlen. *Sitzungsberichte der Königlich Preussischen Akademie der Wissenschaften zu Berlin* (1882) 641–669. Cited: pp. 149, 150 & 216.

[483] G. Kirchhoff, Zur Theorie der Lichtstrahlen. *Annalen der Physik und Chemie* **18** (1883) 663–695. Cited: p. 149.

[484] G. Kirchhoff, *Vorlesungen über Mathematische Physik*, vol. 2: *Mathematische Optik*. Leipzig: Teubner, 1891. Cited: p. 150.

[485] G. Kirchhoff, On the ray theory of light. Translation of [482]. In: *Classical and Modern Diffraction Theory* (ed. K. Klem-Musatov, H.C. Hoeber, T.J. Moser & M.A. Pelissier) pp. 191–203. SEG Geophysics Reprint Ser. No. 29. Tulsa, OK: Society of Exploration Geophysicists, 2016. Cited: p. 150.

[486] A.P. Kiselev & M.V. Perel, Highly localized solutions of the wave equation. *J. Math. Phys.* **41** (2000) 1934–1955. Cited: p. 49.

[487] E. Klaseboer, S. Sepehrirahnama & D.Y.C. Chan, Space-time domain solutions of the wave equation by a non-singular boundary integral method and Fourier transform. *J. Acoust. Soc. Amer.* **142** (2017) 697–707. Cited: pp. 98 & 183.

[488] E. Klaseboer, Q. Sun & D.Y.C. Chan, Field-only integral equation method for time domain scattering of electromagnetic pulses. *Appl. Optics* **56** (2017) 9377–9383. Cited: p. 176.

[489] J.J. Knab, Interpolation of band-limited functions using the approximate prolate series. *IEEE Trans. Information Theory* **IT-25** (1979) 717–720. Cited: pp. 167 & 176.

[490] L. Knopoff, Diffraction of elastic waves. *J. Acoust. Soc. Amer.* **28** (1956) 217–229. Cited: p. 178.

[491] W.L. Ko & T. Karlsson, Application of Kirchhoff's integral equation formulation to an elastic wave scattering problem. *J. Appl. Mech.* **34** (1967) 921–930. Discussion and errata: **35** (1968) 428–430. Cited: p. 178.

[492] S. Kobayashi, Elastodynamics. In: *Boundary Element Methods in Mechanics* (ed. D.E. Beskos) pp. 191–255. Amsterdam: North-Holland, 1987. Cited: pp. 176 & 178.

[493] S. Kobayashi & N. Nishimura, Transient stress analysis of tunnels and caverns of arbitrary shape due to travelling waves. In: *Developments in Boundary Element Methods—2* (ed. P.K. Banerjee & R.P. Shaw) pp. 177–210. Barking: Applied Science Publishers, 1982. Cited: p. 178.

[494] W. Kosiński, *Field Singularities and Wave Analysis in Continuum Mechanics*. Chichester: Ellis Horwood Ltd., 1986. Cited: p. 58.

[495] A.D. Kotsis & J.A. Roumeliotis, Acoustic scattering by a penetrable spheroid. *Acoustical Physics* **54** (2008) 153–167. Cited: p. 128.

[496] O.G. Kozina, G.I. Makarov & N.N. Shaposhnikov, Transient processes in the acoustic fields generated by a vibrating spherical segment. *Soviet Physics-Acoustics* **8** (1962) 53–57. Cited: p. 119.

[497] M. Kranyš, Causal theories of evolution and wave propagation in mathematical physics. *Appl. Mech. Rev.* **42** (1989) 305–322. Cited: p. 22.

[498] H. Kraus, *Thin Elastic Shells*. New York: Wiley, 1967. Cited: p. 82.

[499] H.-O. Kreiss, Initial boundary value problems for hyperbolic systems. *Comm. Pure Appl. Math.* **23** (1970) 277–298. Cited: p. 88.

[500] H.-O. Kreiss, O.E. Ortiz & N.A. Petersson, Initial-boundary value problems for second order systems of partial differential equations. *ESAIM: Mathematical Modelling & Numerical Analysis* **46** (2012) 559–593. Cited: p. 89.

[501] R. Kress, *Linear Integral Equations*, 3rd edition. New York: Springer, 2014. Cited: pp. 139, 140 & 141.

[502] G.A. Kriegsmann, Exploiting the limiting amplitude principle to numerically solve scattering problems. *Wave Motion* **4** (1982) 371–380. Cited: p. 87.

[503] G.A. Kriegsmann, A. Norris & E.L. Reiss, Acoustic scattering by baffled membranes. *J. Acoust. Soc. Amer.* **75** (1984) 685–694. Cited: p. 82.

[504] G.A. Kriegsmann, A.N. Norris & E.L. Reiss, Acoustic pulse scattering by baffled membranes. *J. Acoust. Soc. Amer.* **79** (1986) 1–8. Cited: p. 82.

[505] G. Kristensson, Natural frequencies of circular disks. *IEEE Trans. Antennas & Propag.* **AP-32** (1984) 442–448. Cited: pp. 130 & 133.

[506] G. Kristensson, *Scattering of Electromagnetic Waves by Obstacles*. Edison, NJ: SciTech, 2016. Cited: pp. 24 & 74.

[507] W. Kropp & P.U. Svensson, Application of the time domain formulation of the method of equivalent sources to radiation and scattering problems. *Acustica* **81** (1995) 528–543. Cited: p. 183.

[508] W. Kropp & P.U. Svensson, Time domain formulation of the method of equivalent sources. *Acta Acustica* **3** (1995) 67–73. Cited: p. 183.

[509] K.A. Kuo, H.E.M. Hunt & J.R. Lister, Small oscillations of a pressurized, elastic, spherical shell: model and experiments. *J. Sound Vib.* **359** (2015) 168–178. Cited: p. 84.

[510] Ya.I. Kupets, Diffraction of an acoustic pulse on a soft sphere with a hole. *J. Mathematical Sciences* **96** (1999) 2864–2867. Cited: p. 119.

[511] O.A. Ladyzhenskaya, *The Boundary Value Problems of Mathematical Physics*. New York: Springer, 1985. Cited: pp. 61, 73 & 78.

[512] P. Lalanne, W. Yan. K. Vynck, C. Sauvan & J.-P. Hugonin, Light interaction with photonic and plasmonic resonances. *Laser & Photonics Reviews* **12** (2018) 1700113. Cited: pp. 131, 133, 135 & 136.

[513] A.R. Laliena & F.-J. Sayas, A distributional version of Kirchhoff's formula. *J. Math. Anal. Appl.* **359** (2009) 197–208. Cited: pp. 146 & 149.

[514] H. Lamb, On a peculiarity of the wave-system due to the free vibrations of a nucleus in an extended medium. *Proc. London Math. Soc.*, Ser. 1, **32** (1900) 208–211. Cited: pp. 14 & 135.

[515] H. Lamb, On the diffraction of a solitary wave. *Proc. London Math. Soc.*, Ser. 2, **8** (1910) 422–437. Cited: pp. 50 & 187.

[516] H. Lamb, *Hydrodynamics*, 6th edition. Cambridge: Cambridge University Press, 1932. Cited: pp. 4, 27, 31, 34, 40, 73, 84, 87, 108 & 192.

[517] H. Lamb, *The Dynamical Theory of Sound*, 2nd edition. New York: Dover, 1960. Reprint of 1925 edition. Cited: p. 12.

[518] L.D. Landau & E.M. Lifshitz, *Fluid Mechanics*, 2nd edition. Oxford: Pergamon Press, 1987. Cited: pp. 4, 5, 61 & 73.

[519] J. Larmor, On the mathematical expression of the principle of Huygens. *Proc. London Math. Soc.*, Ser. 2, **1** (1904) 1–13. Cited: p. 156.

[520] I. Lasiecka & R. Triggiani, Recent advances in regularity of second-order hyperbolic mixed problems, and applications. In: *Dynamics Reported: Expositions in Dynamical Systems*, new series, vol. 3 (ed. C.K.R.T. Jones, U. Kirchgraber & H.-O. Walther) pp. 104–162. Berlin: Springer, 1994. Cited: pp. 89, 91 & 92.

[521] M. Lassas, M. Salo & G. Uhlmann, Wave phenomena. In: *Handbook of Mathematical Methods in Imaging*, 2nd edition (ed. O. Scherzer) pp. 1205–1252. New York: Springer, 2015. Cited: p. 5.

[522] V.R. Lauvstad, Transient scattering of a monochromatic acoustical wave by a scatterer fixed in space. *J. Acoust. Soc. Amer.* **38** (1963) 35–46. Cited: pp. 87 & 130.

[523] P. Laven, Time domain analysis of scattering by a water droplet. *Appl. Optics* **50** (2011) F29–F38. Cited: p. 120.

[524] P.D. Lax, *Hyperbolic Partial Differential Equations*. Providence, RI: American Mathematical Society, 2006. Cited: pp. 47, 59, 65 & 71.

[525] P.D. Lax, C.S. Morawetz & R.S. Phillips, Exponential decay of solutions of the wave equation in the exterior of a star-shaped obstacle. *Comm. Pure Appl. Math.* **16** (1963) 477–486. Cited: p. 77.

[526] P.D. Lax & R.S. Phillips, The wave equation in exterior domains. *Bull. Amer. Math. Soc.* **68** (1962) 47–49. Cited: p. 76.

[527] P.D. Lax & R.S. Phillips, *Scattering Theory*. New York: Academic Press, 1967. Cited: pp. 47, 73, 132 & 138.

[528] P.D. Lax & R.S. Phillips, Decaying modes for the wave equation in the exterior of an obstacle. *Comm. Pure Appl. Math.* **22** (1969) 737–787. Cited: p. 138.

[529] P.D. Lax & R.S. Phillips, Scattering theory. *Rocky Mountain J. Math.* **1** (1971) 173–224. Cited: p. 138.

[530] P.D. Lax & R.S. Phillips, On the scattering frequencies of the Laplace operator for exterior domains. *Comm. Pure Appl. Math.* **25** (1972) 85–101. Cited: p. 133.

[531] P.D. Lax & R.S. Phillips, Scattering theory for dissipative hyperbolic systems. *J. Functional Anal.* **14** (1973) 172–235. Cited: p. 80.

[532] P.D. Lax & R.S. Phillips, *Scattering Theory*, revised edition. San Diego, CA: Academic Press, 1989. Cited: pp. 47, 77 & 138.

[533] M. Le Bellac, The Poincaré group. In: *The Scientific Legacy of Poincaré* (ed. É. Charpentier, É. Ghys & A. Lesne) pp. 351–371. Providence, RI: American Mathematical Society, 2010. Cited: p. 23.

[534] P.H. LeBlond & L.A. Mysak, *Waves in the Ocean*. Amsterdam: Elsevier, 1978. Cited: p. 84.

[535] A. Lechleiter & P. Monk, The time-domain Lippmann–Schwinger equation and convolution quadrature. *Numerical Methods for Partial Differential Equations* **31** (2015) 517–540. Cited: p. 171.

[536] S. Lee, Review: the use of equivalent source method in computational acoustics. *J. Computational Acoustics* **25** (2017) 1630001. Cited: p. 183.

[537] S. Lee, K.S. Brentner & P.J. Morris, Acoustic scattering in the time domain using an equivalent source method. *AIAA J.* **48** (2010) 2772–2780. Cited: p. 183.

[538] S. Lee, K.S. Brentner & P.J. Morris, Assessment of time-domain equivalent source method for acoustic scattering. *AIAA J.* **49** (2011) 1897–1906. Cited: p. 183.

[539] Y.W. Lee & D.J. Lee, Derivation and implementation of the boundary integral formula for the convective acoustic wave equation in time domain. *J. Acoust. Soc. Amer.* **136** (2014) 2959–2967. Cited: pp. 125 & 150.

[540] R. Leis, Variations on the wave equation. *Math. Meth. Appl. Sci.* **24** (2001) 339–367. Cited: pp. 1 & 13.

[541] R. Leis, *Initial Boundary Value Problems in Mathematical Physics*. New York: Dover, 2013. Cited: pp. 86 & 88.

[542] J. Lekner, *Theory of Reflection*, 2nd edition. Cham: Springer, 2016. Cited: p. 46.

[543] M. Lenoir, M. Vullierme-Ledard & C. Hazard, Variational formulations for the determination of resonant states in scattering problems. *SIAM J. Math. Anal.* **23** (1992) 579–608. Cited: pp. 133 & 135.

[544] F.G. Leppington & H. Levine, The sound field of a pulsating sphere in unsteady rectilinear motion. *Proc. Roy. Soc.* A **412** (1987) 199–221. Cited: pp. 10, 126 & 127.

[545] M.B. Lesser & D.A. Berkley, Fluid mechanics of the cochlea. Part 1. *J. Fluid Mech.* **51** (1972) 497–512. Cited: p. 81.

[546] H. Levine, *Unidirectional Wave Motions*. Amsterdam: North-Holland, 1978. Cited: pp. 13, 15, 16 & 17.

[547] H. Levine & G.C. Gaunaurd, Energy radiation from point sources whose duration of accelerated motion is finite. *J. Acoust. Soc. Amer.* **110** (2001) 31–36. Cited: p. 57.

[548] H. Levine & F.G. Leppington, The acoustic power from moving and pulsating spheres. *J. Sound Vib.* **146** (1991) 199–210. Cited: p. 126.

[549] J. Li, D. Dault & B. Shanker, A quasianalytical time domain solution for scattering from a homogeneous sphere. *J. Acoust. Soc. Amer.* **135** (2014) 1676–1685. Cited: p. 39.

[550] J. Li, P. Monk & D. Weile, Time domain integral equation methods in computational electromagnetism. In: *Computational Electromagnetism* (ed. A. Bermúdez de Castro & A. Valli) pp. 111–189. *Lect. Notes Math.* **2148**. Cham: Springer, 2015. Cited: p. 176.

[551] J. Li & B. Shanker, Time-dependent Debye–Mie series solutions for electromagnetic scattering. *IEEE Trans. Antennas & Propag.* **63** (2015) 3644–3653. Cited: p. 39.

[552] J. Li & D.S. Weile, Integral accuracy and the stability of two methods for the solution of time-domain integral equations for scattering from perfect conductors. *IEEE Trans. Antennas & Propag.* **67** (2019) 4924–4929. Cited: p. 176.

[553] L.-W. Li, X.-K. Kang & M.-S. Leong, *Spheroidal Wave Functions in Electromagnetic Theory*. New York: Wiley, 2002. Cited: p. 128.

[554] L.-W. Li, M.-S. Leong, T.-S. Yeo, P.-S. Kooi & K.-Y. Tan, Computations of spheroidal harmonics with complex arguments: a review with an algorithm. *Phys. Rev.* E **58** (1998) 6792–6806. Erratum: **71** (2005) 069901. Cited: p. 53.

[555] P. Li & L. Zhang, Analysis of transient acoustic scattering by an elastic obstacle. *Communications in Mathematical Sciences* **17** (2019) 1671–1698. Cited: p. 82.

[556] T. Li, R. Burdisso & C. Sandu, Literature review of models on tire–pavement interaction noise. *J. Sound Vib.* **420** (2018) 357–445. Cited: p. 167.

[557] C. Licht, Évolution d'un système fluide-flotteur. *Journal de Mécanique Théorique et Appliquée* **1** (1982) 211–235. Cited: p. 84.

[558] A. Liénard, Champ électrique et magnétique produit par une charge électrique concentrée en un point et animée d'un mouvement quelconque. *L'Éclairage Électrique* **16** (1898) 5–14, 53–59 & 106–112. Cited: p. 55.

[559] J. Lighthill, *Waves in Fluids*. Cambridge: Cambridge University Press, 1978. Cited: pp. 5, 18, 73, 108 & 127.

[560] M.J. Lighthill, On sound generated aerodynamically I. General theory. *Proc. Roy. Soc.* A **211** (1952) 564–587. Cited: p. 87.

[561] M.J. Lighthill, Viscosity effects in sound waves of finite amplitude. In: *Surveys in Mechanics* (ed. G.K. Batchelor & R.M. Davies) pp. 250–351. Cambridge: Cambridge University Press, 1956. Cited: p. 18.

[562] M.J. Lighthill, *An Introduction to Fourier Analysis and Generalised Functions*. Cambridge: Cambridge University Press, 1958. Cited: p. 90.

[563] M.J. Lighthill, Sound generated aerodynamically. *Proc. Roy. Soc.* A **267** (1962) 147–182. Cited: p. 87.

[564] P.H. Lim & J.M. Ozard, On the underwater acoustic field of a moving point source. I. Range-independent environment. *J. Acoust. Soc. Amer.* **95** (1994) 131–137. Cited: p. 57.

[565] A. Lischke, G. Pang, M. Gulian, F. Song, C. Glusa, X. Zheng, Z. Mao, W. Cai, M.M. Meerschaert, M. Ainsworth & G.E. Karniadakis, What is the fractional Laplacian? A comparative review with new results. *J. Comp. Phys.* **404** (2020) 109009. Cited: p. 19.

[566] S. Litynskyy, Y. Muzychuk & A. Muzychuk, On the numerical solution of the initial-boundary value problem with Neumann condition for the wave equation by the use of the Laguerre transform and boundary elements method. *Acta Mechanica et Automatica* **10** (2016) 285–290. Cited: p. 167.

[567] J.A. Lock & P. Laven, Mie scattering in the time domain. Part 1. The role of surface waves. *J. Opt. Soc. Amer.* A **28** (2011) 1086–1095. Cited: p. 120.

[568] J.A. Lock & P. Laven, Mie scattering in the time domain. Part II. The role of diffraction. *J. Opt. Soc. Amer.* A **28** (2011) 1096–1106. Cited: p. 120.

[569] A.L. Longhorn, The unsteady, subsonic motion of a sphere in a compressible inviscid fluid. *Quart. J. Mech. Appl. Math.* **5** (1952) 64–81. Cited: pp. 10 & 119.

[570] M. Lopez-Fernandez & S. Sauter, Generalized convolution quadrature with variable time stepping. *IMA J. Numer. Anal.* **33** (2013) 1156–1175. Cited: p. 171.

[571] H.A. Lorentz, *The Theory of Electrons*, 2nd edition. New York: Dover, 1952. This edition was first published in 1916. Cited: pp. 11 & 150.

[572] A.E.H. Love, Wave-motions with discontinuities at wave-fronts. *Proc. London Math. Soc.*, Ser. 2, **1** (1904) 37–62. Cited: pp. 63, 68, 108, 146, 149 & 150.

[573] A.E.H. Love, The propagation of wave-motion in an isotropic elastic solid medium. *Proc. London Math. Soc.*, Ser. 2, **1** (1904) 291–344. Cited: p. 176.

[574] A.E.H. Love, Some illustrations of modes of decay of vibratory motions. *Proc. London Math. Soc.*, Ser. 2, **2** (1905) 88–113. Cited: pp. 64, 69, 80, 108 & 135.

[575] M.V. Lowson, The sound field for singularities in motion. *Proc. Roy. Soc.* A **286** (1965) 559–572. Cited: pp. 57 & 155.

[576] J.-y. Lu & J.F. Greenleaf, Nondiffracting X waves—exact solutions to free-space scalar wave equation and their finite aperture realizations. *IEEE Trans. Ultrasonics, Ferroelectrics, & Frequency Control* **39** (1992) 19–31. Cited: p. 51.

[577] C. Lubich, Convolution quadrature and discretized operational calculus. I. *Numer. Math.* **52** (1988) 129–145. Cited: pp. 167 & 168.

[578] Ch. Lubich, On the multistep time discretization of linear initial-boundary value problems and their boundary integral equations. *Numer. Math.* **67** (1994) 365–389. Cited: pp. 89, 92, 93, 167 & 170.

[579] C. Lubich, Convolution quadrature revisited, *BIT Numer. Math.* **44** (2004) 503–514. Cited: p. 170.

[580] D.R. Luke & R. Potthast, The point source method for inverse scattering in the time domain. *Math. Meth. Appl. Sci.* **29** (2006) 1501–1521. Cited: pp. 33 & 99.

[581] R.K. Luneburg, *Mathematical Theory of Optics*. Berkeley: University of California Press, 1964. Reproduced from mimeographed notes issued by Brown University in 1944. Luneburg died in 1949. Cited: pp. 61 & 67.

[582] K.A. Lurie, *An Introduction to the Mathematical Theory of Dynamic Materials*, 2nd edition. Cham: Springer, 2017. Cited: p. 6.

[583] A.S. Lyrintzis, Review: the use of Kirchhoff's method in computational aeroacoustics. *J. Fluids Engineering* **116** (1994) 665–676. Cited: p. 156.

[584] M. Mabrouk & Z. Helali, The scattering theory of C. Wilcox in elasticity. *Math. Meth. Appl. Sci.* **25** (2002) 997–1044. Cited: p. 73.

[585] F. Maestre & P. Pedregal, Dynamic materials for an optimal design problem under the two-dimensional wave equation. *Discrete & Continuous Dynamical Systems*, Ser. A **23** (2009) 973–990. Cited: p. 6.

[586] G. Majda & M. Wei, Relationships between a potential and its scattering frequencies. *SIAM J. Appl. Math.* **55** (1995) 1094–1116. Cited: p. 133.

[587] J. Mäkitalo, M. Kauranen & S. Suuriniemi, Modes and resonances of plasmonic scatterers. *Phys. Rev.* B **89** (2014) 165429. Cited: p. 135.

[588] P. Mann-Nachbar, The interaction of an acoustic wave and an elastic spherical shell. *Quart. Appl. Math.* **15** (1957) 83–93. Cited: p. 119.

[589] G.D. Manolis, A comparative study on three boundary element method approaches to problems in elastodynamics. *Int. J. Numer. Meth. Eng.* **19** (1983) 73–91. Cited: pp. 177 & 178.

[590] W.J. Mansur & C.A. Brebbia, Transient elastodynamics using a time-stepping technique. In: *Boundary Elements* (ed. C.A. Brebbia, T. Futagami & M. Tanaka) pp. 677–698. Berlin: Springer, 1983. Cited: p. 177.

[591] F. Mariani, M.C. Recchioni & F. Zirilli, The use of the Pontryagin maximum principle in a furtivity problem in time-dependent acoustic obstacle scattering. *Waves in Random Media* **11** (2001) 549–575. Cited: pp. 80 & 100.

[592] L. Marin, Natural-mode representation of transient scattered fields. *IEEE Trans. Antennas & Propag.* **AP-21** (1973) 809–818. Cited: p. 137.

[593] L. Marin, Natural-mode representation of transient scattering from rotationally symmetric bodies. *IEEE Trans. Antennas & Propag.* **AP-22** (1974) 266–274. Cited: p. 130.

[594] L. Marin, Transient acoustic scattering from a finite body. *Acustica* **31** (1974) 230–237. Cited: p. 137.

[595] L. Marin, Major results and unresolved issues in singularity expansion method. *Electromagnetics* **1** (1981) 361–373. Cited: p. 130.

[596] L. Marin & R.W. Latham, Representation of transient scattered fields in terms of free oscillations of bodies. *Proc. IEEE* **60** (1972) 640–641. Cited: p. 137.

[597] R.B. Marks, The singular function expansion in time-dependent scattering. *IEEE Trans. Antennas & Propag.* **37** (1989) 1559–1565. Cited: pp. 133 & 142.

[598] J.E. Marsden & T.J.R. Hughes, *Mathematical Foundations of Elasticity*. New York: Dover, 1994. Cited: p. 143.

[599] P.A. Martin, *Multiple Scattering*. Cambridge: Cambridge University Press, 2006. Cited: pp. 33, 40, 85, 98, 128, 130, 133, 134, 140, 147, 161, 166, 174, 177 & 190.

[600] P.A. Martin, Acoustic scattering by a sphere in the time domain. *Wave Motion* **67** (2016) 68–80. Cited: pp. 112, 117 & 121.

[601] P.A. Martin, The pulsating orb: solving the wave equation outside a ball. *Proc. Roy. Soc.* A **472** (2016) 20160037. Cited: pp. 58, 94 & 108.

[602] P.A. Martin, Asymptotic approximations for radial spheroidal wavefunctions with complex size parameter. *Stud. Appl. Math.* **140** (2018) 255–269. Cited: p. 130.

[603] P.A. Martin, On in-out splitting of incident fields and the far-field behaviour of Herglotz wavefunctions. *Math. Meth. Appl. Sci.* **41** (2018) 2961–2970. Cited: p. 100.

[604] P.A. Martin, Acoustics and dynamic materials. *Mechanics Research Comm.* **105** (2020) 103502. Cited: p. 9.

[605] T. Maruyama, T. Saitoh, T.Q. Bui & S. Hirose, Transient elastic wave analysis of 3-D large-scale cavities by fast multipole BEM using implicit Runge–Kutta convolution quadrature. *Comput. Meth. Appl. Mech. Eng.* **303** (2016) 231–259. Cited: p. 178.

[606] E. Marx, Electromagnetic pulse scattered by a sphere. *IEEE Trans. Antennas & Propag.* **35** (1987) 412–417. Cited: p. 120.

[607] D. Mavaleix-Marchessoux, M. Bonnet, S. Chaillat & B. Leblé, A fast boundary element method using the Z-transform and high-frequency approximations for large-scale three-dimensional transient wave problems. *Int. J. Numer. Meth. Eng.* **121** (2020) 4734–4767. Cited: pp. 108 & 171.

[608] J. McCully, The Laguerre transform. *SIAM Rev.* **2** (1960) 185–191. Cited: p. 101.

[609] M. McIver & P. McIver, Water waves in the time domain. *J. Engng. Math.* **70** (2011) 111–128. Cited: p. 84.

[610] E. Mecocci, L. Misici, M.C. Recchioni & F. Zirilli, A new formalism for time-dependent wave scattering from a bounded obstacle. *J. Acoust. Soc. Amer.* **107** (2000) 1825–1840. Cited: pp. 98 & 100.

[611] L. Mees, G. Gouesbet & G. Gréhan, Scattering of laser pulses (plane wave and focused Gaussian beam) by spheres. *Appl. Optics* **40** (2001) 2546–2550. Cited: p. 120.

[612] J. Meixner & F.W. Schäfke, *Mathieusche Funktionen und Sphäroidfunktionen*. Berlin: Springer, 1954. Cited: p. 55.

[613] J. Meixner, F.W. Schäfke & G. Wolf, *Mathieu Functions and Spheroidal Functions and Their Mathematical Foundations: Further Studies. Lect. Notes Math.* **837**. Berlin: Springer, 1980. Cited: p. 53.

[614] R.B. Melrose, *Geometric Scattering Theory*. Cambridge: Cambridge University Press, 1995. Cited: p. 132.

[615] B.L. Merchant, P.J. Moser, A. Nagl & H. Überall, Complex pole patterns of the scattering amplitude for conducting spheroids and finite-length cylinders. *IEEE Trans. Antennas & Propag.* **36** (1988) 1769–1778. Cited: p. 130.

[616] B.L. Merchant, A. Nagl & H. Überall, Resonance frequencies of conducting spheroids and the phase matching of surface waves. *IEEE Trans. Antennas & Propag.* **AP-34** (1986) 1464–1467. Cited: p. 136.

[617] B.L. Merchant, A. Nagl & H. Überall, A method for calculating eigenfrequencies of arbitrarily shaped convex targets: eigenfrequencies of conducting spheroids and their relation to helicoidal surface wave paths. *IEEE Trans. Antennas & Propag.* **37** (1989) 629–634. Cited: p. 136.

[618] M. Messner & M. Schanz, An accelerated symmetric time-domain boundary element formulation for elasticity. *Eng. Anal. Bound. Elem.* **34** (2010) 944–955. Cited: p. 178.

[619] M.H. Meylan & C.J. Fitzgerald, The singularity expansion method and near-trapping of linear water waves. *J. Fluid Mech.* **755** (2014) 230–250. Cited: p. 133.

[620] M.H. Meylan & C. Fitzgerald, Computation of long lived resonant modes and the poles of the S-matrix in water wave scattering. *J. Fluids & Struct.* **76** (2018) 153–165. Cited: p. 133.

[621] K.A. Michalski, Bibliography of the singularity expansion method and related topics. *Electromagnetics* **1** (1981) 493–511. Cited: p. 204.

[622] B.L. Michielsen, G.C. Herman, A.T. de Hoop & D. de Zutter, Three-dimensional relativistic scattering of electromagnetic waves by an object in uniform translational motion. *J. Math. Phys.* **22** (1981) 2716–2722. Cited: p. 125.

[623] J. Miklowitz, Modern corner, edge, and crack problems in linear elastodynamics involving transient waves. *Advances in Applied Mechanics* **25** (1987) 47–181. Cited: p. 120.

[624] V. Milenkovic & S. Raynor, Reflection of a plane acoustic step wave from an elastic spherical membrane. *J. Acoust. Soc. Amer.* **39** (1966) 556–563. Cited: p. 119.

[625] K. Miller, Stabilized numerical analytic prolongation with poles. *SIAM J. Appl. Math.* **18** (1970) 346–363. Cited: p. 133.

[626] T. Miloh, A note on impulsive sphere motion beneath a free-surface. *J. Engng. Math.* **41** (2001) 1–11. Cited: p. 84.

[627] R. Misawa, K. Niino & N. Nishimura, Boundary integral equations for calculating complex eigenvalues of transmission problems. *SIAM J. Appl. Math.* **77** (2017) 770–788. Cited: pp. 134 & 135.

[628] R. Mittra, Integral equation methods for transient scattering. In: *Transient Electromagnetic Fields* (ed. L.B. Felsen) pp. 73–128. *Topics in Applied Physics* **10**. Berlin: Springer, 1976. Cited: pp. 175 & 176.

[629] K.M. Mitzner, Numerical solution for transient scattering from a hard surface of arbitrary shape—retarded potential technique. *J. Acoust. Soc. Amer.* **42** (1967) 391–397. Comments by R.P. Shaw: [761]. Cited: pp. 165 & 229.

[630] G. Monegato, L. Scuderi & M.P. Stanić, Lubich convolution quadratures and their application to problems described by space-time BIEs. *Numerical Algorithms* **56** (2011) 405–436. Cited: pp. 168, 170 & 171.

[631] C.S. Morawetz, The limiting amplitude principle. *Comm. Pure Appl. Math.* **15** (1962) 349–361. Cited: p. 87.

[632] C.S. Morawetz, The limiting amplitude principle for arbitrary finite bodies. *Comm. Pure Appl. Math.* **18** (1965) 183–189. Cited: p. 87.

[633] C.S. Morawetz, Exponential decay of solutions of the wave equation. *Comm. Pure Appl. Math.* **19** (1966) 439–444. Cited: pp. 47 & 77.

[634] C.S. Morawetz, Energy flow: wave motion and geometrical optics. *Bull. Amer. Math. Soc.* **76** (1970) 661–674. Cited: pp. 47, 73 & 77.

[635] C.S. Morawetz, *Notes on Time Decay and Scattering for Some Hyperbolic Problems*. Philadelphia: SIAM, 1975. Cited: pp. 24, 73 & 108.

[636] C.S. Morawetz, J.V. Ralston & W.A. Strauss, Decay of solutions of the wave equation outside nontrapping obstacles. *Comm. Pure Appl. Math.* **30** (1977) 447–508. Cited: p. 77.

[637] C.L. Morfey, Rotating blades and aerodynamic sound. *J. Sound Vib.* **28** (1973) 587–617. Cited: p. 87.

[638] M.A. Morgan, Singularity expansion representations of fields and currents in transient scattering. *IEEE Trans. Antennas & Propag.* **AP-32** (1984) 466–473. Cited: p. 138.

[639] W.R. Morgans, The Kirchhoff formula extended to a moving surface. *Phil. Mag.*, Ser. 7, **9** (1930) 141–161. Cited: p. 156.

[640] L. Morino, B.K. Bharadvaj, M.I. Freedman & K. Tseng, Boundary integral equation for wave equation with moving boundary and applications to compressible potential aerodynamics of airplanes and helicopters. *Comput. Mech.* **4** (1989) 231–243. Cited: p. 156.

[641] T. Morley, A simple proof that the world is three-dimensional. *SIAM Rev.* **27** (1985) 69–71. Errata: **28** (1986) 229. Cited: p. 32.

[642] P.M. Morse, *Vibration and Sound*. New York: McGraw-Hill, 1936. Cited: p. 108.

[643] P.M. Morse & H. Feshbach, *Methods of Theoretical Physics*. New York: McGraw-Hill, 1953. Cited: pp. 11, 12, 19, 55, 86, 151 & 158.

[644] P.M. Morse & K.U. Ingard, *Theoretical Acoustics*. New York: McGraw-Hill, 1968. Reprint: Princeton, NJ: Princeton University Press, 1986. Cited: pp. 1, 4, 5, 12, 19, 57 & 81.

[645] H.E. Moses, The time-dependent inverse source problem for the acoustic and electromagnetic equations in the one- and three-dimensional cases. *J. Math. Phys.* **25** (1984) 1905–1923. Cited: p. 88.

[646] H.E. Moses & R.T. Prosser, Propagation of an electromagnetic field through a planar slab. *SIAM Rev.* **35** (1993) 610–620. Cited: p. 18.

[647] C.C. Mow, Transient response of a rigid spherical inclusion in an elastic medium. *J. Appl. Mech.* **32** (1965) 637–642. Cited: p. 120.

[648] N. Mujica, R. Wunenburger & S. Fauve, Scattering of a sound wave by a vibrating surface. *European Physical Journal B–Condensed Matter & Complex Systems* **33** (2003) 209–213. Cited: p. 128.

[649] M.K. Myers, On the acoustic boundary condition in the presence of flow. *J. Sound Vib.* **71** (1980) 429–434. Cited: p. 126.

[650] M.K. Myers & J.S. Hausmann, Computation of acoustic scattering from a moving rigid surface. *J. Acoust. Soc. Amer.* **91** (1992) 2594-2605. Cited: p. 125.

[651] A. Najafi-Yazdi, G.A. Brès & L. Mongeau, An acoustic analogy formulation for moving sources in uniformly moving media. *Proc. Roy. Soc.* A **467** (2011) 144–165. Cited: p. 154.

[652] L. Nannen & M. Wess, Computing scattering resonances using perfectly matched layers with frequency dependent scaling functions. *BIT Numer. Math.* **58** (2018) 373–395. Cited: p. 133.

[653] F. Natterer, Sonic imaging. In: *Handbook of Mathematical Methods in Imaging*, 2nd edition (ed. O. Scherzer) pp. 1253–1278. New York: Springer, 2015. Cited: p. 5.

[654] H.C. Neilson, Y.P. Lu & Y.F. Wang, Transient scattering by arbitrary axisymmetric surfaces. *J. Acoust. Soc. Amer.* **63** (1978) 1719–1726. Cited: p. 166.

[655] J.A. Neubert & J.L. Lumley, Derivation of the stochastic Helmholtz equation for sound propagation in a turbulent fluid. *J. Acoust. Soc. Amer.* **48** (1970) 1212–1218. Cited: p. 5.

[656] J.N. Newman, A simplified derivation of the ordinary differential equations for the free-surface Green functions. *Appl. Ocean Res.* **94** (2020) 101973. Cited: p. 181.

[657] R.G. Newton, Analytic properties of radial wave functions. *J. Math. Phys.* **1** (1960) 319–347. Cited: p. 133.

[658] R.G. Newton, *Scattering Theory of Waves and Particles*, 2nd edition. New York: Springer, 1982. Cited: p. 133.

[659] G. Ni, S.J. Elliott, M. Ayat & P.D. Teal, Modelling cochlear mechanics. *BioMed Research Int.* **2014** (2014) 150637. Cited: p. 81.

[660] A. Nisbet, Electromagnetic potentials in a heterogeneous non-conducting medium. *Proc. Roy. Soc.* A **240** (1957) 375–381. Cited: p. 25.

[661] NIST Digital Library of Mathematical Functions, dlmf.nist.gov. Cited: pp. 34, 35, 36, 44, 51, 52, 53, 54, 55, 101, 114, 116, 117, 118, 133, 167, 171 & 172.

[662] Y. Niwa, T. Fukui, S. Kato & K. Fujiki, An application of the integral equation method to two-dimensional elastodynamics. *Theoretical and Applied Mechanics: Proc. 28th Japan National Congress for Applied Mechanics, 1978*, pp. 281–290. Tokyo: University of Tokyo Press, 1980. Cited: p. 177.

[663] Y. Niwa, S. Hirose & M. Kitahara, Application of the boundary integral equation (BIE) method to transient response analysis of inclusions in a half space. *Wave Motion* **8** (1986) 77–91. Cited: p. 178.

[664] A.N. Norris, Acoustic integrated extinction. *Proc. Roy. Soc.* A **471** (2015) 20150008. Cited: p. 23.

[665] A.N. Norris & D.A. Rebinsky, Acoustic coupling to membrane waves on elastic shells. *J. Acoust. Soc. Amer.* **95** (1994) 1809–1829. Cited: p. 82.

[666] A.N. Norris & D.A. Rebinsky, Membrane and flexural waves on thin shells. *J. Vibration & Acoustics* **116** (1994) 457–467. Cited: p. 82.

[667] J.D. Norton, Is there an independent principle of causality in physics? *British J. Philosophy of Science* **60** (2009) 475–486. Cited: p. 22.

[668] F.R. Norwood & J. Miklowitz, Diffraction of transient elastic waves by a spherical cavity. *J. Appl. Mech.* **34** (1967) 735–744. Cited: p. 120.

[669] S.K. Numrich & H. Überall, Scattering of sound pulses and the ringing of target resonances. *Physical Acoustics* **21** (1992) 235–318. Cited: p. 138.

[670] H.M. Nussenzveig, *Causality and Dispersion Relations*. New York: Academic Press, 1972. Cited: pp. 22, 23 & 135.

[671] H.L. Oestreicher, Field of a spatially extended moving sound source. *J. Acoust. Soc. Amer.* **29** (1957) 1223–1232. Cited: p. 125.

[672] T. Oguchi, Eigenvalues of spheroidal wave functions and their branch points for complex values of propagation constants. *Radio Sci.* **5** (1970) 1207–1214. Cited: p. 53.

[673] F.W.J. Olver, The asymptotic expansion of Bessel functions of large order. *Phil. Trans. Roy. Soc.* A **247** (1954) 328–368. Cited: p. 114.

[674] A. Osipov, V. Rokhlin & H. Xiao, *Prolate Spheroidal Wave Functions of Order Zero: Mathematical Tools for Bandlimited Approximation*. New York: Springer, 2013. Cited: p. 54.

[675] V.E. Ostashev, *Acoustics in Moving Inhomogeneous Media*. London: Spon, 1997. Cited: pp. 1 & 6.

[676] B. Osting & M.I. Weinstein, Long-lived scattering resonances and Bragg structures. *SIAM J. Appl. Math.* **73** (2013) 827–852. Cited: p. 136.

[677] L. Page & N.I. Adams, Jr, The electrical oscillations of a prolate spheroid. Paper I. *Phys. Rev.* **53** (1938) 819–831. Cited: p. 130.

[678] R.Y.S. Pak & X. Bai, A regularized boundary element formulation with weighted-collocation and higher-order projection for 3D time-domain elastodynamics. *Eng. Anal. Bound. Elem.* **93** (2018) 135–142. Cited: p. 177.

[679] C.G. Panagiotopoulos & G.D. Manolis, Three-dimensional BEM for transient elastodynamics based on the velocity reciprocal theorem. *Eng. Anal. Bound. Elem.* **35** (2011) 507–516. Cited: p. 177.

[680] W.K.H. Panofsky & M. Phillips, *Classical Electricity and Magnetism*, 2nd edition. Reading, MA: Addison-Wesley, 1962. Cited: pp. 49 & 57.

[681] Y.-H. Pao & C.-C. Mow, *Diffraction of Elastic Waves and Dynamic Stress Concentrations*. New York: Crane, Russak & Co., 1973. Cited: p. 120.

[682] Y.-H. Pao & V. Varatharajulu, Huygens' principle, radiation conditions, and integral formulas for the scattering of elastic waves. *J. Acoust. Soc. Amer.* **59** (1976) 1361–1371. Cited: pp. 176 & 178.

[683] M. Parmar, A. Haselbacher & S. Balachandar, On the unsteady inviscid force on cylinders and spheres in subcritical compressible flow. *Phil. Trans. Roy. Soc.* A **366** (2008) 2161–2175. Cited: p. 10.

[684] J.-M. Parot, C. Thirard & C. Puillet, Elimination of a non-oscillatory instability in a retarded potential integral equation. *Eng. Anal. Bound. Elem.* **31** (2007) 133–151. Cited: p. 192.

[685] V.Z. Parton & V.G. Boriskovsky, *Dynamic Fracture Mechanics*, vol. 1: *Stationary Cracks*, revised edition. New York: Hemisphere, 1989. Cited: p. 192.

[686] L.W. Pearson, Present thinking on the use of the singularity expansion in electromagnetic scattering computation. *Wave Motion* **5** (1983) 355–368. Cited: p. 138.

[687] L.W. Pearson, A note on the representation of scattered fields as a singularity expansion. *IEEE Trans. Antennas & Propag.* **AP-32** (1984) 520–524. Cited: p. 138.

[688] A. Peirce & E. Siebrits, Stability analysis and design of time-stepping schemes for general elastodynamic boundary element models. *Int. J. Numer. Meth. Eng.* **40** (1997) 319–342. Cited: p. 177.

[689] Z. Peng, K.-H. Lim & J.-F. Lee, A discontinuous Galerkin surface integral equation method for electromagnetic wave scattering from nonpenetrable targets. *IEEE Trans. Antennas & Propag.* **61** (2013) 3617–3628. Cited: p. 176.

[690] R. Penrose, Solutions of the zero-rest-mass equations. *J. Math. Phys.* **10** (1969) 38–39. Cited: p. 46.

[691] R. Penrose, On the origins of twistor theory. In: *Gravitation and Geometry* (ed. W. Rindler & A. Trautman) pp. 341–361. Naples: Bibliopolis, 1987. Cited: p. 46.

[692] B.A. Peterson, V.V. Varadan & V.K. Varadan, T-matrix approach to study the vibration frequencies of elastic bodies in fluids. *J. Acoust. Soc. Amer.* **74** (1983) 1051–1056. Cited: p. 130.

[693] S. Petropavlovsky, S. Tsynkov & E. Turkel, A method of boundary equations for unsteady hyperbolic problems in 3D. *J. Comp. Phys.* **365** (2018) 294–323. Cited: p. 151.

[694] S.V. Petropavlovsky & S.V. Tsynkov, Method of difference potentials for evolution equations with lacunas. *Computational Math. & Math. Phys.* **60** (2020) 711–722. Cited: p. 151.

[695] P.G. Petropoulos, Reflectionless sponge layers as absorbing boundary conditions for the numerical solution of Maxwell equations in rectangular, cylindrical, and spherical coordinates. *SIAM J. Appl. Math.* **60** (2000) 1037–1058. Cited: p. 42.

[696] A.D. Pierce, *Acoustics*. New York: Acoustical Society of America, 1989. Cited: pp. 1, 2, 5, 7, 9, 31, 73, 81, 108 & 119.

[697] A.D. Pierce, Wave equation for sound in fluids with unsteady inhomogeneous flow. *J. Acoust. Soc. Amer.* **87** (1990) 2292–2299. Cited: pp. 1, 6 & 7.

[698] J.C. Piquette & A.L. Van Buren, Nonlinear scattering of acoustic waves by vibrating surfaces. *J. Acoust. Soc. Amer.* **76** (1984) 880–889. Cited: p. 128.

[699] J.C. Piquette, A.L. Van Buren & P.H. Rogers, Censor's acoustical Doppler effect analysis—Is it a valid method? *J. Acoust. Soc. Amer.* **83** (1988) 1681–1682. Cited: p. 128.

[700] A.J. Poggio & E.K. Miller, Integral equation solution of three-dimensional scattering problems. In: *Computer Techniques for Electromagnetics* (ed. R. Mittra) pp. 159–264. Oxford: Pergamon Press, 1973. Cited: pp. 174, 175 & 176.

[701] H. Poincaré, *Électricité et optique*, 2nd edition. Paris: Gauthier-Villars, 1901. Cited: p. 11.

[702] D. Pölz & M. Schanz, Space-time discretized retarded potential boundary integral operators: quadrature for collocation methods. *SIAM J. Sci. Comput.* **41** (2019) A3860–A3886. Cited: p. 167.

[703] A.J. Pray, Y. Beghein, N.V. Nair, K. Cools, H. Bağcı & B. Shanker, A higher order space-time Galerkin scheme for time domain integral equations. *IEEE Trans. Antennas & Propag.* **62** (2014) 6183–6191. Cited: p. 176.

[704] J. Prieur & G. Rahier, Aeroacoustic integral methods, formulation and efficient numerical implementation. *Aerospace Science & Technology* **5** (2001) 457–468. Addendum: **6** (2002) 323. Cited: p. 156.

[705] A.C. Prunty & R.K. Snieder, Theory of the linear sampling method for time-dependent fields. *Inverse Prob.* **35** (2019) 055003. Cited: p. 31.

[706] T. Qiu & F.J. Sayas, The Costabel–Stephan system of boundary integral equations in the time domain. *Mathematics of Computation* **85** (2016) 2341–2364. Cited: p. 171.

[707] A.G. Ramm, Mathematical foundations of the singularity and eigenmode expansion methods (SEM and EEM). *J. Math. Anal. Appl.* **86** (1982) 562–591. Cited: pp. 137, 138 & 140.

[708] S.M. Rao & D.R. Wilton, Transient scattering by conducting surfaces of arbitrary shape. *IEEE Trans. Antennas & Propag.* **39** (1991) 56–61. Cited: p. 176.

[709] J. Rauch, *Hyperbolic Partial Differential Equations and Geometric Optics*. Providence, RI: American Mathematical Society, 2012. Cited: pp. 17, 59 & 61.

[710] J.B. Rauch & F.J. Massey III, Differentiability of solutions to hyperbolic initial-boundary value problems. *Trans. Amer. Math. Soc.* **189** (1974) 303–318. Cited: p. 78.

[711] Lord Rayleigh (J.W. Strutt), *The Theory of Sound*, vol. 2, 2nd edition. London: Macmillan & Co., 1896. Reprint: New York: Dover, 1945. Cited: pp. 1, 18, 31, 87 & 108.

[712] M. Reed & B. Simon, *Methods of Modern Mathematical Physics*, vol. 1: *Functional Analysis*. New York: Academic Press, 1972. Cited: pp. 135 & 139.

[713] R.C. Restrick, III, Electromagnetic scattering by a moving conducting sphere. *Radio Sci.* **3** (1968) 1144–1154. Cited: p. 125.

[714] L. Rhaouti, A. Chaigne & P. Joly, Time-domain modeling and numerical simulation of a kettledrum. *J. Acoust. Soc. Amer.* **105** (1999) 3545–3562. Cited: p. 82.

[715] J. Rheinstein, Backscatter from spheres: a short pulse view. *IEEE Trans. Antennas & Propag.* **AP-16** (1968) 89–97. Cited: p. 120.

[716] D.C. Rizos & D.L. Karabalis, An advanced direct time domain BEM formulation for general 3-D elastodynamic problems. *Comput. Mech.* **15** (1994) 249–269. Cited: p. 177.

[717] D.C. Rizos & S. Zhou, An advanced direct time domain BEM for 3-D wave propagation in acoustic media. *J. Sound Vib.* **293** (2006) 196–212. Cited: p. 167.

[718] G.F. Roach, *Wave Scattering by Time-Dependent Perturbations*. Princeton, NJ: Princeton University Press, 2007. Cited: pp. 13 & 24.

[719] P.H. Rogers, Comments on Censor [150] and author's reply. *J. Sound Vib.* **28** (1973) 764–768. Cited: pp. 128 & 200.

[720] J.H. Rose, Phase retrieval for the variable velocity classical wave equation. *Inverse Prob.* **2** (1986) 219–228. Cited: p. 5.

[721] T.E. Roth & W.C. Chew, Development of stable A-Φ time-domain integral equations for multiscale electromagnetics. *IEEE J. Multiscale & Multiphysics Computational Techniques* **3** (2018) 255–265. Cited: p. 176.

[722] T.E. Roth & W.C. Chew, Stability analysis and discretization of **A**-Φ time domain integral equations for multiscale electromagnetics. *J. Comp. Phys.* **408** (2020) 109102. Cited: pp. 91 & 176.

[723] M. Rousseau, G.A. Maugin & M. Berezovski, Elements of study on dynamic materials. *Archive of Applied Mechanics* **81** (2011) 925–942. Cited: pp. 6 & 7.

[724] A.J. Rudgers, Acoustic pulses scattered by a rigid sphere immersed in a fluid. *J. Acoust. Soc. Amer.* **45** (1969) 900–910. Cited: p. 119.

[725] A.J. Rudgers, Separation and analysis of the acoustic field scattered by a rigid sphere. *J. Acoust. Soc. Amer.* **52** (1972) 234–246. Cited: p. 119.

[726] B. Russell, On the notion of cause. In: *Mysticism and Logic and Other Essays*, pp. 180–208. London: George Allen & Unwin Ltd., 1917. Cited: p. 22.

[727] B.P. Rynne, Stability and convergence of time marching methods in scattering problems. *IMA J. Appl. Math.* **35** (1985) 297–310. Cited: p. 166.

[728] B.P. Rynne, Instabilities in time marching methods for scattering problems. *Electromagnetics* **6** (1986) 129–144. Cited: p. 166.

[729] B.P. Rynne, The well-posedness of the electric field integral equation for transient scattering from a perfectly conducting body. *Math. Meth. Appl. Sci.* **22** (1999) 619–631. Cited: pp. 175 & 176.

[730] B.P. Rynne & P.D. Smith, Stability of time marching algorithms for the electric field integral equation. *J. Electro. Waves Appl.* **4** (1990) 1181–1205. Cited: p. 176.

[731] A. Sadigh & E. Arvas, Treating the instabilities in marching-on-in-time method from a different perspective. *IEEE Trans. Antennas & Propag.* **41** (1993) 1695–1702. Cited: p. 176.

[732] T. Saitoh & S. Hirose, Parallelized fast multipole BEM based on the convolution quadrature method for 3-D wave propagation problems in time-domain. *IOP Conference Series: Materials Science and Engineering* **10** (2010) 012242. Cited: p. 171.

[733] R. Sakamoto, Mixed problems for hyperbolic equations. II. Existence theorems with zero initial datas and energy inequalities with initial datas. *Journal of Mathematics of Kyoto University* **10** (1970) 403–417. Cited: p. 88.

[734] R. Sakamoto, *Hyperbolic Boundary Value Problems*. Cambridge: Cambridge University Press, 1982. Cited: pp. 16, 78 & 88.

[735] T. Sakurai & H. Sugiura, A projection method for generalized eigenvalue problems using numerical integration. *J. Comp. Appl. Math.* **159** (2003) 119–128. Cited: p. 135.

[736] J. Salo, J. Fagerholm, A.T. Friberg & M.M. Salomaa, Unified description of nondiffracting *X* and *Y* waves. *Phys. Rev.* E **62** (2000) 4261–4275. Cited: p. 51.

[737] K. Sandberg & G. Beylkin, Full-wave-equation depth extrapolation for migration. *Geophysics* **74** (2009) WCA121–WCA128. Cited: p. 5.

[738] S.A. Sauter & M. Schanz, Convolution quadrature for the wave equation with impedance boundary conditions. *J. Comp. Phys.* **334** (2017) 442–459. Cited: p. 171.

[739] S. Sauter & A. Veit, A Galerkin method for retarded boundary integral equations with smooth and compactly supported temporal basis functions. *Numer. Math.* **123** (2013) 145–176. Cited: p. 167.

[740] S. Sauter & A. Veit, Retarded boundary integral equations on the sphere: exact and numerical solution. *IMA J. Numer. Anal.* **34** (2014) 675–699. Cited: p. 167.

[741] S. Sauter & A. Veit, Adaptive time discretization for retarded potentials. *Numer. Math.* **132** (2016) 569–595. Cited: p. 167.

[742] F.-J. Sayas, *Retarded Potentials and Time Domain Boundary Integral Equations: A Road Map*. Cham: Springer, 2016. Cited: pp. 24, 89, 91, 156, 159, 168, 170 & 176.

[743] M. Schanz, *Wave Propagation in Viscoelastic and Poroelastic Continua: A Boundary Element Approach*. Berlin: Springer, 2001. Cited: pp. 168 & 178.

[744] M. Schanz & H. Antes, Application of 'operational quadrature methods' in time domain boundary element methods. *Meccanica* **32** (1997) 179–186. Cited: p. 168.

[745] E. Schlemmer, W.M. Rucker & K.R. Richter, Boundary element computations of 3D transient scattering from lossy dielectric objects. *IEEE Trans. Magnetics* **29** (1993) 1524–1527. Cited: p. 176.

[746] E. Schmidt, Zur Theorie der linearen und nichtlinearen Integralgleichungen. I. Teil: Entwicklung willkürlicher Funktionen nach Systemen vorgeschriebener. *Mathematische Annalen* **63** (1907) 433–476. Cited: p. 140.

[747] G. Schmidt, Spectral and scattering theory for Maxwell's equations in an exterior domain. *Arch. Rational Mech. Anal.* **28** (1968) 284–322. Cited: p. 73.

[748] G.A. Schott, *Electromagnetic Radiation*. Cambridge: Cambridge University Press, 1912. Cited: p. 57.

[749] L.S. Schulman, *Time's Arrows and Quantum Measurement*. Cambridge: Cambridge University Press, 1997. Cited: pp. 22 & 24.

[750] L. Schwartz, *Mathematics for the Physical Sciences*. Reading, MA: Addison-Wesley, 1966. Cited: pp. 90, 97 & 152.

[751] J. Schwinger, On the classical radiation of accelerated electrons. *Phys. Rev.* **75** (1949) 1912–1925. Cited: p. 57.

[752] T.B.A. Senior & P.L.E. Uslenghi, The prolate spheroid. Chapter 11 (pp. 416–471) in [121]. Cited: p. 128.

[753] T.B.A. Senior & P.L.E. Uslenghi, The oblate spheroid. Chapter 13 (pp. 503–527) in [121]. Cited: p. 52.

[754] J. Sesma, A naive procedure for computing angular spheroidal functions. arXiv: 1606.00149, April 2018. Cited: pp. 53 & 55.

[755] B. Seymour & E. Varley, Exact representations for acoustical waves when the sound speed varies in space and time. *Stud. Appl. Math.* **76** (1987) 1–35. Cited: pp. 6 & 12.

[756] A.M. Shaarawi, R.W. Ziolkowski & I.M. Besieris, On the evanescent fields and the causality of the focus wave modes. *J. Math. Phys.* **36** (1995) 5565–5587. Cited: p. 49.

[757] B. Shanker, A.A. Ergin, K. Aygün & E. Michielssen, Analysis of transient electromagnetic scattering phenomena using a two-level plane wave time-domain algorithm. *IEEE Trans. Antennas & Propag.* **48** (2000) 510–523. Errata: **49** (2001) 1243. Cited: p. 176.

[758] B. Shanker, A.A. Ergin, K. Aygün & E. Michielssen, Analysis of transient electromagnetic scattering from closed surfaces using a combined field integral equation. *IEEE Trans. Antennas & Propag.* **48** (2000) 1064–1074. Cited: p. 176.

[759] B. Shanker, A.A. Ergin, M. Lu & E. Michielssen, Fast analysis of transient electromagnetic scattering phenomena using the multilevel plane wave time domain algorithm. *IEEE Trans. Antennas & Propag.* **51** (2003) 628–641. Cited: p. 176.

[760] R.P. Shaw, Diffraction of acoustic pulses by obstacles of arbitrary shape with a Robin boundary condition. *J. Acoust. Soc. Amer.* **41** (1967) 855–859. Cited: p. 165.

[761] R.P. Shaw, Comments on Mitzner [629]. *J. Acoust. Soc. Amer.* **43** (1968) 638–639. Cited: pp. 165 & 223.

[762] R.P. Shaw, Retarded potential approach to the scattering of elastic pulses by rigid obstacles of arbitrary shape. *J. Acoust. Soc. Amer.* **44** (1968) 745–748. Cited: p. 178.

[763] R.P. Shaw, Diffraction of plane acoustic pulses by obstacles of arbitrary cross section with an impedance boundary condition. *J. Acoust. Soc. Amer.* **44** (1968) 1062–1068. Cited: p. 165.

[764] R.P. Shaw, Integral equation formulation of dynamic acoustic fluid–elastic solid interaction problems. *J. Acoust. Soc. Amer.* **53** (1973) 514–520. Cited: p. 165.

[765] R.P. Shaw, A history of boundary elements. In: *Boundary Elements XV* (ed. C.A. Brebbia & J.J. Rencis) pp. 265–280. Southampton: Computational Mechanics Publications Ltd., 1993. Cited: p. 165.

[766] R.P. Shaw & J.A. English, Transient acoustic scattering by a free (pressure release) sphere. *J. Sound Vib.* **20** (1972) 321–331. Cited: p. 165.

[767] R.P. Shaw & M.B. Friedman, Diffraction of pulses by deformable cylindrical obstacles of arbitrary cross section. *Proc. 4th U.S. National Congress of Applied Mechanics*, vol. 1, pp. 371–379. New York: ASME, 1962. Cited: p. 165.

[768] Y. Shi, H. Bağcı & M. Lu, On the internal resonant modes in marching-on-in-time solution of the time domain electric field integral equation. *IEEE Trans. Antennas & Propag.* **61** (2013) 4389–4392. Cited: p. 166.

[769] Y. Shibata & H. Soga, Scattering theory for the elastic wave equation. *Publications of the Research Institute for Mathematical Sciences, Kyoto University* **25** (1989) 861–887. Cited: p. 73.

[770] K.S. Shifrin & I.G. Zolotov, Quasi-stationary scattering of electromagnetic pulses by spherical particles. *Appl. Optics* **33** (1994) 7798–7804. Cited: p. 120.

[771] Y. Shindo, Axisymmetric elastodynamic response of a flat annular crack to normal impact waves. *Engineering Fracture Mech.* **19** (1984) 837–848. Cited: p. 192.

[772] T. Shiozawa, Electromagnetic scattering by a moving small particle. *J. Appl. Phys.* **39** (1968) 2993–2997. Cited: p. 125.

[773] A. Shlivinski, E. Heyman & A.J. Devaney, Time domain radiation by scalar sources: plane wave to multipole transform. *J. Math. Phys.* **42** (2001) 5915–5919. Cited: p. 36.

[774] R.D. Sidman, Scattering of acoustical waves by a prolate spheroidal obstacle. *J. Acoust. Soc. Amer.* **52** (1972) 879–883. Cited: p. 130.

[775] A.J.F. Siegert, On the derivation of the dispersion formula for nuclear reactions. *Phys. Rev.* **56** (1939) 750–752. Cited: p. 133.

[776] G.C. Sih & G.T. Embley, Sudden twisting of a penny-shaped crack. *J. Appl. Mech.* **39** (1972) 395–400. Cited: p. 192.

[777] G.C. Sih, G.T. Embley & R.S. Ravera, Impact response of a finite crack in plane extension. *Int. J. Solids Struct.* **8** (1972) 977–993. Cited: p. 192.

[778] A. Silbiger, Comments on Chertock [174] and author's reply. *J. Acoust. Soc. Amer.* **33** (1961) 1630. Cited: pp. 55 & 201.

[779] A.G. Sitenko, *Scattering Theory*. Berlin: Springer, 1991. Cited: p. 133.

[780] S.L. Skorokhodov & D.V. Khristoforov, Calculation of the branch points of the eigenfunctions corresponding to wave spheroidal functions. *Computational Math. & Math. Phys.* **46** (2006) 1132–1146. Cited: p. 53.

[781] S. Smale, Smooth solutions of the heat and wave equations. *Commentarii Mathematici Helvetici* **55** (1980) 1–12. Cited: p. 78.

[782] V.I. Smirnov, La solution d'un problème aux limites pour l'équation des ondes dans le cas du cercle et de la sphère. *Comptes Rendus (Doklady) de l'Académie des Sciences de l'URSS*, New Series, **14** (1937) 13–16. Cited: p. 37.

[783] P.D. Smith, Instabilities in time marching methods for scattering: cause and rectification. *Electromagnetics* **10** (1990) 439–451. Cited: pp. 166, 175 & 176.

[784] F. Smithies, *Integral Equations*. Cambridge: Cambridge University Press, 1958. Cited: pp. 139 & 140.

[785] I.N. Sneddon, *Fourier Transforms*. New York: McGraw-Hill, 1951. Cited: pp. 13, 84 & 99.

[786] K. Sobczyk, *Stochastic Wave Propagation*. Amsterdam: Elsevier, 1985. Cited: p. 5.

[787] C. Sohl, M. Gustafsson & G. Kristensson, The integrated extinction for broadband scattering of acoustic waves. *J. Acoust. Soc. Amer.* **122** (2007) 3206–3210. Cited: p. 23.

[788] A. Sommerfeld, Theoretisches über die Beugung der Röntgenstrahlen. *Zeitschrift für Mathematik und Physik* **46** (1901) 11–97. Cited: p. 187.

[789] A. Sommerfeld, *Optics*. New York: Academic Press, 1954. Cited: p. 136.

[790] C.R. Steele, Behavior of the basilar membrane with pure-tone excitation. *J. Acoust. Soc. Amer.* **55** (1974) 148–162. Cited: p. 81.

[791] P.D. Stefanov, Inverse scattering problem for moving obstacles. *Mathematische Zeitschrift* **207** (1991) 461–480. Cited: p. 125.

[792] O. Steinbach & G. Unger, Combined boundary integral equations for acoustic scattering-resonance problems. *Math. Meth. Appl. Sci.* **40** (2017) 1516–1530. Cited: pp. 134 & 135.

[793] S. Steinberg, Meromorphic families of compact operators. *Arch. Rational Mech. Anal.* **31** (1968) 372–379. Cited: p. 135.

[794] G.W. Stewart, On the early history of the singular value decomposition. *SIAM Rev.* **35** (1993) 551–566. Cited: p. 140.

[795] J.J. Stoker, *Water Waves*. New York: Interscience, 1957. Cited: pp. 27 & 84.

[796] G.G. Stokes, On the theories of the internal friction of fluids in motion, and of the equilibrium and motion of elastic solids. *Transactions of the Cambridge Philosophical Society* **8** (1849) 287–319. Also: *Mathematical and Physical Papers*, vol. 1, pp. 75–129. Cambridge: Cambridge University Press, 1880. Cited: p. 18.

[797] R.H. Stolt & A.B. Weglein, *Seismic Imaging and Inversion*. Cambridge: Cambridge University Press, 2012. Cited: pp. 4 & 5.

[798] M. Strasberg, Gas bubbles as sources of sound in liquids. *J. Acoust. Soc. Amer.* **28** (1956) 20–26. Cited: p. 127.

[799] J.A. Stratton, *Electromagnetic Theory*. New York: McGraw-Hill, 1941. Cited: pp. 19, 24, 25, 57, 137, 146 & 150.

[800] H.C. Strifors, G.C. Gaunaurd, B. Brusmark & S. Abrahamson, Transient interactions of an EM pulse with a dielectric spherical shell. *IEEE Trans. Antennas & Propag.* **42** (1994) 453–462. Cited: p. 120.

[801] D.J. Struik (ed.), *A Source Book in Mathematics, 1200–1800*. Cambridge: Harvard University Press, 1969. Cited: p. 12.

[802] D.J. Struik, *Lectures on Classical Differential Geometry*, 2nd edition. New York: Dover, 1988. Reprint of 1961 edition. Cited: p. 136.

[803] Y. Su, W. Sheng & Y. Han, An accurate time-domain algorithm using high-order spatial basis functions for bodies of revolution. *J. Electro. Waves Appl.* **29** (2015) 574–588. Cited: p. 176.

[804] A. Sym, Solitons of wave equation. *J. Nonlinear Math. Phys.* **12** (2005) 648–659. Cited: pp. 30 & 47.

[805] J.L. Synge, Hamilton's method in geometrical optics. *J. Opt. Soc. Amer.* **27** (1937) 75–82. Cited: p. 61.

[806] T. Tada, Boundary integral equation method for earthquake rupture dynamics. *International Geophysics* **94** (2009) 217–267. Cited: p. 192.

[807] T. Tada, E. Fukuyama & R. Madariaga, Non-hypersingular boundary integral equations for 3-D non-planar crack dynamics. *Comput. Mech.* **25** (2000) 613–626. Cited: p. 192.

[808] T. Tada & R. Madariaga, Dynamic modelling of the flat 2-D crack by a semi-analytic BIEM scheme. *Int. J. Numer. Meth. Eng.* **50** (2001) 227–251. Cited: p. 192.

[809] A.M. Tagirdzhanov & A.P. Kiselev, Complexified spherical waves and their sources. A review. *Optics & Spectroscopy* **119** (2015) 257–267. Cited: p. 49.

[810] T. Takahashi, An interpolation-based fast-multipole accelerated boundary integral equation method for the three-dimensional wave equation. *J. Comp. Phys.* **258** (2014) 809–832. Cited: pp. 166 & 167.

[811] T. Takahashi, N. Nishimura & S. Kobayashi, A fast BIEM for three-dimensional elastodynamics in time domain. *Eng. Anal. Bound. Elem.* **28** (2004) 165–180. Cited: p. 177.

[812] S.-C. Tang & D.H.Y. Yen, Interaction of a plane acoustic wave with an elastic spherical shell. *J. Acoust. Soc. Amer.* **47** (1970) 1325–1333. Cited: p. 119.

[813] I. Tardy, G.-P. Piau, P. Chabrat & J. Rouch, Computational and experimental analysis of the scattering by rotating fans. *IEEE Trans. Antennas & Propag.* **44** (1996) 1414–1421. Cited: p. 125.

[814] V.I. Tatarski, *Wave Propagation in a Turbulent Medium*. New York: McGraw-Hill, 1961. Cited: p. 6.

[815] J.R. Taylor, *Scattering Theory: The Quantum Theory on Nonrelativistic Collisions*. New York: Wiley, 1972. Cited: p. 133.

[816] K. Taylor, Acoustic generation by vibrating bodies in homentropic potential flow at low Mach number. *J. Sound Vib.* **65** (1979) 125–136. Cited: pp. 125 & 127.

[817] M.E. Taylor, *Partial Differential Equations II: Qualitative Studies of Linear Equations*, 2nd edition. New York: Springer, 2011. Cited: pp. 73, 132 & 135.

[818] F.L. Teixeira & W.C. Chew, PML-FDTD in cylindrical and spherical grids. *IEEE Microwave & Guided Wave Lett.* **7** (1997) 285–287. Cited: p. 42.

[819] R. Temam, Suitable initial conditions. *J. Comp. Phys.* **218** (2006) 443–450. Cited: p. 78.

[820] S.A. Thau & T.-H. Lu, Diffraction of transient horizontal shear waves by a finite crack and a finite rigid ribbon. *Int. J. Eng. Sci.* **8** (1970) 857–874. Cited: p. 187.

[821] S.A. Thau & T.-H. Lu, Transient stress intensity factors for a finite crack in an elastic solid caused by a dilatational wave. *Int. J. Solids Struct.* **7** (1971) 731–750. Cited: p. 192.

[822] J.H. Thompson, Closed solutions for wedge diffraction. *SIAM J. Appl. Math.* **22** (1972) 300–306. Cited: p. 187.

[823] J.J. Thomson, On electrical oscillations and the effects produced by the motion of an electrified sphere. *Proc. London Math. Soc.*, Ser. 1, **15** (1884) 197–218. Cited: p. 135.

[824] A.G. Tijhuis, Toward a stable marching-on-in-time method for two-dimensional transient electromagnetic scattering problems. *Radio Sci.* **19** (1984) 1311–1317. Cited: p. 176.

[825] A.G. Tijhuis, *Electromagnetic Inverse Profiling*. Utrecht: VNU Science Press, 1987. Cited: p. 166.

[826] L. Ting & M.J. Miksis, Exact boundary conditions for scattering problems. *J. Acoust. Soc. Amer.* **80** (1986) 1825–1827. Cited: p. 150.

[827] T. Tokita, Exponential decay of solutions for the wave equation in the exterior domain with spherical boundary. *Journal of Mathematics of Kyoto University* **12** (1972) 413–430. Cited: p. 120.

[828] J.S. Toll, Causality and the dispersion relation: logical foundations. *Phys. Rev.* **104** (1956) 1760–1770. Cited: p. 22.

[829] B.E. Treeby, J. Budisky, E.S. Wise, J. Jaros & B.T. Cox, Rapid calculation of acoustic fields from arbitrary continuous-wave sources. *J. Acoust. Soc. Amer.* **143** (2018) 529–537. Cited: p. 86.

[830] B.E. Treeby & B.T. Cox, Modeling power law absorption and dispersion for acoustic propagation using the fractional Laplacian. *J. Acoust. Soc. Amer.* **127** (2010) 2741–2748. Erratum: **130** (2011) 610. Cited: p. 19.

[831] L.N. Trefethen, Pseudospectra of matrices. In: *Numerical Analysis 1991* (ed. D.F. Griffiths & G.A. Watson) pp. 234–266. Harlow: Longman Scientific & Technical, 1992. Cited: p. 140.

[832] L.N. Trefethen & M. Embree, *Spectra and Pseudospectra. The Behavior of Nonnormal Matrices and Operators*. Princeton, NJ: Princeton University Press, 2005. Cited: p. 140.

[833] F. Treves, *Basic Linear Partial Differential Equations*. New York: Academic Press, 1975. Cited: pp. 29, 89, 90, 91 & 97.

[834] C. Truesdell, Outline of the history of flexible or elastic bodies to 1788. *J. Acoust. Soc. Amer.* **32** (1960) 1647–1656. Cited: p. 12.

[835] C. Truesdell & R. Toupin, The classical field theories. In: *Handbuch der Physik*, vol. III/1 (ed. S. Flügge) pp. 226–858. Berlin: Springer, 1960. Cited: p. 69.

[836] L.W. Tu, *An Introduction to Manifolds*, 2nd edition. New York: Springer, 2011. Cited: pp. 186 & 187.

[837] G.E. Tupholme, Generation of an axisymmetrical acoustic pulse by a deformable sphere. *Proc. Camb. Phil. Soc.* **63** (1967) 1285–1308. Cited: p. 119.

[838] G.E. Tupholme, Elastic pulse generation by tractions applied to a spherical cavity. *Applied Scientific Research* **40** (1983) 299–325. Cited: p. 120.

[839] M. Tygel & P. Hubral, *Transient Waves in Layered Media*. Amsterdam: Elsevier, 1987. Cited: pp. 24 & 45.

[840] H. Überall, P.P. Delsanto, J.D. Alemar, E. Rosario & A. Nagl, Application of the singularity expansion method to elastic wave scattering. *Appl. Mech. Rev.* **43** (1990) 235–249. Cited: p. 138.

[841] H. Überall, L.R. Dragonette & L. Flax, Relation between creeping waves and normal modes of vibration of a curved body. *J. Acoust. Soc. Amer.* **61** (1977) 711–715. Cited: p. 136.

[842] H. Überall, G.C. Gaunaurd & E. Tanglis, Interior and exterior resonances in acoustic scattering. II. – Targets of arbitrary shape (T-matrix approach). *Il Nuovo Cimento B* **77** (1983) 73–86. Cited: p. 136.

[843] H. Überall & H. Huang, Acoustical response of submerged elastic structures obtained through integral transforms. *Physical Acoustics* **12** (1976) 217–275. Cited: p. 119.

[844] H. Überall, P.J. Moser, B.L. Merchant, A. Nagl, K.B. Yoo, S.H. Brown, J.W. Dickey & J.M. D'Archangelo, Complex acoustic and electromagnetic resonance frequencies of prolate spheroids and related elongated objects and their physical interpretation. *J. Appl. Phys.* **58** (1985) 2109–2124. Cited: p. 130.

[845] H.A. Ülkü, H. Bağcı & E. Michielssen, Marching on-in-time solution of the time domain magnetic field integral equation using a predictor-corrector scheme. *IEEE Trans. Antennas & Propag.* **61** (2013) 4120–4131. Cited: pp. 171 & 175.

[846] G. Unger, A. Trügler & U. Hohenester, Novel modal approximation scheme for plasmonic transmission problems. *Phys. Rev. Lett.* **121** (2018) 246802. Cited: p. 135.

[847] P.L.E. Uslenghi (ed.), *Electromagnetic Scattering*. New York: Academic Press, 1978. Cited: pp. 196 & 204.

[848] J. Van Bladel, *Relativity and Engineering*. Berlin: Springer, 1984. Cited: pp. 49 & 125.

[849] A.L. Van Buren, Prolate spheroidal functions for complex *c*. www.mathieuandspheroidalwavefunctions.com Cited: p. 55.

[850] E. van 't Wout, D.R. van der Heul, H. van der Ven & C. Vuik, Design of temporal basis functions for time domain integral equation methods with predefined accuracy and smoothness. *IEEE Trans. Antennas & Propag.* **61** (2013) 271–280. Cited: p. 176.

[851] E. van 't Wout, D.R. van der Heul, H. van der Ven & C. Vuik, Stability analysis of the marching-on-in-time boundary element method for electromagnetics. *J. Comp. Appl. Math.* **294** (2016) 358–371. Cited: p. 176.

[852] D.A. Vechinski & S.M. Rao, A stable procedure to calculate the transient scattering by conducting surfaces of arbitrary shape. *IEEE Trans. Antennas & Propag.* **40** (1992) 661–665. Comments by B.P. Rynne and authors' reply: **41** (1993) 517–520. Cited: p. 176.

[853] S.R. Vechinski & T.H. Shumpert, Natural resonances of conducting bodies of revolution. *IEEE Trans. Antennas & Propag.* **38** (1990) 1133–1136. Cited: p. 135.

[854] A. Veit, M. Merta, J. Zapletal & D. Lukáš, Efficient solution of time-domain boundary integral equations arising in sound-hard scattering. *Int. J. Numer. Meth. Eng.* **107** (2016) 430–449. Cited: p. 167.

[855] J.V. Venås & T. Jenserud, Exact 3D scattering solutions for spherical symmetric scatterers. *J. Sound Vib.* **440** (2019) 439–479. Corrigendum: **474** (2020) 115270. Cited: p. 119.

[856] E. Ventsel & T. Krauthammer, *Thin Plates and Shells*. New York: Marcel Dekker, 2001. Cited: p. 82.

[857] A.D. Venttsel', On boundary conditions for multidimensional diffusion processes. *Theory of Probability & Its Applications* **4** (1959) 164–177. Cited: p. 80.

[858] J.D. Victor, Temporal impulse responses from flicker sensitivities: causality, linearity, and amplitude data do not determine phase. *J. Opt. Soc. Amer.* A **6** (1989) 1302–1303. Cited: p. 24.

[859] V. Volterra, Sur la théorie des ondes liquides et la méthode de Green. *Journal de Mathématiques Pures et Appliquées*, Ser. 9, **13** (1934) 1–18. Cited: pp. 26 & 180.

[860] O. von Estorff & H. Antes, On FEM–BEM coupling for fluid–structure interaction analyses in the time domain. *Int. J. Numer. Meth. Eng.* **31** (1991) 1151–1168. Cited: p. 82.

[861] J.R. Wait, Diffraction of a spherical wave pulse by a half-plane screen. *Canadian J. Phys.* **35** (1957) 693–696. Cited: p. 187.

[862] S.P. Walker, M.J. Bluck & I. Chatzis, The stability of integral equation time-domain computations for three-dimensional scattering; similarities and differences between electrodynamic and elastodynamic computations. *Int. J. Numerical Modelling: Electronic Networks, Devices & Fields* **15** (2002) 459–474. Cited: pp. 176 & 177.

[863] G.C. Wan & M.S. Tong, Refining transient electromagnetic scattering analysis: a new approach based on the magnetic field integral equation. *IEEE Antennas & Propagation Magazine* **59**, No. 1 (2017) 66–73. Cited: p. 176.

[864] H. Wang, D.J. Henwood, P.J. Harris & R. Chakrabarti, Concerning the cause of instability in time-stepping boundary element methods applied to the exterior acoustic problem. *J. Sound Vib.* **305** (2007) 289–297. Cited: p. 166.

[865] M. Wang, J.B. Freund & S.K. Lele, Computational prediction of flow-generated sound. *Ann. Rev. Fluid Mech.* **38** (2006) 483–512. Cited: pp. 87 & 156.

[866] X. Wang, R.A. Wildman, D.S. Weile & P. Monk, A finite difference delay modeling approach to the discretization of the time domain integral equations of electromagnetics. *IEEE Trans. Antennas & Propag.* **56** (2008) 2442–2452. Cited: pp. 171 & 176.

[867] Z.-H. Wang, S.W. Rienstra, C.-X. Bi & B. Koren, An accurate and efficient computational method for time-domain aeroacoustic scattering. *J. Comp. Phys.* **412** (2020) 109442. Cited: p. 183.

[868] K. Watanabe, *Integral Transform Techniques for Green's Function*, 2nd edition. Cham: Springer, 2015. Cited: p. 187.

[869] E.J. Watson, *Laplace Transforms and Applications*. New York: Van Nostrand Reinhold, 1981. Cited: pp. 17, 21 & 116.

[870] J.C. Wawa & F.L. DiMaggio, Dynamic response of a submerged prolate spheroidal shell to a longitudinal shock wave. *Computers & Struct.* **20** (1985) 975–989. Cited: p. 130.

[871] A.G. Webster, *Partial Differential Equations of Mathematical Physics*, 2nd edition. New York: Dover, 1955. Cited: pp. 12, 85 & 146.

[872] J.V. Wehausen, Initial-value problem for the motion in an undulating sea of a body with fixed equilibrium position. *J. Engng. Math.* **1** (1967) 1–17. Cited: pp. 84 & 180.

[873] J.V. Wehausen, The motion of floating bodies. *Ann. Rev. Fluid Mech.* **3** (1971) 237–268. Cited: pp. 27, 84 & 180.

[874] J.V. Wehausen, Causality and the radiation condition. *J. Engng. Math.* **26** (1992) 153–158. Cited: p. 180.

[875] M. Wei, G. Majda & W. Strauss, Numerical computation of the scattering frequencies for acoustic wave equations. *J. Comp. Phys.* **75** (1988) 345–358. Cited: p. 133.

[876] D.S. Weile, G. Pisharody, N.-W. Chen, B. Shanker & E. Michielssen, A novel scheme for the solution of the time-domain integral equations of electromagnetics. *IEEE Trans. Antennas & Propag.* **52** (2004) 283–295. Cited: p. 176.

[877] P.J. Westervelt, Scattering of sound by sound. *J. Acoust. Soc. Amer.* **29** (1957) 199–203. Cited: p. 88.

[878] V.H. Weston, Pulse return from a sphere. *IRE Trans. Antennas & Propagation* **7** (1959) S43–S51. Cited: pp. 114 & 120.

[879] G.F. Wheeler & W.P. Crummett, The vibrating string controversy. *Amer. J. Phys.* **55** (1987) 33–37. Cited: pp. 12 & 13.

[880] L.T. Wheeler & E. Sternberg, Some theorems in classical elastodynamics. *Arch. Rational Mech. Anal.* **31** (1968) 51–90. Cited: p. 176.

[881] G.B. Whitham, *Linear and Nonlinear Waves*. New York: Wiley, 1974. Cited: pp. 31, 59, 73 & 108.

[882] E.T. Whittaker, On the partial differential equations of mathematical physics. *Mathematische Annalen* **57** (1903) 333–355. Cited: p. 45.

[883] E. Wiechert, Elektrodynamische Elementargesetze. *Archives Néerlandaises des Sciences Exactes et Naturelles*, Ser. 2, **5** (1900) 549–573. Also: *Annalen der Physik*, 4 Folge, **4** (1901) 667–689. Cited: p. 55.

[884] A.C. Wijeyewickrema & L.M. Keer, Transient elastic wave scattering by a rigid spherical inclusion. *J. Acoust. Soc. Amer.* **86** (1989) 802–809. Cited: p. 120.

[885] C.H. Wilcox, The initial-boundary value problem for the wave equation in an exterior domain with spherical boundary. *Notices Amer. Math. Soc.* **6** (1959) 869–870. Cited: pp. 115 & 119.

[886] C.H. Wilcox, Initial-boundary value problems for linear hyperbolic partial differential equations of the second order. *Arch. Rational Mech. Anal.* **10** (1962) 361–400. Cited: p. 73.

[887] C.H. Wilcox, Scattering theory for the d'Alembert equation in exterior domains. *Lect. Notes Math.* **442**. Berlin: Springer, 1975. Cited: pp. 73, 74 & 208.

[888] C.H. Wilcox, Spectral and asymptotic analysis of acoustic wave propagation. In: [328], pp. 385–473. Cited: p. 73.

[889] R.A. Wildman, G. Pisharody, D.S. Weile, S. Balasubramaniam & E. Michielssen, An accurate scheme for the solution of the time-domain integral equations of electromagnetics using higher order vector bases and bandlimited extrapolation. *IEEE Trans. Antennas & Propag.* **52** (2004) 2973–2984. Cited: p. 176.

[890] E.B. Wilson & G.N. Lewis, The space-time manifold of relativity. The non-Euclidean geometry of mechanics and electromagnetics. *Proc. American Academy of Arts & Sciences* **48** (1912) 389–507. Cited: p. 11.

[891] D.W. Wright & R.S.C. Cobbold, Acoustic wave transmission in time-varying phononic crystals. *Smart Materials & Structures* **18** (2009) 015008. Cited: p. 6.

[892] S.F. Wu, *The Helmholtz Equation Least Squares Method.* New York: Springer, 2015. Cited: p. 120.

[893] S.F. Wu, H. Lu & M.S. Bajwa, Reconstruction of transient acoustic radiation from a sphere. *J. Acoust. Soc. Amer.* **117** (2005) 2065–2077. Cited: p. 120.

[894] X.-F. Wu & A. Akay, Sound radiation from vibrating bodies in motion. *J. Acoust. Soc. Amer.* **91** (1992) 2544–2555. Cited: p. 156.

[895] R. Wunenburger, N. Mujica & S. Fauve, Experimental study of the Doppler shift generated by a vibrating scatterer. *J. Acoust. Soc. Amer.* **115** (2004) 507–514. Cited: p. 128.

[896] M.Y. Xia, G.H. Zhang, G.L. Dai & C.H. Chan, Stable solution of time domain integral equation methods using quadratic B-spline temporal basis functions. *J. Comp. Math.* **25** (2007) 374–384. Cited: p. 176.

[897] P. Yang, J. Li, X. Gu & D. Wu, Application of the 3D time-domain Green's function for finite water depth in hydroelastic mechanics. *Ocean Engng.* **189** (2019) 106386. Cited: p. 179.

[898] M. Yazdani, J.R. Mautz, J.K. Lee & E. Arvas, Transient electromagnetic scattering by a radially uniaxial dielectric sphere: the generalized Debye and Mie series solutions. *IEEE Trans. Antennas & Propag.* **64** (2016) 1039–1046. Cited: p. 120.

[899] A.E. Yılmaz, J.-M. Jin & E. Michielssen, Time domain adaptive integral method for surface integral equations. *IEEE Trans. Antennas & Propag.* **52** (2004) 2692–2708. Cited: p. 167.

[900] G. Yu, W.J. Mansur, J.A.M. Carrer & L. Gong, Stability of Galerkin and collocation time domain boundary element methods as applied to the scalar wave equation. *Computers & Struct.* **74** (2000) 495–506. Cited: p. 166.

[901] A.H. Zemanian, *Distribution Theory and Transform Analysis*. New York: McGraw-Hill, 1965. Cited: pp. 90 & 97.

[902] S.H. Zemell, New derivation of the exact solution for the diffraction of a cylindrical or spherical pulse by a wedge. *Int. J. Eng. Sci.* **14** (1976) 845–851. Cited: p. 187.

[903] A. Zenginoğlu, Hyperboloidal layers for hyperbolic equations on unbounded domains. *J. Comp. Phys.* **230** (2011) 2286–2302. Cited: pp. 42 & 44.

[904] Ch. Zhang & D. Gross, A non-hypersingular time-domain BIEM for 3-D transient elastodynamic crack analysis. *Int. J. Numer. Meth. Eng.* **36** (1993) 2997–3017. Cited: p. 192.

[905] H.L. Zhang, Y.X. Sha, X.Y. Guo, M.Y. Xia & C.H. Chan, Efficient analysis of scattering by multiple moving objects using a tailored MLFMA. *IEEE Trans. Antennas & Propag.* **67** (2019) 2023–2027. Cited: p. 125.

[906] P. Zhang & T.L. Geers, Excitation of a fluid-filled, submerged spherical shell by a transient acoustic wave. *J. Acoust. Soc. Amer.* **93** (1993) 696–705. Cited: pp. 119, 120 & 121.

[907] Y. Zhao, D. Ding & R. Chen, A discontinuous Galerkin time-domain integral equation method for electromagnetic scattering from PEC objects. *IEEE Trans. Antennas & Propag.* **64** (2016) 2410–2417. Cited: p. 176.

[908] M.-D. Zhu, T.K. Sarkar & H. Chen, A stabilized marching-on-in-degree scheme for the transient solution of the electric field integral equation. *IEEE Trans. Antennas & Propag.* **67** (2019) 3232–3240. Cited: p. 176.

[909] M.-D. Zhu, T.K. Sarkar, H. Chen & Y. Wu, On the stability of time-domain magnetic field integral equation using Laguerre functions. *IEEE Trans. Antennas & Propag.* **67** (2019) 3939–3947. Cited: p. 176.

[910] M. Zworski, Resonances in physics and geometry. *Notices Amer. Math. Soc.* **46** (1999) 319–328. Cited: pp. 28 & 133.

Citation index

Index

Printed in the United States
by Baker & Taylor Publisher Services